AF782463

Microgravity Research in Support of Technologies for the Human Exploration and Development of Space and Planetary Bodies

Committee on Microgravity Research
Space Studies Board
Commission on Physical Sciences, Mathematics, and Applications
National Research Council

NOTICE: The project that is the subject of this report was approved by the Governing Board of the National Research Council, whose members are drawn from the councils of the National Academy of Sciences, the National Academy of Engineering, and the Institute of Medicine. The members of the committee responsible for the report were chosen for their special competences and with regard for appropriate balance.

Support for this project was provided by Contract NASW 96013 between the National Academy of Sciences and the National Aeronautics and Space Administration. Any opinions, findings, conclusions, or recommendations expressed in this material are those of the authors and do not necessarily reflect the views of the sponsor.

International Standard Book Number 0-309-06491-0

Cover design by Penny Margolskee.

Copies of this report are available from

Space Studies Board
National Research Council
2101 Constitution Avenue, NW
Washington, DC 20418

Printed in the United States of America

THE NATIONAL ACADEMIES

National Academy of Sciences
National Academy of Engineering
Institute of Medicine
National Research Council

The **National Academy of Sciences** is a private, nonprofit, self-perpetuating society of distinguished scholars engaged in scientific and engineering research, dedicated to the furtherance of science and technology and to their use for the general welfare. Upon the authority of the charter granted to it by the Congress in 1863, the Academy has a mandate that requires it to advise the federal government on scientific and technical matters. Dr. Bruce M. Alberts is president of the National Academy of Sciences.

The **National Academy of Engineering** was established in 1964, under the charter of the National Academy of Sciences, as a parallel organization of outstanding engineers. It is autonomous in its administration and in the selection of its members, sharing with the National Academy of Sciences the responsibility for advising the federal government. The National Academy of Engineering also sponsors engineering programs aimed at meeting national needs, encourages education and research, and recognizes the superior achievements of engineers. Dr. William A. Wulf is president of the National Academy of Engineering.

The **Institute of Medicine** was established in 1970 by the National Academy of Sciences to secure the services of eminent members of appropriate professions in the examination of policy matters pertaining to the health of the public. The Institute acts under the responsibility given to the National Academy of Sciences by its congressional charter to be an adviser to the federal government and, upon its own initiative, to identify issues of medical care, research, and education. Dr. Kenneth I. Shine is president of the Institute of Medicine.

The **National Research Council** was organized by the National Academy of Sciences in 1916 to associate the broad community of science and technology with the Academy's purposes of furthering knowledge and advising the federal government. Functioning in accordance with general policies determined by the Academy, the Council has become the principal operating agency of both the National Academy of Sciences and the National Academy of Engineering in providing services to the government, the public, and the scientific and engineering communities. The Council is administered jointly by both Academies and the Institute of Medicine. Dr. Bruce M. Alberts and Dr. William A. Wulf are chairman and vice chairman, respectively, of the National Research Council.

COMMITTEE ON MICROGRAVITY RESEARCH

RAYMOND VISKANTA, Purdue University, *Chair*
ROBERT A. ALTENKIRCH, Mississippi State University
ROBERT L. ASH, Old Dominion University
ROBERT J. BAYUZICK, Vanderbilt University
CHARLES W. CARTER, JR., University of North Carolina at Chapel Hill
GRETCHEN J. DARLINGTON,* Baylor College of Medicine
RICHARD T. LAHEY, JR., Rensselaer Polytechnic Institute
RALPH A. LOGAN, AT&T Bell Laboratories (retired)
FRANKLIN K. MOORE, Cornell University
WILLIAM W. MULLINS, Carnegie Mellon University (emeritus)
ROSALIA N. SCRIPA,* University of Alabama at Birmingham
FORMAN A. WILLIAMS, University of California at San Diego

SANDRA J. GRAHAM, Study Director
ANNE K. SIMMONS, Senior Project Assistant
CATHY GRUBER, Senior Project Assistant (through May 1998)

---

*Former member.

SPACE STUDIES BOARD

---

*Former member.

COMMISSION ON PHYSICAL SCIENCES, MATHEMATICS, AND APPLICATIONS

Preface

The study that is the subject of this report was initiated in early 1996 by a request to the Committee on Microgravity Research (CMGR) from the leadership of NASA's Microgravity Science and Applications Division[1] to perform an assessment of scientific and related technological issues facing NASA's Human Exploration and Development of Space (HEDS) endeavor. The committee agreed to consider mission enabling and enhancing technologies that, for development, would require an improved understanding of fluid and material behavior in a reduced-gravity environment. The committee would then identify opportunities for microgravity research to contribute to the understanding of fundamental scientific questions underlying exploration technologies and make recommendations for some areas of directed research. The study was to be carried out in two phases. The phase I report, *An Initial Review of Microgravity Research in Support of Human Exploration and Development of Space*, was published in 1997 (National Academy Press, Washington, D.C.). That first report represented a preliminary look at broad categories of HEDS technologies and contained primarily programmatic recommendations. For the second phase of the study, the committee undertook a more in-depth examination of a wide range of specific technologies that might be applicable to human exploration. As no single office at NASA had assembled a list of critical technologies needed for HEDS, the committee has included the results of its own technology survey in this report. This survey was carried out by canvassing the available literature, participating in relevant workshops, and receiving extensive briefings from experts in NASA, industry, and academia. The goal of this phase II report was to provide specific recommendations for areas of research on fundamental phenomena. The phenomena recommended for study would be those that had the potential to significantly affect the operation of future exploration technologies and that needed to be better understood to enable the optimization or eventual development of those technologies. Since the time frame for technology development from fundamental research is generally quite long, the committee chose to focus, in this phase II report, primarily on those technology areas that might be important for space exploration one to three decades into the future.

In its study, the committee utilized a large number of past reports from various sources. Among the previous National Research Council reports relevant to this study, the committee took particular note of the following:

---

[1]Now the Microgravity Research Division (MRD).

• *Microgravity Research Opportunities for the 1990s*, Space Studies Board, National Research Council (National Academy Press, Washington, D.C., 1995), reviewed the various research topics currently studied within the different scientific disciplines of NASA's microgravity research program and suggested research and programmatic priorities and recommendations. The report focused on fundamental research that could contribute to basic advances within individual disciplines.

• *Space Technology for the New Century*, Aeronautics and Space Engineering Board, National Research Council (National Academy Press, Washington, D.C., 1998), examined space technology needs in the post-2000 time frame and identified a few high-risk, high-payoff areas where research investments might benefit a range of future missions.

• *Advanced Technology for Human Support in Space*, Aeronautics and Space Engineering Board, National Research Council (National Academy Press, Washington, D.C., 1997), reviewed the NASA programs that support development of technologies for human life support and recommended improved strategies for managing the development process.

• *Space Technology to Meet Future Needs*, Aeronautics and Space Engineering Board, National Research Council (National Academy Press, Washington, D.C., 1987), evaluated national advanced space technology requirements and recommended a long-term technology program focus for NASA.

Acknowledgment of Reviewers

This report has been reviewed by individuals chosen for their diverse perspectives and technical expertise, in accordance with procedures approved by the National Research Council's (NRC's) Report Review Committee. The purpose of this independent review is to provide candid and critical comments that will assist the authors and the NRC in making the published report as sound as possible and to ensure that the report meets institutional standards for objectivity, evidence, and responsiveness to the study charge. The contents of the review comments and draft manuscript remain confidential to protect the integrity of the deliberative process. We wish to thank the following individuals for their participation in the review of this report:

Rino Buonamici, Westinghouse Hanford Company (retired),
Daniel C. Drucker, University of Florida (emeritus),
Jerry P. Gollub, Haverford College,
Lionel Isenberg, Jet Propulsion Laboratory (retired),
Joseph Miller, TRW Space and Electronics Group (retired),
Simon Ostrach, Case Western Reserve University,
Julio M. Ottino, Northwestern University,
Frederick G. Pohland, University of Pittsburgh,
William C. Reynolds, Stanford University, and
William A. Sirignano, University of California at Irvine.

Although the individuals listed above have provided many constructive comments and suggestions, responsibility for the final content of this report rests solely with the authoring committee and the NRC.

Contents

Executive Summary

CHARGE TO THE COMMITTEE AND REPORT ORGANIZATION

The primary charge to the Committee on Microgravity Research (CMGR) from the Microgravity Science and Applications Division (MSAD)[1] of NASA reads:

> CMGR will undertake an assessment of scientific and related technological issues facing NASA's Human Exploration and Development of Space (HEDS) endeavor. The committee will look specifically at mission enabling and enhancing technologies which, for development, require an improved understanding of fluid and material behavior in a reduced gravity environment. These might range from construction assembly techniques such as welding in space, to chemical processing of extraterrestrially derived fuels and oxygen. The committee will identify opportunities which exist for microgravity research to contribute to the understanding of fundamental science questions underlying exploration technologies and make recommendations for some areas of directed research.

In addition to the above charge, which is stated in full in Appendix A, the committee was asked to give some consideration to radiation hazards and shielding.

The committee and MSAD mutually interpreted the main thrust of the charge to be the determination of the gravity-related physicochemical phenomena most relevant to HEDS technology needs and the recommendation of fundamental research on those phenomena. The technologies considered were those judged to be relevant in the next one to three decades.

The organization of this report reflects the committee's interpretation of the charge. Following the introduction and brief descriptions of relevant phenomena and concepts, the report surveys a set of selected HEDS-enabling technologies, classified according to function. The survey is intended not to be comprehensive but to identify those underlying scientific phenomena that are vital to the technologies, that are gravity related, and that present a compelling need for research. The committee defines a gravity-related phenomenon as a phenomenon that is either directly affected by reduced gravity or that becomes significant as gravity level is reduced. A

[1] Now the Microgravity Research Division.

phenomenon of the latter type may sometimes be used to compensate for the loss of gravity (an example is surface-tension-driven flow in wicks and heat pipes in the absence of gravity-induced convection). Selected phenomena and their dependence on, or importance in, reduced gravity are then discussed, along with the research needed to develop predictive models and better databases for characterizing the phenomena. The remainder of the report deals with other gravity-related features of HEDS technologies, discusses microgravity countermeasures (e.g., artificial gravity), and offers research and programmatic recommendations.

TECHNOLOGIES SURVEYED

The selected technologies are discussed according to their functions: (1) power generation and storage, (2) space propulsion, (3) life support, (4) hazard control, (5) materials production and storage, and (6) construction and maintenance. They were examined for their dependence on gravity level by considering the gravity dependence of the components (subsystems) or processes of which they are made up. In many instances, the subsystems (e.g., pumps or boilers) or processes (e.g., electrolysis) are common to many technologies, so the phenomena underlying them were recognized as especially important for HEDS. An example of a strongly gravity-dependent subsystem is a heat exchanger subsystem, such as a boiler or a condensation-based space radiator, that uses a two-phase fluid (i.e., liquid and vapor). Its operation is radically affected by microgravity because the phenomena of buoyant convection and density stratification are absent. An example of a strongly gravity-dependent process is liquid electrolysis, common to life support and fuel production systems. The phenomenon of buoyancy-driven migration of the gases (i.e., bubbles) in the liquid does not occur in microgravity, so phase separation of the product gases from the liquid must be accomplished by other means.

Although the charge to the committee did not specifically include an evaluation of technologies, a remark on this point seems in order. Now and in the past NASA has chosen not to use active, high-power-density systems that involve heat transfer by phase change (e.g., condensation and boiling) to meet energy needs but has chosen instead to use lower-power-density, passive systems such as solar collectors, fuel cells, and radio isotope generators. This approach has been motivated by the requirement to reduce risk and to ensure reliability, since the performance of multiphase (two or more phases) flow and heat transfer processes in reduced gravity is not well understood and was therefore considered risky. Unfortunately, however, the lower-power-density systems will not be able to supply enough energy for proposed long-duration, crewed space and interplanetary missions. The high efficiency and high power-to-weight ratio of closed-cycle multiphase systems, based on the use of the latent heat of phase change (i.e., condensation and evaporation) to transfer energy, are so attractive that the committee believes it is imperative for NASA to undertake a directed research program on multiphase flow and heat transfer that will enable it to decide if systems dependent on these processes can be successfully controlled and utilized in space. Accordingly, one of the higher-priority recommendations in this report proposes this research.

PHENOMENA IDENTIFIED AS AFFECTED BY OR DOMINANT IN REDUCED GRAVITY

Phenomena that are identified as underlying HEDS-enabling technologies and that either are directly affected by gravity level or emerge as dominant factors in reduced gravity are generally organized in Chapter IV of this report as follows: (1) surface and interfacial phenomena, referring to effects stemming from surface wetting and interfacial tension; (2) multiphase flow and heat transfer, referring to the flow of more than one fluid phase in pipes, pumps, and phase-change components, and flow in porous media, exemplified by the flow of fluids in the packed and fluidized particulate beds used in chemical reactors; (3) multiphase system dynamics, which deals with the global instabilities that may occur in multiphase systems; (4) solidification, referring to the phase change of a liquid to a solid, as occurs in casting or welding; (5) fire phenomena and combustion, used in some power generation and propulsion systems and occurring in accidental fires; and (6) granular mechanics, referring to such topics as the response of granular media and soils to geotechnical loads and the flow of granular materials in chutes and hoppers.

RECOMMENDED HIGHER-PRIORITY RESEARCH ON FUNDAMENTAL PHENOMENA

Of the specific areas recommended for research in this report, those discussed below were considered by the committee to have higher priority based on their potential to affect a wide range of HEDS technologies that are mission enabling. In each area, the technological importance of the phenomena is briefly explained first.

Surface or Interfacial Phenomena

Surface tension effects are of critical importance in such diverse HEDS technologies as welding, liquid-phase sintering, the operation of wicks in heat pipes for thermal management, the use of capillary vanes (wet by the liquid) in cryogenic storage tanks to control the position and movement of liquids, lubrication, and boiling/condensation heat transfer, including the rewetting of hot surfaces. A special (Marangoni) effect occurs when the surface tension varies over the surface of a liquid (or the interface between two liquids) because of thermal and composition gradients. Marangoni effects can produce strong gravity-independent convection, which may be beneficial, as in the stirring of weld pools and the enhancement of the critical heat flux in multicomponent boiling, or detrimental, as in the migration of fluid in a thermal gradient to unwanted locations.

The committee's recommendations, which are based on the critical issues underlying the technologies, call for research on the following topics:

- *The physics of wetting*, with an emphasis on hysteresis effects, the dynamics of the wetting process, and the molecular basis of wetting, to elucidate the wetting of both solid and porous media (e.g., wicks and nanoscale media) and to provide a basis for the choice of material combinations and conditions for optimal wetting and wetting agents; and
- *Capillary-driven flows,* with modeling of the flows induced by the Marangoni effect, which are complicated because of the feedback between the flow and the surface temperature and composition gradients that drive the flow.

Multiphase Flow and Heat Transfer

Multiphase flow and heat transfer are the fundamental processes in systems using a fluid of two or more phases (e.g., liquid and vapor) to transport mass, momentum, and energy. They are critical to the operation of many power production and utilization systems and other systems that require high energy-transport efficiency and high power-to-weight ratios. Multiphase systems have these characteristics because they are able to utilize the latent heat of evaporation/condensation to efficiently transfer energy. Their successful operation in Earth's gravity often depends on buoyancy-driven convection and density-induced stratification of the phases, processes that are reduced or absent in microgravity. Moreover, new flow regimes may occur in which the spatial distribution of the phases reflects the absence of a gravitational force. Therefore, to exploit the attractive advantages of multiphase systems under microgravity conditions, it is imperative to determine how they can be used and controlled in the absence of gravity.

It is recognized that there will probably be a continuing need for experimental microgravity data and appropriate empirical correlations, since some physical phenomena and HEDS design issues go beyond current, and anticipated near-term, computational capabilities. Nevertheless, the primary objective of the proposed research is the development of a reliable, physically based,[2] multidimensional two-fluid model for the computational fluid dynamics (CFD) analysis of multiphase flow and heat transfer phenomena of importance to the HEDS program. Indeed, the following recommended research is aimed at developing predictive models of multiphase flow and heat transfer and testing these models against reduced-scale data taken in microgravity environments:

---

[2]In the context of this report, a physically based model is one that is developed from fundamental principles and physical mechanisms, as opposed to an empirical model. See the glossary, Appendix C.

• *The development of physically based models to predict the flow regimes, flow regime transitions, and the multiphase flow and heat transfer that occur in fractional gravity and microgravity environments.* These models should include the effects of two-phase turbulence, surface-tension-induced forces, and the axial and lateral interfacial and wall forces on the flowing phases (i.e., the flow-regime-specific interfacial and wall constitutive laws). They should be suitable for use in three-dimensional CFD solvers.

• *Assessment of the predictive capabilities of these models by comparing them with the results of reduced-scale and separate-effects experiments performed under microgravity conditions.* In particular, detailed data are needed on flow-regime-specific phenomena in simple and complex geometry conduits (where gravity dependence may occur even at high flow rates); the distribution and separation of the phases for the various flow regimes, including the effect of phase separations induced by swirl; and the local velocity, temperature, void fraction, and turbulence fields.

• *A program parallel to the one described above for assessing the effect of gravity level on forced convective flows*, especially the forced-flow boiling curve for different boiling regimes (e.g., nucleate and film). Such a program is essential, since forced flow can compensate for some of the problems arising from the loss of buoyancy.

Multiphase System Dynamics

In systems using multiphase flow, effects on a global scale may emerge from the interaction of the components. In particular, phase or time lags in feedback loops can cause potentially dangerous instabilities that are revealed only by analysis of the system as a whole. The following research is recommended:

• *The development of models and the collection and analysis of stability data on boiling and condensing systems at fractional gravity and microgravity levels.* In particular, the effect of gravity level on excursive instabilities, as well as on those dynamic instabilities that can be induced by compressibility and lags in the propagation of density waves around closed loops in multiphase systems, needs to be investigated and analyzed.

Fire Phenomena

Accidental fires are a major hazard in the confined quarters of spacecraft. The structure and dynamics of fires and flames are drastically altered in microgravity, primarily because there is no buoyancy-driven convection or sedimentation (e.g., of smoke particles). Accordingly, the following research is recommended:

• *Experimental, theoretical, and computational studies of flame spread over surfaces of solid materials in microgravity and fractional gravity.* These studies should focus on generic materials, both cellulose and synthetic polymers, and should include ignition requirements, flame-spread rates, and flame structure. Parallel studies on the production of gaseous fuel from solid-fuel pyrolysis are needed.

• *Gravity effects in smoldering, as in the case of electrical cable fires.* In particular, the research should look at the initiation and termination of smoldering, propagation rates of smoldering fronts, and the production of hazardous or flammable products from smoldering, including conditions for transition from smoldering to flaming combustion.

Granular Mechanics

The granular materials encountered in lunar and Martian soils will serve both as the physical foundation that supports people, equipment, and buildings and as a raw material to be mined and used for construction and for the extraction of valuable resources. Granular material in the form of dust is expected to be a serious environmental problem on the Moon and Mars. The behavior of granular materials in response to loads and digging, with respect to their flow in chutes and hoppers, or in atmospheric transport (i.e., dust) and adhesion to surfaces, is affected by gravity level. Accordingly, the following research is recommended:

- *The response of granular material to applied stress.* The knowledge gained will allow researchers to examine separately the effects of gravity and shearing using both analytical and experimental studies. Predictive models of granular deformation and flow under reduced gravity need to be developed that include the effects of particle size and shape, the effects of particle constitution, and the effects of particle agitation and of electrostatic charge, especially at low pressures.
- *Predictive models of the behavior of dust in spacecraft and extraterrestrial environments.* An understanding of this behavior will permit the reliable prediction of dust transport and deposition. An understanding is also needed of the cohesion and adhesion mechanisms that control dust attachment, where the attraction mechanism appears to be electrostatic.

OTHER CONCERNS

Reduced-Gravity Countermeasures

Because reduced or variable gravity is generally a troublesome complication of system design for HEDS (and has harmful consequences for human health), research should be carried out on means to counter the adverse effects. Such means would probably be mechanical in nature, involving rotation or vibration, and could be implemented at a range of levels, from that of the whole spacecraft down to the level of small but critical components. Design studies of structural and system problems would be required to establish technical practicality and costs for large-scale artificial-gravity concepts.

Applied research looking toward economic and effective artificial gravity should emphasize solutions that would apply to both technical and biological systems.

Research on and development of reduced-gravity countermeasures are given high priority in the report and must obviously proceed hand in hand with the microgravity research recommended elsewhere in this report, because the latter will establish the target gravity levels desired for various components and systems. In turn, the specific benefits of an artificial gravity system must be understood and weighed against the penalties (e.g., weight and cost) so that design trade-offs can be made. In other words, it is to be expected that artificial gravity will be part of integrated system designs for HEDS.

Indirect Effects of Reduced Gravity

Reduced gravity will have indirect effects on systems and components, necessitating designs different from the corresponding, more familiar ones on Earth. For example, seemingly mundane components such as piping, valves, and bearings will have to be adapted to the altered structural forces and loads in reduced- and variable-gravity environments. Then, too, products of wear and decay are presumably less easily managed in microgravity. Such concerns are additional elements in a central HEDS issue, namely the effect of reduced and variable gravity on system reliability and safety.

RECOMMENDED RESEARCH WITH A LOWER PRIORITY

The committee also recommends research on other fundamental phenomena in addition to those described above. These phenomena, listed below in no particular order, were judged to be somewhat less critical to mission success, so the research has a lower priority.

- Marangoni material parameters;
- Static equilibrium capillary shapes;
- The effect of gravity on convective condensation heat transfer;
- The effect of gravity on the heat transfer characteristics of fluid flow in porous media;
- The transport of flame suppressant to fires in reduced gravity;
- Diffusion-flame structure of fuels and flame products as affected by gravity levels;

- Flammability and flame behavior of gaseous combustible mixtures, sprays, and dust clouds; and
- The effect of gravity on nucleation and growth of solid from the melt.

PROGRAMMATIC RECOMMENDATIONS

It should be clearly understood that the committee's research recommendations deal with fundamental phenomena associated with fluid and material behavior rather than with the direct development of subsystems and their integration into technologies operable in a reduced-gravity environment. However, the committee recognizes that blending conceptual design needs and phenomenological research findings requires a great deal of communication, coordination, and interdisciplinary collaboration among designers and researchers. For this reason, it makes recommendations in this report concerning the goals, research planning, and programmatic activities of NASA that support gravity-related research for HEDS. Similar recommendations made in the phase I report (NRC, 1997) are reflected in this more extensive study as well. It was thought then, and is still believed, that in view of the long time scale needed for the evolution of basic scientific concepts into practical applications, the suggested research programs will require a sustained commitment on the part of NASA to achieve an understanding of gravity-related issues.

A Research Approach for the Development of Multiphase Flow and Heat Transfer Technology

For NASA to be able to decide whether multiphase and phase-change systems can be used and controlled in future HEDS missions, a well-focused experimental and analytical research program will be needed to develop an understanding of how multiphase systems and processes behave in reduced gravity. Since parametric full-scale testing in space is not feasible, NASA should consider developing a reliable three-dimensional, two-fluid CFD model that can be used to help design and analyze multiphase systems and subsystems for HEDS missions. The approach that has been recommended is that a reliable, physically based analytical model be developed and qualified against appropriate terrestrial and microgravity data. The resulting computational model could then be used to analyze and optimize final designs and to scale up the reduced-scale data obtained in space. While this is expected to be the most reliable, least expensive, and quickest means of developing the potentially enabling technology required by HEDS, programmatic changes would be required to accomplish this goal. In particular, it would be necessary for NASA to refocus its multiphase fluid physics research program and to be much more proactive than it has been in defining and managing the research needed to develop predictive capabilities for multiphase flow and heat transfer. In this context, NASA should investigate the possibility of consulting the U.S. Department of Energy-Naval Reactors program for help in designing research programs aimed at developing the required multivariate, physically based computational models.

Coordination of Research and Design

The NASA office responsible for microgravity research should diligently inform NASA at large about the issues of reduced gravity that are foreseen for space hardware design, so that such considerations may enter design thinking at the concept stage. It should also apprise the microgravity research community of design issues relevant to microgravity research, and NASA should encourage the blending of conceptual design and phenomenological research. This will require active communication and coordination among basic researchers and system designers and users, which should be specifically encouraged by such means as regular workshops and study groups in which both mission technologists and microgravity scientists participate.

Microgravity Research and the International Space Station

It is expected that the International Space Station (ISS) will provide a unique platform for conducting long-duration microgravity scientific research and assessing the efficiency and long-term suitability of many of the technical systems important to HEDS. The committee reiterates a recommendation from its phase I report (NRC, 1997, p. 39): in addition to carrying out basic research aboard the ISS, NASA should take advantage of the station and its subsystems, using them for testbed studies of scientific and engineering concepts applicable to HEDS technologies. In particular, the ISS can play an important role in the multiphase flow and heat transfer research program recommended above.

Peer Review for Reduced-Gravity Research

The NASA Research Announcement process and its peer review system have greatly enhanced the productivity and quality of NASA's gravity-related research. These mechanisms should be maintained as steps are taken to develop areas of science affecting HEDS technologies.

REFERENCE

National Research Council (NRC), Space Studies Board. 1997. An Initial Review of Microgravity Research in Support of Human Exploration and Development of Space. Washington, D.C.: National Academy Press.

I

Introduction

OBJECTIVES

The frontier represented by the near solar system confronts humanity with intriguing challenges and opportunities. With the inception of the Human Exploration and Development of Space (HEDS) enterprise in 1995, NASA has acknowledged the opportunities and has accepted the very significant challenges. This report was commissioned by NASA to assist it in coordinating the scientific information relevant to anticipating, identifying, and solving the technical problems that must be addressed throughout the HEDS program over the coming decades. Specifically, the committee was asked to ". . . undertake an assessment of scientific and related technological issues facing NASA's Human Exploration and Development of Space endeavor," to ". . . look specifically at mission enabling and enhancing technologies which, for development, require an improved understanding of fluid and material behavior in a reduced gravity environment . . . [and which] might range from construction assembly techniques such as welding in space, to chemical processing of extraterrestrially derived fuels and oxygen," and to ". . . identify opportunities which exist for microgravity research to contribute to the understanding of fundamental science questions underlying exploration technologies and make recommendations for some areas of directed research" (see Appendix A). This report therefore sets research priorities for that portion of microgravity research that NASA may direct toward contributing to the long-term goal of HEDS technology development. In a previous report (NRC, 1995), the committee set priorities for the microgravity research that was directed primarily at increasing basic knowledge within each discipline. It should be noted that the relative balance of resources devoted to these two categories of research will be determined by NASA priorities and funding availability and that this report does not attempt to make a recommendation in that regard.

In the current HEDS enterprise there are crucial technological challenges associated with travel within the solar system and with the long-term survival and productivity of missions and, ultimately, extraterrestrial colonies in environments quite different from those found on Earth. The similarities between these challenges and those faced in the sixteenth century by the explorers and colonizers who expanded the horizons of the world known to western Europe derive from what unifies all physical exploration. Participants must be able to carry out certain *functions* that are inevitably similar: power generation, propulsion, life support, hazards management, the ability to exploit resources encountered along the way, and so on. These functions provide us with the framework for considering the technological requirements for HEDS activities.

THE EXPLORATION ENVIRONMENT

The differences between intercontinental and extraterrestrial exploration arise from the differences in how the environment affects the systems and people required for the central functions. Whereas the environmental characteristics of a space environment may be tolerated by spacecraft, they can be lethal to humans. For example, since our nominal body temperature is 37 °C and water makes up a significant part of our mass, we cannot survive in environments with ambient pressures below 62 mb, because water would begin to boil at our body temperature. It is certainly known that humans cannot survive in the vacuum of space, and that pressure limit also means that humans cannot survive on Mars by simply wearing a breathing mask.

The successful Apollo missions to the Moon have demonstrated that humans can live and work on another planetary body, but for destinations beyond the Moon, it must be remembered that more time will likely be spent travelling in space during a round-trip mission than will be spent exploring the planetary body. Except for the period of acceleration during trajectory corrections, humans and their machines will spend hundreds of days in the microgravity environment around the Sun. Furthermore, the 1.28-second delay in communications signals between the Moon and Earth was an irritation, but the up-to-21-minute communication delay for radio transmissions between Earth and Mars makes two-way human conversation impossible and more fundamentally means that systems being operated—either in the vicinity of Mars or on its surface—must be operated autonomously.

The free-space environment in the inner solar system has already been characterized, along with many of the aspects of other planetary bodies that could be involved in future HEDS missions. Some of the important environmental characteristics are discussed below.

Radiation in Space

The National Research Council has published two studies that address the space radiation aspects of human interplanetary space travel (NRC, 1996, 1998). Without question, it will be necessary to shield human space travelers from specific types of radiation events. Shield designs must be based on detailed knowledge of the radiation characteristics and of how the various types of radiation or particle beams interact with the shielding material and with human tissue. During low solar flare activity, the space environment is characterized by the relatively constant flux of galactic cosmic radiation. Galactic cosmic radiation can emanate from any direction and consists of approximately 87 percent protons, 12 percent helium ions, and 1 percent heavier ions, with energies ranging from 100 MeV per nucleon to 10 GeV per nucleon. In addition to galactic cosmic radiation, solar particle events associated with solar flares or solar storms can produce orders-of-magnitude increases in energetic protons. The proton fluxes can be lethal to humans who are not appropriately shielded (Parker, 1997).

Eckart (1996) has discussed the circumstances under which serious radiation hazards can be produced by major solar flare events, prolonged exposure in the Van Allen belt around Earth, and prolonged exposure to (the omnidirectional) galactic cosmic radiation outside Earth's atmospheric and magnetic shields. Research is ongoing to characterize the various hazards, to assess the cumulative effects of combinations of hazards, and to develop appropriate shielding systems.

The hazards resulting from micrometeorite impacts and from collisions with space debris (in Earth orbit) should not be overlooked, but they have been studied elsewhere and do not have a major bearing on microgravity-specific research; therefore they are not covered in this report.

Planetary Bodies

To date, the vast majority of the design studies for human exploration missions have focused on the Moon and Mars. Missions to the Moon will probably involve longer stays than the Apollo missions and may include the development of a permanent human presence. Facilities that can support indefinite human stays will need to be much more reliable and will demand extra protection against the lunar vacuum conditions, as well as shielding against the major solar radiation events and meteorite impacts that must be anticipated for long-duration missions. For HEDS purposes a much wider range of systems must be operated for extended periods of time in the reduced

TABLE I.1 Characteristics of Planetary Bodies in the Solar System

Planetary Body	Mean Solar Distance	Communication Delay	Gravity Level ($\times g_0$)	Escape Velocity (km/s)	Surface Pressure (mbar)	Surface Temperature (K)
Venus	0.723 AU	2.3-14.3 min	0.880	10.3	92,000	733
Earth	(149.6 M km)	(499.0 s Sun)	(9.8066 m/s^2)	11.2	1,013.25	288 (avg)
Moon		1.28 s	0.169	2.37	Negligible	80-390
Mars	1.524 AU	4.4-21.0 min	0.380	5.0	6-10	130-300
Phobos			0.0008-0.002	0	Negligible	
Deimos			Variable	Variable	Negligible	
Jupiter	5.203 AU	35-52 min	2.640	61	Uncertain	124
Io			0.180	2.56	Negligible	
Europa			0.140	2.06	Negligible	
Ganymede			0.150	2.75	Negligible	
Callisto			0.130	2.45	Negligible	
Saturn	9.539 AU	1.18-1.46 h	1.150	37	Uncertain	95
Titan			0.140	2.64	1,496	94
Uranus	19.18 AU	2.52-2.80 h	1.170	22	Uncertain	58
Neptune	30.06 AU	4.02-4.31 h	1.180	25	Uncertain	56
Triton			0.081	0.31	0.02	38

SOURCE: Data from Chamberlin and Hunten (1987).

gravitational environment of the Moon, where the surface gravity is 1.66 m/s^2 (0.169 g_0) and the Sun shines continuously for almost 14 terrestrial days.

Mars has an atmosphere, and its day length is almost identical to that of Earth. However, it does not have a strong enough magnetic field to shield its surface from harmful solar ionizing radiation, and it is so far from Earth that two-way interactive communication and control are impractical. Its surface gravity is 3.72 m/s^2 (0.380 g_0), and in many respects, Mars would be much more habitable during extended human stays than the Moon. A variety of other planetary bodies are of interest to future HEDS missions, including Jupiter's Galilean satellites, Europa and Callisto, which are now believed to contain liquid water oceans beneath their very cold water-ice shells. Titan has a surface pressure that is greater than 1 atmosphere, but its orbit around Saturn means that communications are delayed by more than 1 hour. Geysers have been observed on Triton (Soderblom et al., 1990), but it is so far from Earth that any sort of round-trip mission will require decades for completion using current technologies. Possible planetary bodies that can be considered as targets of opportunity are listed in Table I.1, along with their average distances from the Sun, average communications delay, gravitational accelerations, and surface pressure and temperature characteristics. Current knowledge of atmospheric compositions on most of those bodies has been summarized by Chamberlin and Hunten (1987).

Asteroids and comets are very different targets of opportunity because they are potential sources of raw material for space construction and for consumables, such as water and rocket propellant, and because some of these objects are potential impact threats to Earth. The satellites of Mars (Phobos and Deimos) are thought to be captured asteroids, and it is possible that some of the satellites orbiting other planets are also captured asteroids. Saturn's rings and the more tenuous rings surrounding Jupiter and Neptune are probably composed of captured asteroid and cometary debris. Excluding these captured asteroids and comets, which are less attractive sources of raw material and which pose no threat to Earth, there remain a huge number of objects that could be studied and exploited as part of a long-term HEDS program.

Gradie et al. (1989) have presented an overview of the location, size, and compositional distribution of the known asteroids. The vast majority of observed asteroids have orbital semimajor axes between the orbits of Mars and Jupiter, and the main asteroid belt is considered to be between 2.1 and 3.3 AU. There are 14 classes of asteroids, and although classifications are based on combinations of albedo and spectral features (Tholen and Barucci, 1989) and do not lead directly to compositional or evolutional conclusions, they do correlate strongly with orbital characteristics and size. Asteroids with diameters of 25 km or less are an order of magnitude more prevalent than asteroids with diameters of 200 km or more. Davis et al. (1989) have argued convincingly that the larger asteroids will be covered with rubble resulting from the gravitational reaccumulation of debris, after collisions, whereas the smaller asteroids lack the gravitational pull needed to recapture ejected material. Furthermore, the vast majority of asteroids are rotating (tumbling) at three or four revolutions per day. Lewis et al. (1993) have examined the near-Earth asteroids (NEAs), and they estimate that there are more than 70,000 NEAs with diameters greater than 100 m. Since these objects are irregular in shape, they have variable surface gravities (assuming Phobos and Deimos are indicative of NEA irregularities, since those bodies have mean ellipsoidal radii of $13.3 \times 11.1 \times 9.3$ km and $7.5 \times 6.2 \times 5.4$ km, respectively, and their surface gravitational accelerations vary locally between 0.0008 g_0 and 0.002 g_0). The variety of NEA types range from ordinary chondrites, which are silicate-dominated, primitive, unmelted and undifferentiated materials (approximately 88 percent of meteorite falls) to stony irons and nearly pure iron-nickel-cobalt alloys (see Lewis and Hutson, 1993).

Comets are the most numerous objects in the solar system, with more than 10^{11} nuclei residing in the Oort cloud, whose aphelia are located between 20,000 and 70,000 AU from the Sun (Oort, 1990), and in all probability they represent the most primitive compositions in the solar system. Based on observations of the Giacobini-Zinner and Halley comets, it is now believed that most comets have nuclei composed of between 80 and 90 percent mixtures of (ortho- and para-) water ice, confirming the model of Whipple (1950, 1951). Comet nuclei are not spherical and may tend more toward prolate shapes with aspect ratios of 2:1; however, the surveyed population is too small to permit generalization. The nuclei are known to be inhomogeneous because of the jetlike particle ejections that occur during close approach to the Sun. Huebner and McKay (1990) have discussed the high concentrations of organic molecules and their implications for the formation of life. The chemical reactions that are sustained by cometary trajectories passing near the Sun appear to explain why there is a higher concentration of organic molecules in the outer solar system than in the inner solar system. The majority of short-period comets (orbital periods of less than 200 years) that have been studied (Halley, Arend-Rigaux, Neujmin 1, Schwassmann-Wachmann 1, Tempel 2, Encke, IRAS-Araki-Alcock, and Chiron) have effective diameters on the order of 10 km, and their estimated densities are somewhat less than water ice, suggesting that they are "fluffy."

Although Mars and the Moon are the focal points of current HEDS mission design studies, this study has attempted to include other targets of opportunity such as those identified in Table I.1 and certain comets and asteroids that might be selected in the future. This complementary set of planetary bodies can be explored or exploited as part of an expanded HEDS program, but more importantly it provides guidance on the range of gravitational environments that will be encountered. The influences of reduced gravity on systems that have been designed and tested on Earth to be operated on the Moon or Mars are often subtle but in many cases are predictable using straightforward scaling parameters. On the other hand, systems that are to be operated aboard manned spacecraft or on surfaces such as those of the moons of Mars, an asteroid, or a comet will be subjected to gravitational forces that are often highly variable and where steady acceleration levels are so low that the importance of the basic transport processes controlling comparable terrestrial hardware designs can be either negated or overwhelmed by other phenomena.

Microgravity refers to acceleration environments that are small compared with those found on Earth's surface. This definition includes the steady gravitational environments found on the surfaces of planetary bodies such as the Moon and Mars, where the gravity level is a significant fraction of Earth's gravity, along with the highly unsteady, near-zero acceleration environments that exist on spacecraft. Both types of microgravity environment are integral to HEDS missions involving extraterrestrial bodies, but they represent very different areas of microgravity research. Furthermore, the word *microgravity* in the literature refers mainly to gravity levels that are less than 0.01 g_0, even though the formal definition is broader. The purpose of this report is to identify and determine priorities for microgravity research, in its broader sense, that support the HEDS program. However, since it is not possible to use the broad definition of microgravity in the body of this report without generating confusion, the term *fractional gravity* is used to refer to gravity environments that are less than 1 g_0 but greater than the much lower gravitational levels found on spacecraft or on the surfaces of small planetary bodies such as Phobos and Deimos. The term *microgravity* is used to refer to these near-zero gravitational levels.

REPORT ORGANIZATION AND DEVELOPMENT

The overriding challenge posed by the charge to the committee is twofold. Two diverse groups of objects—the engineering subsystems necessary for HEDS functions and the microgravity-specific physical phenomena that affect them—must first be identified. Then, dependencies between the two groups must be adduced, and questions derived about the fundamental scientific obstacles to the enabling technologies. Once these questions have been derived, the task boils down to one of enumerating as many as possible of the known (and possibly unknown) problem areas in which further scientific inquiry can be expected to produce solutions.

Previous NRC reports have dealt with exploiting microgravity as a parameter in pursuit of new basic science (NRC, 1992, 1995). Priorities were assigned to microgravity research topics primarily on the basis of their potential to expand basic knowledge within a given discipline. To meet the goals of this current study an additional step was needed: the committee first examined the technological barriers and only then asked what scientific research would level them.

The organization of this report reflects this approach (Figure I.1). The committee first lists in some detail the technological requirements for HEDS (Chapter III), coming to the underlying scientific phenomena only after assembling this list. Chapters III and IV constitute the heart of the report. They are essentially parallel constructions: the first (Chapter III) identifies the HEDS technologies and describes their microgravity-dependent *subsystems*, and the second (Chapter IV) identifies the physical bases for microgravity sensitivity and describes the associated *phenomena*. A logical symmetry between Chapters III and IV arises from the links between them: microgravity sensitivities generate potential failure modes in the subsystems; the need to improve subsystem efficiency drives the generation of relevant new scientific questions associated with microgravity-induced sensitivity. Several important issues that lie outside the purview of this parallelism are developed in Chapter V. Finally, having derived the scientific questions in this manner, the committee makes research recommendations in Chapter VI and programmatic recommendations in Chapter VII.

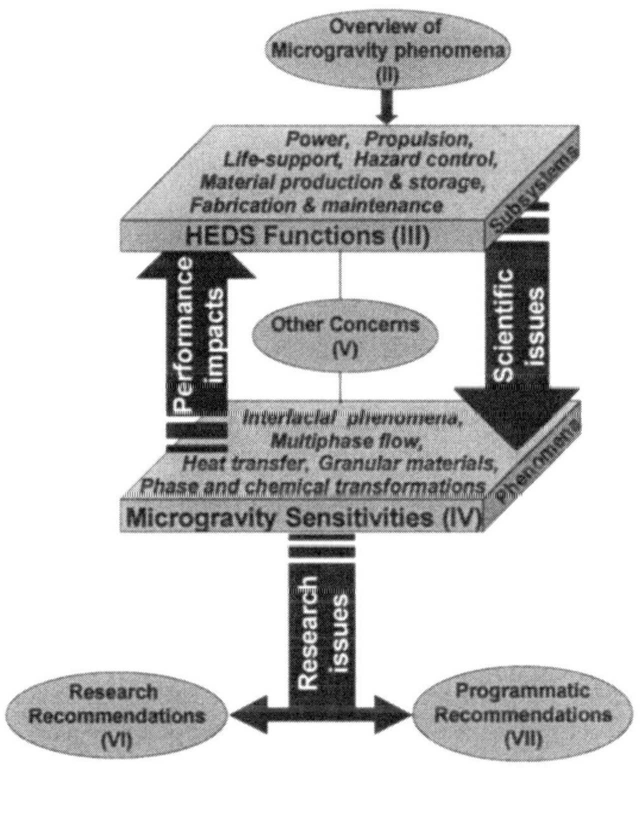

FIGURE I.1 Structured map of the logical dependencies between material presented in different chapters of this report.

In developing the material for this report the committee first considered the broad capabilities that the HEDS program must have in order to succeed. These were judged to be the following:

- Power generation and storage,
- Space propulsion,
- Life support,
- Hazard control,
- Material production and storage, and
- Construction and maintenance.

Certain important functions, such as communications, were excluded because gravity plays no significant role. Within each of the above functions, various enabling technologies were considered, and these are discussed in Chapter III. Some of the technologies were highly speculative since the committee did not rule out NASA's far-term needs from its considerations. These technology systems were then broken down into *subsystems* (or in some cases *processes*), which is the level at which most microgravity effects are expected to intervene. This last was a critical step because it was impractical for the committee to consider in detail all of the systems that have been suggested for use in HEDS programs or to predict the new systems that might be developed in the next few decades. However, many basic subsystems, such as radiators, show up repeatedly in a wide range of technologies, and their use is likely to continue in future technologies. Therefore, by examining a diverse, but not comprehensive, set of systems, the committee was able to identify the most commonly occurring subsystems and consider how their performance might be affected by reduced gravity.

Of the various phenomena known to become problematic in microgravity and considered in Chapter IV, one phenomenon underlies nearly all the issues the committee identified as problem areas worthy of research investment. This phenomenon is the profound effect of gravitation on the separation of distinct phases, particularly fluid phases. Terrestrial gravity fields provide reliable separations based on density differences; in the near absence of gravitational forces, phases of different density do not spontaneously separate. The lack of phase separation in microgravity has severely compromised a range of promising technologies associated with all HEDS functions, from propulsion to sanitation. An inability to reliably predict multiphase behavior in microgravity has led NASA to utilize single-phase systems even though functioning two-phase systems would have substantially enhanced efficiency.

Chapter V addresses the indirect effects of microgravity on system design, various ways to counter microgravity, and also certain program issues such as experimental design matrices and the development of physics-based predictive models and probabilistic risk assessment.

The concerns developed in Chapters III through V are integrated into a discussion of research recommendations and priorities in Chapter VI. There, the committee assesses how the technical problems described in Chapter III can best be solved by addressing the underlying physical phenomena outlined in Chapter IV.

Finally, in Chapter VII, important programmatic issues are addressed that are critical to the conduct of the recommended research.

REFERENCES

Chamberlin, J.W., and D.M. Hunten. 1987. Theory of Planetary Atmospheres, 2nd Ed. New York: Academic Press.

Davis, D.R., S.J. Weidenschilling, P. Farinella, P. Paolicchi, and R.P. Binzel. 1989. Asteroid collisional history: Effects on sizes and spins. Pp. 805-826 in Asteroids II. R.P. Binzel, T. Gehrels, and M.S. Matthews, eds. Tucson: University of Arizona Press.

Eckart, P. 1996. Spaceflight Life Support and Biospherics. Torrance, Calif.: Microcosm Press, and Dordrecht, Netherlands: Kluwer Academic Publishers.

Gradie, J.C., C.R. Chapman, and E.F. Tedesco. 1989. Distribution of taxonomic classes and the compositional structure of the asteroid belt. Pp. 316-335 in Asteroids II. R.P. Binzel, T. Gehrels, and M.S. Matthews, eds. Tucson: University of Arizona Press.

Huebner, W.F., and C.P. McKay. 1990. Implications of comet research. Pp. 305-331 in Physics and Chemistry of Comets. New York: Springer-Verlag.

Lewis, J.S., and M.L. Hutson. 1993. Asteroidal resource opportunities suggested by meteorite data. Pp. 523-542 in Resources of Near-Earth Space. J. Lewis, M.S. Matthews, and M.L. Guerrieri, eds. Tucson and London: University of Arizona Press.

Lewis, J.S., M.S. Matthews, and M.L. Guerrieri, eds. 1993. Resources of Near-Earth Space. Tucson and London: University of Arizona Press.

National Research Council (NRC), Space Studies Board (SSB). 1992. Toward a Microgravity Research Strategy. Washington, D.C.: National Academy Press.

NRC, SSB. 1995. Microgravity Research Opportunities for the 1990s. Washington, D.C.: National Academy Press.

NRC, SSB. 1996. Radiation Hazards to Crews of Interplanetary Missions: Biological Issues and Research Strategies. Washington, D.C.: National Academy Press.

NRC, SSB. 1998. A Strategy for Research in Space Biology and Medicine in the New Century. Washington, D.C.: National Academy Press.

Oort, J.H. 1990. Orbital distribution of comets. Pp. 235-244 in Physics and Chemistry of Comets. New York: Springer-Verlag.

Parker, E.N. 1997. Mass ejection and a brief history of the solar wind concept. Pp. 3-27 in Cosmic Winds and the Heliosphere. J.R. Jokipii, C.P. Sonett, and M.S. Giampapa, eds. Tucson: University of Arizona Press.

Soderblom, L.A., S.W. Kieffer, T.L. Becker, R.H. Brown, A.F. Cook II, C.J. Hansen, T.V. Johnson, R.L. Kirk, and E.M. Shoemaker. 1990. Triton's geyser-like plumes: Discovery and basic characterization. Science 250:410-415.

Tholen, D.J., and M.A. Barucci. 1989. Asteroid taxonomy. Pp. 298-315 in Asteroids II. R.P. Binzel, T. Gehrels, and M.S. Matthews, eds. Tucson: University of Arizona Press.

Whipple, F.L. 1950. A comet model I: The acceleration of comet Encke. Astrophys. J. 111:375-394.

Whipple, F.L. 1951. A comet model II: Physical relations for comets and meteors. Astrophys. J. 113:464-474.

II

Brief Descriptions of Phenomena Important in Reduced Gravity

Chapter III surveys a set of HEDS-enabling technologies for the purpose of identifying critical underlying physical phenomena that are affected by, or that play a dominant role in, reduced gravity. These phenomena, and the research needed to extend our knowledge of them, are discussed in Chapter IV. The purpose of Chapter II is to briefly introduce these phenomena, along with associated concepts such as scaling, so that reference to them in the following chapters will be intelligible to a wider range of readers.

Because of the equivalence between force and mass times acceleration, many of the dimensionless groups used to generalize design data may be expressed as force ratios that contain gravitational acceleration. The consequences of local variations in Earth's gravitational acceleration can be accounted for easily through the scaling that is inherent in the relationships among the various dimensionless groups, and it is expected that for the gravitational accelerations that exist on the surfaces of the Moon and Mars, such known scaling laws are adequate for describing a number of phenomena. (For this reason most HEDS technology problems that need to be addressed by fundamental research on basic transport phenomena are likely to occur at near zero gravity rather than at fractional gravity levels.) As the gravitational level decreases below that of Earth, the Moon, and Mars, as may obtain in the space environment, the magnitudes of the forces associated with differences in density may become small compared with the forces associated with other phenomena, which then may become significant and even dominant. In some cases, the emergence of other phenomena is predictable as gravitational levels are reduced, while in other cases, the influence of competing phenomena may be unexpected, and the engineering systems needed for successful HEDS developments may then be compromised. There are a number of phenomena that may play a significant role in such systems.

Interfacial phenomena refer to effects caused by the presence of an interface or bounding surface between two different thermodynamic phases, i.e., a solid, liquid, or gas. The defining characteristic of an interface is the work required to create a unit area of the interface, which is referred to as the free energy per unit area or (for a liquid) the surface tension of the interface. Many of the static forms assumed by liquids on a small scale reflect the tendency of the multiphase system to minimize the total interfacial free energy, which in simple cases is equivalent to minimizing the total interfacial area. This accounts for the spherical shape of isolated droplets and bubbles and other more complicated configurations of liquids in contact with a solid of prescribed shape, as occurs in the storage of liquid fuels. These shapes, both static and dynamic (e.g., ripples on a liquid surface), are discussed in Section IV.B. In addition, surface tension regulates wetting phenomena, as illustrated by the flow of solder on a work piece. Also, variations in surface tension over a liquid surface, arising from temperature or composition gradients, drive special flows referred to as Marangoni flows. Surface tension is not affected by gravity level, but surface tension as a driving force for the configuration and motion of liquids becomes increasingly important as the gravity level is reduced. For this reason, understanding and utilizing interfacial phenomena are key to the control and management of liquids in reduced gravity.

Multiphase flow, described in Section IV.C, refers to the flow of more than one fluid phase in a system of pipes, pumps, and phase-change components; an example is the flow of vapor and liquid in a power-generating system. In Earth gravity, phases tend to stratify depending on their density, with the denser phase flowing or collecting in the bottom of a pipe or component, whereas in microgravity there is no driving force for stratification. Various flow regimes occur corresponding to different configurations of the two phases, including bubbles of gas in a liquid, droplets of liquid in a gas, and the annular flow of a film of liquid along the walls of a conduit with gas and entrained liquid droplets flowing in a central core. Flow-regime-specific predictive models of multiphase flow are essential for the design of multiphase systems for operation in microgravity, including power-generating systems that depend on the flow of more than one phase, as in the case of a Rankine cycle power plant (typified by a steam power plant on Earth). Rankine-cycle-based systems are attractive because of the efficient heat transfer due to phase change during evaporation and condensation.

Flow in porous media is exemplified by the movement of liquids in the wicks used in heat pipes and other thermal management devices based on capillarity, the flow of fluids in packed and fluidized beds used in chemical reactors, and the flow of nutrients and gases in soils used for plant cultivation in space. Gravity, pressure gradients, and capillarity (i.e., surface tension forces) are the primary driving forces for flow, and gravity becomes less important as it is reduced. The behavior of the flow under these conditions is poorly understood, as discussed in Section IV.C.

Heat transfer refers to the flux of heat across a boundary or interface by conduction, convection, or radiation and is central to the operation of power and propulsion systems and to the management of the energy budget in a spacecraft. Of these processes, convection (with or without phase change) driven by thermally induced density differences is strongly dependent on gravity level and hence is the focus of Section IV.D. In this case, the transferred heat is conveyed to (or from) the bounding surface from (or to) another part of the system by a moving fluid or fluids. If the fluid is single-phase, as in the case of a Brayton cycle, then thermally induced buoyancy convection is diminished in reduced gravity and absent in microgravity. If the fluid is two-phase, as in the case of a Rankine cycle, then all the issues of multiphase flow and heat transfer referred to above are present. In addition, the process of heat transfer at the walls of a condenser or evaporator, where the phase change occurs, is critically affected by gravity level. In microgravity, surface tension or capillary forces play a dominant role in the degree and dynamics of the wetting of the walls by the liquid phase.

In systems using multiphase flow, system effects on a global scale may emerge from the interaction among the components of the network. In particular, the phase or time lags in feedback loops can cause potentially dangerous instabilities that are only revealed by analysis of the system as a whole. These effects are discussed in Section IV.C.

Solidification of a melt occurs in such processes as casting, welding, and liquid-phase sintering. It is affected by gravity level, as discussed in Section IV.E, because of buoyancy-induced convection in the liquid phase that, in turn, affects the distribution of solutes, foreign particles, and bubbles in the liquid as it freezes. This in turn affects the grain structure, porosity, and impurity distribution (i.e., the microstructure) of the solidified or cast material. The microstructure and properties of a material solidified in microgravity differ markedly from those of a material solidified in Earth gravity (Curreri and Stefanescu, 1988).

Combustion, either intentional in power-generating and propulsion systems or unintentional in accidental fires, is influenced by gravity level because of the large density differences between the ambient gas and the hot product gases. Flame structure is drastically altered from the upward convective form in Earth gravity to a quiescent spherical form controlled by diffusion in microgravity. The effects of gravity on such combustion characteristics as flammability, smoldering, and flame spread are discussed in Section IV.F.

Granular mechanics includes such topics as the angle of repose of piles of granular material, the response of granular media and soils to loads and to digging, and the flow of granular materials in chutes and hoppers. Gravity affects the degree of compaction of soils and provides the driving force for flow. In addition, the behavior of granular materials is affected by the size distribution, shape, and physical constitution of the grains and by environmental factors such as a hard vacuum, which favors electrostatic charge buildup on grains in a deformed or flowing bed. These and other topics are discussed in Section IV.G.

REFERENCE

Curreri, P.A., and D.M. Stefanescu. 1988. Low-gravity effects during solidification. P. 147 in Metals Handbook: Casting, 9th Ed., Vol. 15. Metals Park, Ohio: ASM International.

III

Survey of Technologies for the Human Exploration and Development of Space

This chapter provides the essential foundation for subsequent discussion of microgravity phenomena and the determination of related research needs important for HEDS. That foundation consists of rather detailed descriptions and assessments of the various technologies that seem most important for HEDS systems. In the committee's view, these are the types of technologies that must operate reliably and efficiently in the various space environments of interest.

Two important considerations were involved in the selection of technologies for discussion. First, it was clear that this report could not usefully embark on studies of systems and mission architectures, with their specific design problems, based on premature mission assumptions. Secondly, the range of technologies identified needed to be broad enough to cover reasonable possibilities for incorporation into future systems but did not have to cover all conceivable possibilities. Therefore, this report emphasizes technologies that have a wide range of potential applications and that are expected to be significantly influenced by gravity level.

In this chapter, the technologies selected for discussion are grouped according to their probable functions in the HEDS program. Since some are quite well developed already, while others exist only as concepts, the level of detail varies considerably. Systems to serve HEDS functions are identified first, followed by their components or subsystems. Especially at the subsystem level, microgravity concerns are identified and summarized in a table for each function.

Tables III.G.1 to III.G.3 at the end of this chapter relate microgravity phenomena to the various subsystems and processes. However, the phenomena are not treated in detail, nor are relevant research issues described. Rather, in Chapter IV, the identified microgravity concerns are related to physical phenomena, and physical research areas are identified that may provide the knowledge base needed to design systems and components that will be reliable and effective in the microgravity environments of interest.

III.A POWER GENERATION AND STORAGE

Introduction

Future HEDS missions for the exploration and colonization of the solar system will require enabling technologies for adequate, reliable electrical power generation and storage. Advanced, high-efficiency power generation and storage will be required for deep-space missions, lunar and planetary bases, and extended human exploration.

Extensive up-to-date discussions are available (NRC, 1987, 1998; Bennett et al., 1996; Brandhorst et al., 1996; Detwiler et al., 1996; Bennett, 1998) and there is no need to repeat them here. Because the electrical power technology requirements for spacecraft are similar to the requirements for the extended human occupation of the Moon or Mars (when energy demand is not constant) they are discussed together.

Space propulsion is, of course, the dominant and limiting power-generation requirement for HEDS. However, due to the wide range of systems that must be considered, propulsion is discussed separately in Section III.B. Many of the means of power generation applicable to spacecraft and station power discussed in this section are also applicable to propulsion.

For the purpose of the present discussion, the primary energy sources for conversion to electrical power on a spacecraft are the following: (1) solar radiation, (2) chemical and electrochemical, and (3) nuclear (radioisotope thermoelectric generators (RTGs), dynamic isotope power (DIP) sources and fission and fusion power). The choice of energy source and power-generation system and subsystem is dictated largely by the mission requirements. These energy sources can be utilized in open or closed thermodynamic systems. A closed-cycle system is one in which a working fluid is heated, does work, and is recycled (Figure III.A.1); an open-cycle system is one in which a working fluid is heated, does work, and is discharged, carrying waste heat with it. The electric power generated requires a power management and distribution system that includes regulators, converters, control circuits, etc. (Figure III.A.2). Energy storage devices may also be required, since some energy sources (e.g., solar radiation) are not continuous.

The power-to-mass ratio (in kW_e/kg) of the power system is an important consideration for space missions. Small versus large power needs and autonomous versus manual control are additional factors. The fact that electrical power generation and onboard propulsion subsystems can account for one-half to three-fourths of the mass of the typical Earth-orbiting satellite or planetary spacecraft provides the motivation to reduce their mass, which would allow more of the spacecraft's mass to be devoted to payload. The desire to reduce costs and maintain reliable performance has led to the consideration of both some old and some new technologies for electric power generation; these technologies are reviewed and discussed in detail in Brandhorst et al. (1996), Detwiler et al. (1996), and Landis et al. (1996).

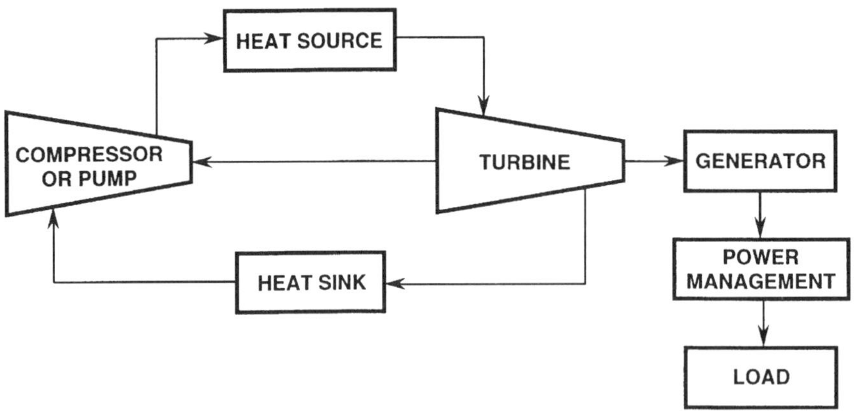

FIGURE III.A.1 Generic diagram of a simplified closed-cycle space power system. For a Brayton cycle, the heat source is a heat exchanger where heat from the source is added to the working gas, and the heat sink is also a heat exchanger (i.e., space radiator) where the working fluid is cooled. For a Rankine cycle, the heat source is a boiler where the working fluid is boiled, and the heat sink is a heat exchanger (i.e., space radiator) where the working fluid is condensed.

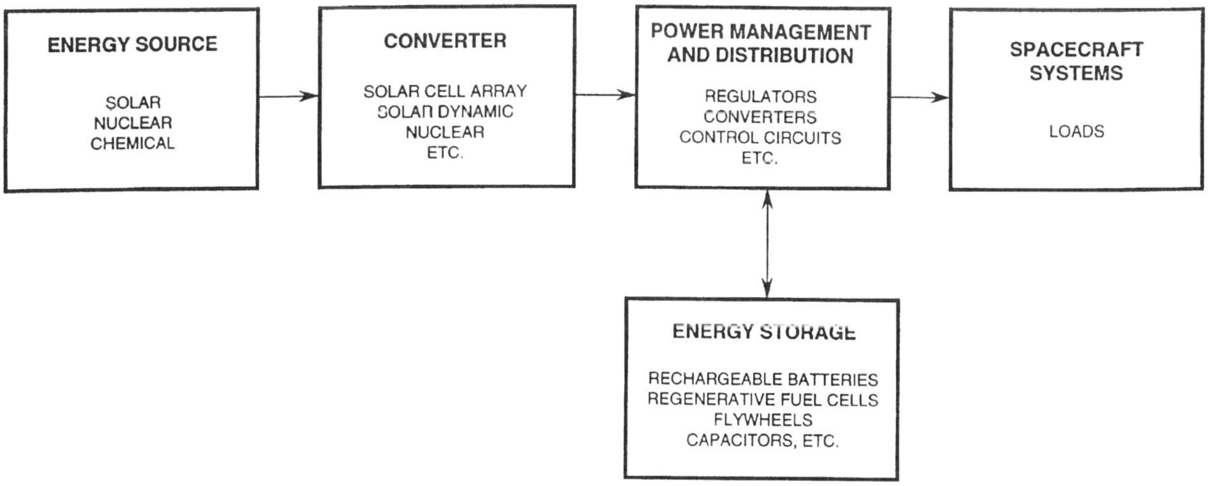

FIGURE III.A.2 Schematic of a generic electric power system. Based on Bennett (1998).

Gravity is an important consideration in active (thermal) subsystems for power generation. Many of these subsystems involve single and/or multiphase fluid and thermal management. Such important subsystems as boilers, condensers, evaporators, heat exchangers, normal and cryogenic fluid storage units, fuel cells, radiators, and heat pipes involve fluid flow and/or transport phenomena, including heat and mass transfer, phase separation, and others. Because fluid flow and transport phenomena are affected by gravity, a full understanding of the phenomena is needed for the design of the systems and for their safe and efficient operation in microgravity or reduced-gravity environments.

Power Generation Systems

Solar Power Systems

The principal solar/electric power systems are of two types: passive (i.e., photovoltaic or photoelectric) and active (thermal). In the literature (Bennett, 1998) the former is referred to as static and the latter as dynamic. The solar cell arrays used on most spacecraft usually consist of a large number of cells that convert a fraction of the solar radiation incident on them to electricity by means of the photoelectric effect. The solar cells are connected into appropriate series/parallel circuits to produce needed power at required voltages and currents. On the recent Mars Pathfinder mission, the lander, the Sojourner, and the cruise system were all powered by gallium-arsenide solar cells. This was the first use by NASA of solar/photovoltaic power on the surface of Mars. Recent discussions of the advances in solar cell technology are available (Landis et al., 1996; Bennett et al., 1996; Bennett, 1998). There are two ways of using solar energy: directly, by thermoelectric means, or indirectly, using solar radiation to heat a working fluid, which then drives a turbine/alternator (generator). The former method simply requires thermoelectric elements placed at the focus of a concentrator. While simple to construct, the power density is low and the system has not been used by NASA to power a spacecraft. Moreover, solar power will be of limited value for deep space missions and may be unreliable at extraterrestrial sites, e.g., Mars.

Indeed the use of photovoltaic solar cell arrays will be restricted to spacecraft that do not travel beyond the Mars orbit. This is the case primarily because the solar irradiation (insolation) decreases as a square of the distance from the Sun. The collectors/concentrators would have to be larger, and this would prohibitively increase the mass of the propulsion subsystem. Ionizing radiation, low intensity, and low- and high-temperature (for the solar probe) degradation effects are other problems that must be addressed in the use of solar cell arrays for photovoltaic electric power generation (Bennett, 1998). Solar/thermal dynamic systems can overcome some of the problems, such as radiation damage to solar cells and the limited life of chemical energy storage systems, but there are issues with the technology that have not yet been solved. The need for moving components and the reliability of parts of solar dynamic systems are some of the viability issues that have not yet been addressed.

The active method of using solar radiation to produce electric power is to heat a working fluid that can drive a turbine/generator in much the same way as is done in the electric power industry. To this end, a large number of solar collectors can be used to convert a fraction of the incident solar radiation to sensible and/or latent energy of the fluid. Heat is transferred from the structural collector elements to the fluid by forced convection and/or flow boiling. The energy collected can then be stored in a thermal energy storage device for use when insolation is not available. The component technologies for active power conversion for both the Brayton and Rankine cycles are fully mature. However, waste heat rejection in Brayton and Rankine thermodynamic power-generation cycles is of major concern, because space radiators represent a significant portion of the weight of the system. NASA's Glenn Research Center has performed the first full-scale demonstration of a complete space-configured 2-kW solar active system based on the Brayton cycle in a relevant space environment. However, one of the criticisms that has been leveled at solar/thermal active systems is that the conversion systems depend on moving parts that are considered to be intrinsically less reliable and shorter-lived than those in photovoltaic conversion systems. In summary, solar power systems may not be feasible for many deep-space missions, lunar and planetary bases, and extended human exploration missions or for powering high-thrust, high-efficiency propulsion systems (NRC, 1998).

Chemical Power Systems

Power systems based on chemical energy sources include batteries and fuel cells. Storable chemical reactants (e.g., nitrogen tetroxide, mixed amines, hydrogen, oxygen, and other chemicals) can be stored aboard spacecraft or on the surface of Mars for power generation using a mass open or closed thermodynamic system. Considerable technology relevant to this application is available from the Apollo program. The principal unknown in using chemical reactants to produce electric power is the ability of the spacecraft systems to tolerate the effects of any chemical effluents that are released. Although chemical energy sources appear attractive because they offer rapid response, as Figure III.A.3 illustrates, chemical energy sources for electric power generation are suitable only for short-duration functions and/or missions. Also, as can be seen from Figure III.A.3, a fundamental shortcoming of the chemical energy sources for power generation is that the mass of the chemical reactants becomes prohibitive for burns and/or missions of long durations (NRC, 1987; Bennett, 1998). Stored chemical energy could be used to meet short-duration peak or emergency power demand, but for long-duration missions, solar or nuclear energy sources with energy conversion technologies based on a Brayton or Rankine cycle power-generation system would be the most suitable.

Fuel cells are electrochemical devices that convert the chemical energy of a fuel directly into electrical and thermal energy (Blomen and Mugerwa, 1993). In a typical fuel cell, gaseous fuels are fed continuously to the anode (negative electrode) compartment and an oxidant, e.g., oxygen or air, is fed continuously to the cathode (positive electrode) compartment. The electrochemical reactions take place at the electrodes to produce an electric (direct) current. The fuel cell theoretically has the capability of producing electrical energy for as long as the fuel and oxidant are fed to the electrodes. In reality, degradation or malfunction of components limits the practical operating life of fuel cells. Besides directly producing electricity and having the capacity to serve as energy storage devices, fuel cells also produce heat and water. The heat can be utilized effectively for the generation of additional electricity or for other purposes, depending on the temperature. A practical consideration for fuel cells is their compatibility with the available fuels and oxidants. For HEDS missions, at least four applications of fuel cells are possible: (1) electric power generation in a space vehicle or at an extraterrestrial site, (2) surface transport on Mars or the Moon, (3) production of oxygen (O_2) from carbon dioxide (CO_2) on Mars, and (4) production of potable water for life support.

One of the main attractive features of fuel cell systems is the expected high fuel-to-electricity efficiency and the fact that they can also be used as storage devices. This efficiency, which runs from 40 to 60 percent based on

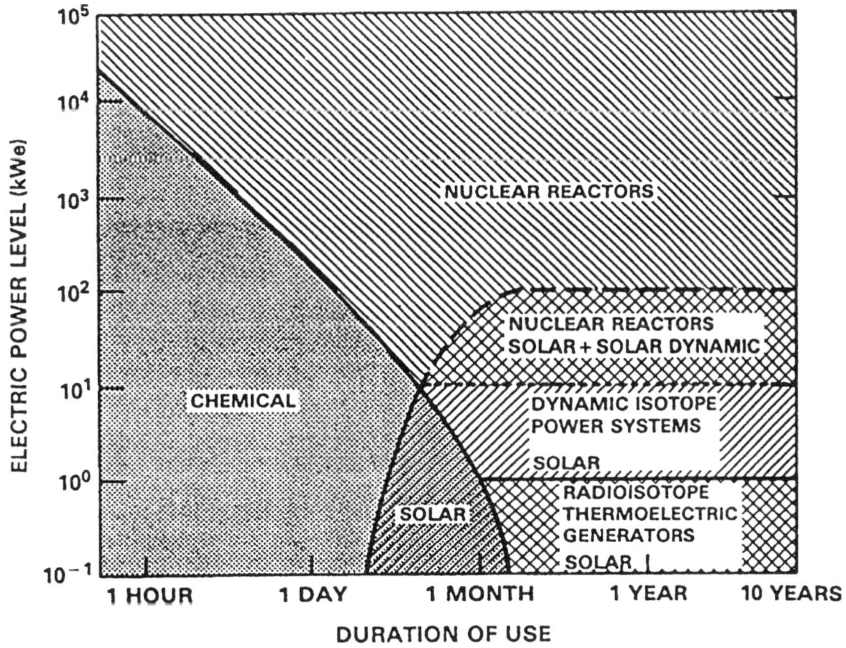

FIGURE III.A.3 Qualitative diagram illustrating the regimes of applicability of various space power systems. Courtesy of Gary L. Bennett, Metaspace Enterprises.

the lower heating value (LHV) of the fuel, is higher than that of almost all other energy conversion systems. In addition, high-temperature fuel cells produce high-grade heat, which is available for cogeneration applications. If waste heat is utilized, the theoretical efficiency can reach 80 percent (Klaiber, 1996). Because fuel cells operate at near-constant efficiency, independent of size, small fuel cells are nearly as efficient as large ones. Thus, fuel cell power plants can be configured in a wide range of electrical power levels from watts to megawatts. Fuel cells are quiet and operate with virtually no noxious emissions, but they are sensitive to certain fuel contaminants, e.g., carbon monoxide (CO), hydrogen sulfide (H_2S), ammonia (NH_3), and halides, depending on the type of fuel cell. Thus, the contaminants must be minimized in the fuel gas.

Fuel cells have been identified by the National Critical Technologies Panel as one of the 22 key technologies the United States must develop and implement in order to achieve economic prosperity and maintain national security (National Critical Technologies Panel, 1993). A variety of fuel cells have been developed for terrestrial and space applications (Blomen and Mugerwa, 1993). Fuel cells are usually classified according to the type of electrolyte used in the cell: alkaline fuel cells (AFCs), phosphoric acid fuel cells (PAFCs), molten carbonate fuel cells (MCFCs), solid oxide fuel cells (SOFCs), and proton-exchange membrane fuel cells (PEMFCs). The operating temperature ranges from ~80 °C for PEMFCs to ~1000 °C for SOFCs (Kroschwitz and Bickford, 1994). The physicochemical and thermomechanical properties of materials used for the cell components (e.g., electrodes, electrolyte, bipolar separator, and current collector) determine the practical operating temperature and useful life of the cells. The properties of the electrolyte are especially important. Solid polymer and aqueous electrolytes can be used only at ~200 °C or lower because of high water-vapor pressure and/or rapid degradation at higher temperatures. The operating temperature of high-temperature fuel cells is determined by the melting point for MCFCs or the ionic-conductivity requirements for SOFCs of the electrolyte. The operating temperature dictates the type of fuel that can be utilized. Interfacial and transport (flow, heat, mass, charge) phenomena in the membranes (porous media) under reduced or microgravity conditions are also important issues in the design and safe operation of fuel cell systems.

Nuclear Power Systems

Nuclear energy sources for power generation come in three types: radioisotope, fission reactor, and fusion reactor. Since sustained fusion has not yet been demonstrated in a laboratory and no reactors are likely to be available—even for terrestrial applications—until well into the twenty-first century, this type of nuclear reactor is not considered. An up-to-date discussion of nuclear power technology for spacecraft applications is available (Bennett, 1998; NRC, 1987). Suffice it to summarize that since 1961, the United States has flown 44 radioisotope thermoelectric generators (RTGs) and one nuclear fission reactor (see below) using thermoelectric conversion to provide power for 25 space systems. The Galileo mission to Jupiter, the Ulysses mission to explore the polar regions of the Sun, and the Saturn-bound Cassini mission are powered by RTGs operating at 1000 °C (Bennett et al., 1995; Bennett, 1998). For example, the Cassini spacecraft was developed and launched in October 1997 on a mission to investigate Saturn and its rings, satellites, and magnetosphere. It is powered by three RTGs. RTGs have been used by NASA for many years, and this technology is mature and reliable. It is not sensitive to gravity; however, it is currently limited to relatively low power levels (see Figure III.A.3).

The United States has flown one space nuclear fission reactor (SNAP-10A), which was launched in 1965 and provided 500 W of electric power. SNAP-10A was a liquid-metal-cooled nuclear reactor with a thermoelectric conversion. A ground-test twin of the flight version of SNAP-10A operated unattended for over a year, demonstrating the feasibility of the fission nuclear reactor. According to reports (Bennett, 1998), the former Soviet Union launched perhaps as many as 33 low-power (~1 to 2 kW_e) nuclear fission reactors from 1967 to 1988 to power its radar ocean reconnaissance satellites. All of the reactors used thermoelectric elements to convert thermal energy to electricity.

Fission nuclear reactors can be characterized as having a very good power-to-weight ratio. An example of a reactor designed for space use is the reactor that was being worked on for SP-100. Jointly undertaken by the Department of Energy (DOE), the Department of Defense (DOD), and NASA, the SP-100 program had the goal of developing a space nuclear reactor technology that could support a range of projected missions, including nuclear electric propulsion and planetary surface operations. Several systems were designed for a power output of 100 kW_e and incorporated a high-temperature, liquid-metal-cooled reactor. One design concept used an inert-gas Brayton cycle with a turbine generator, while another was designed for use with an advanced thermoelectric converter. Most of the nuclear component development had been completed on SP-100 before the project was cancelled in 1994. Currently the United States has no useful space nuclear reactor program, even though recent studies continue to show (AIAA, 1995; NRC, 1998; Friedensen, 1998) that human exploration of the Moon and Mars will require this technology. To quote a recent National Research Council report (1998, p. 19): "The committee is well aware that political constraints may make R&D on advanced space nuclear power systems unpopular. However, the committee could not ignore the fact that space nuclear power will be a key enabling technology for future space activities that will not be able to rely on solar power."

Energy Storage

Reliance on intermittent (i.e., solar) energy sources as a primary means of electrical power generation requires some method of energy storage. The storage methods may be chemical (primary and secondary—rechargeable—batteries, and primary and regenerative fuel cells), electrical (capacitors), mechanical (flywheels, gravitational liquid or solid), or thermal (latent or sensible heat). The design and performance of storage systems are judged on lifetime, reliability, safety, efficiency, and specific energy. For example, the two principal systems that are being used or being considered are nickel-based and lithium-based batteries. An up-to-date review of storage systems used or under development for spacecraft applications is available (Bennett, 1998). In summary, great progress is being made by both NASA and DOD on a range of battery technologies that promise improvements by a factor of 10 in specific energy over the old nickel-cadmium batteries. Lithium-based batteries (i.e., rechargeable lithium ion batteries) have a potential to achieve a specific storage of up to 200 W_e-h/kg.

Fuel cells have been used by NASA to power the internal systems of Gemini and Apollo spacecraft and are currently used to power the space shuttle. The lower operating temperatures and higher efficiencies (which

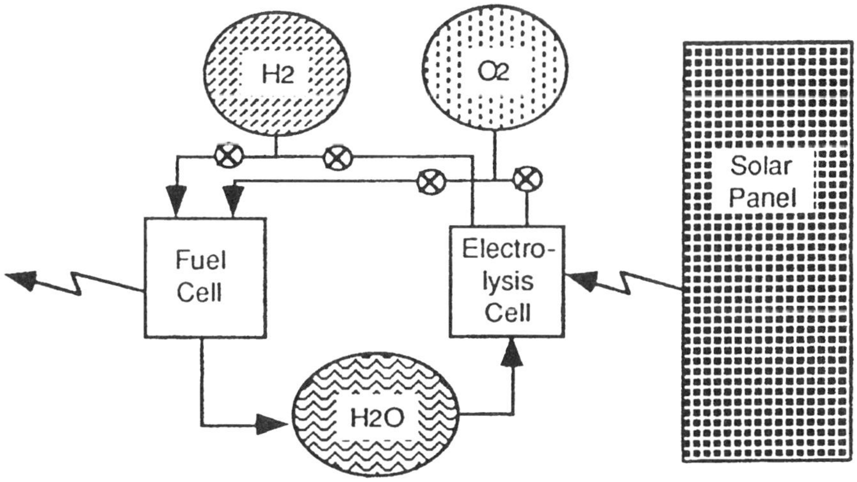

FIGURE III.A.4 Schematic diagram of the fuel cell-electrolysis cycle. SOURCE: Mayer (1992). Reproduced with permission of the American Society of Civil Engineers.

translate to reduced weight) of PEMFCs[1] make them attractive options for planetary missions. Their superior performance and longer life have led NASA to look closely at PEMFCs. The first PEMFC used in an operational space system was the GE-built, 1-kW Gemini power plant. Two 1-kW modules provided primary power for each of the seven Gemini spacecraft in the early 1960s. Each module could provide the full mission power requirements. The performance and life of the Gemini fuel cells were limited by the polystyrene sulfonic acid membrane used at that time. In 1968 an improved Nafion membrane was introduced, significantly improving performance and life. New fuel cell technologies for electric power generation continue to be developed, and there are about 200 units being used for Earth applications in about 15 countries (Hirschenhofer, 1996). As discussed previously, four types of fuel cells (classified primarily on the basis of electrolyte and ranging in operating temperature from about 100 to 1000 °C) are being developed for terrestrial applications in North America. To reduce risk due to unreliability of fuel cells, NASA has decided to use batteries on the International Space Station (ISS), but these batteries are heavy and may not be the most efficient or cost-effective means of power generation or storage on the surfaces of the Moon or Mars.

An attractive alternative for power storage is the regenerative fuel cell or, even more simply, separate electrolysis and fuel cell systems (Figure III.A.4). During the day, excess solar power may be used to electrolyze water and generate hydrogen and oxygen. At night these gases are recombined in a fuel cell to produce electricity and water. Electrolysis and fuel cell systems require a supply of oxygen and hydrogen, and platinum (wire) is used for electrochemically active surfaces. Potassium salts serve as the electrolyte. Fuel cell technology is mature and the efficiency of electrolysis is also high. One very significant advantage of the regenerative H_2-O_2 fuel cell is that it can have a dual function—it can be used not only for energy storage but also for life support on a spacecraft (Eckart, 1996). The electrochemical reactions involving hydrogen and oxygen are the only practical ones at the present time. The oxygen is usually derived from air, but it can be produced on Mars using solid oxide electrolysis (Sridhar and Vaniman, 1997). Hydrogen may be obtained from several fuel sources, e.g., steam-reformed Earth fossil fuels. Other fuels such as methanol can also be used (Klaiber, 1996).

[1]Proton-exchange membrane fuel cells (PEMFCs) are also referred to as solid polymer fuel cells (SPFCs). By virtue of its intrinsic simplicity and high power density, the PEMFC/SPFC has a distinct advantage over other fuel cell technologies. In this discussion the name PEMFC is used.

The intermittent nature of solar energy availability (when, for example, spacecraft or satellites enter into planetary shadow) for low-Earth-orbit or other applications presents a particular challenge for space power management schemes. One alternative to photovoltaic (PV) cells with battery storage is a solar dynamic system with latent heat thermal energy storage (LHTES) via solar heat receivers. During the charging phase heat is stored in the phase-change material while it is melted; during discharge the latent heat is released as the material is solidified. Solar receivers with integral LHTES are needed for generating electric power in space when using solar energy in conjunction with a Brayton cycle (Shaltens and Mason, 1996). A eutectic mixture of LiF-CaF_2 salts, which has a melting temperature of 1413 °F (767 °C), is used as a phase-change material in this application.

Electrical (capacitor, superconducting magnet), mechanical (flywheel), and thermal (latent and sensible heat) storage systems also have potential for use on specific missions but have not yet been developed for spacecraft applications (Bennett, 1998). For example, studies have shown that a solar active system can produce a unit of power from a smaller collector area than is required for an array of solar cells. The improvement in system efficiency results from the increased conversion efficiency of a solar active power cycle compared with solar cells and from the higher specific (kW_e-h/kg) energy storage capacity compared with batteries.

Some Selected Subsystem Technologies

There are many passive and active electric power-generation systems and subsystems. The conversion system that changes the thermal power into electric power distinguishes passive from active power-generation systems. If the conversion system does not use a working fluid it is considered to be passive, but if it employs a working fluid then it is considered to be active. For example, the alkali metal thermal-to-electric conversion (AMTEC), which utilizes high-pressure sodium vapor supplied to one side of a solid electrolyte of beta-alumina, causing sodium vapor to be removed from the other side, is considered to be an active system in spite of the fact that it does not have any rotating parts. The more important subsystems are identified in Table III.A.1, along with the potential of reduced gravity to affect their operation. It is beyond the scope of this report to identify and discuss all of the subsystems in detail; rather, some selected ones are mentioned. Since some of the components of space propulsion systems are the same as those of power-generation and storage systems, reference is made to Section III.B of this report for a discussion of the subsystems common to both.

Passive power systems can use solar (photovoltaic), nuclear-radioisotope (thermoelectric, thermoionic, and thermophotovoltaic), and chemical (fuel cells) energy sources. In the past, static electric power generation systems have enabled, or enhanced, some of the most challenging and exciting space missions, including NASA missions such as the Pioneer flights to Jupiter and Saturn, the Voyager flights to Jupiter, Saturn, Uranus, and Neptune, and the Galileo mission to orbit Jupiter (Bennett et al., 1996). The main disadvantage of the passive power generation systems is that they are much less efficient than the active systems and therefore have a significant weight disadvantage vis-à-vis those systems.

There are various designs for closed-cycle, active power-generation systems. The three most important types are the Brayton, Rankine, and Stirling cycles. It should be mentioned that solar, nuclear, and chemical energy can be used as a source of heat for all cycles. A Brayton cycle is a conventional closed-cycle system that employs a gas turbine and in which the working fluid is a gas flowing throughout the power-generating loop. A Rankine cycle is like a conventional steam cycle in which the vapor is produced in a boiler, does work in a turbine, and condenses in a condenser (radiator). A Stirling cycle is a closed-cycle reciprocating engine whose working fluid is a high-pressure gas, either helium (He) or hydrogen (H_2). For space power systems the use of both solar and nuclear energy conversion systems based on the Stirling cycle engines has been considered, and a recent discussion that cites a large number of relevant references is available (de Monte and Benvenuto, 1998).

A schematic of a closed-cycle Brayton space power generation system is shown in Figure III.A.1. The energy source could be solar, nuclear, or chemical. The technology for the conventional Brayton cycle is well established. The main advantages of the cycle are its high efficiency and very good specific power (in kW_e/kg). As already discussed, a 100 kW_e power unit based on nuclear energy (SP-100 program) was designed but never tested (Bennett, 1998). The problem of rejecting heat from the Brayton cycle space power-generation system is the main concern. The large mass and size of the radiator make it a dominant component of the overall power system. The choice of optimum temperature level for power conversion depends on the compromise between materials limitations and thermal performance. Low heat rejection (radiator) temperature improves thermal performance but results in large, massive radiators.

The block diagram for a closed-cycle Rankine system is the same as that illustrated for a Brayton cycle (see Figure III.A.1). The main difference between the two is that the Rankine cycle involves liquid/vapor mixtures. Boiling takes place in the heat source (boiler, nuclear reactor core) and condensation of the vapor occurs in the radiator. A dynamic system based on the Rankine cycle, which is expected to be more efficient and lighter in weight per unit power generated, has not been designed and operated. The primary reason for NASA's lack of interest in the Rankine cycle is that it involves two-phase flow and boiling/condensation heat transfer in some of its components (i.e., boiler, condenser, separator piping, etc.) and these processes are not sufficiently well understood in microgravity or fractional gravity environments to allow for designing active electric power-generation systems for spacecraft. Nevertheless, the Rankine cycle for space power generation is very attractive because of its relatively high efficiency and the lower mass of the conversion system compared with the Brayton cycle (Gilland and George, 1992). It should be noted that for use in space the boiler, condenser, piping, valves, pump, and thermal management systems need to be designed for safe, efficient, and long-life operation. However, until there is a better understanding of how multiphase systems behave in space, NASA will not be in a position to utilize them.

Boiler for the Rankine Cycle

As noted above, the Rankine cycle is quite efficient for electric power generation and has a higher power-to-weight ratio than a Brayton cycle. However, as noted previously, the cycle (see Figure III.A.1) has components (heat exchanger-boiler, condenser-radiator, phase separator, etc.) in which the working fluid has both liquid and vapor phases. A boiler is an essential subsystem for electric power generation using a Rankine cycle for either spacecraft or stationary power at extraterrestrial sites. However, the operation of a heat exchanger in which the working fluid is boiled (evaporated) will be greatly affected by gravity.

The two-phase flow and heat transfer processes and the flow separation processes in microgravity (near zero) environments are significantly different from those on Earth or on the Moon or Mars. Predictive models for two-phase transport developed for Earth applications are often empirically based and are inadequate for a microgravity environment. Thus, designers of space power-generation systems will be challenged to develop reliable subsystems and technologies that involve two-phase flow and transport phenomena in reduced-gravity environments. The theoretical models and computer codes need to be capable of modeling two-phase flow, boiling and condensation heat transfer, and flow separation and distribution phenomena for all gravity levels. This would permit simulation of microgravity in a continuously variable manner and would not only lead to an increased understanding of, and insight into, the fundamental multiphase phenomena but would also allow NASA engineers to design and evaluate the performance of multiphase systems for use in HEDS missions.

Radiators

Radiators are the only effective means of rejecting heat in space without altering the mass of the spacecraft. Of course, heat can be stored in mass that is ejected from the spacecraft, but this method is not practical for missions of long duration. A comparison of different space power systems has been made (NRC, 1990), and it was found that depending on the type of system and power capability, the radiator can account for between 35 and 60 percent of the total system mass. Radiators that take advantage of a two-phase working fluid are more efficient and are relatively lightweight, but they are subject to possible two-phase flow instabilities, freezing, structural damage caused by oscillatory forces due to periodic or condensation-induced loads, damage by meteorites, and other phenomena. Innovative radiator systems based on moving-belt, liquid-droplet, liquid-sheet, bubble membrane, heat pipes, and other concepts have been proposed (Massardo et al., 1997; Ohtani et al., 1998), but neither the fundamental physical processes of two-phase flow and heat transfer nor the proposed concepts appear to have been studied in sufficient detail to determine their practical feasibility. Space system designers continue to demand design-specific data because they do not understand two-phase flow and phase-change heat-transfer phenomena in microgravity. In addition, the radiators must be robust and reliable, since it is difficult to repair them in space and their impact on life support, mission success, and cost can be very large.

Liquid-droplet radiators and liquid-sheet radiators are among the most promising technologies for achieving lightweight heat exchangers for space applications. In such radiator concepts, neither flow affected by surface tension and thermocapillary forces, nor radiation heat transfer from, say, a cloud of small droplets to the ambient surroundings, has been studied in long-duration microgravity environments, so neither is fully understood. Optimization of the liquid-droplet radiator has revealed that the minimum specific mass is estimated to be 27 percent less than the specific mass of the system with a heat-pipe radiator (Massardo et al., 1997). A recent experimental study of the liquid-droplet radiator has been performed, and the rate at which energy is radiated by a cloud of droplets as a function of droplet velocity and frequency has been measured (Ohtani et al., 1998). At the droplet velocities being considered, there do not appear to be any major microgravity issues associated with fluid dynamics for the system, but heat transfer may be affected by atomic oxygen, and in the space environment micrometeoroids and space debris are of concern.

Proton-Exchange Membrane Fuel Cells

Of the various existing fuel cell systems, the proton-exchange membrane fuel cell (PEMFC) is the most promising, especially for space power-generation and transportation, because of the simplicity of its design and its low-temperature operations. Today's PEMFC membranes are solid, hydrated sheets of a sulfonated fluoropolymer similar to Teflon. The acid concentration of the membrane is fixed and cannot be diluted by product or process water. The acid concentration of a particular membrane is characterized by equivalent weight, EW (grams dry polymer/mole ion exchange sites). This number is the reciprocal of the ion-exchange capacity in moles per gram. Generally, a lower EW and thinner membranes result in higher cell performance. However, thinner membranes also result in higher parasitic cross-diffusion of reactant gas.

As noted above, fuel cells have been used and tested in microgravity. PEMFCs have limited life, and the STS-84 mission in April 1997 was terminated after only 3 days due to problems with an onboard PEMFC. The cell showed low voltage output, and there was concern on the part of the space shuttle crew and NASA ground personnel that the H_2 and O_2 in the cell could cause an explosion. No formal report on the cause for the low cell voltage appears to have been released by NASA or its contractors. NASA has decided to use batteries on the ISS to reduce risk from potentially unreliable fuel cells, but the batteries are relatively heavy and may not be the most cost-effective means of power generation and transportation on the surface of the Moon or Mars.

Like all fuel cells, PEMFCs can serve the dual purpose of generating electric power as well as producing water for human consumption. A PEMFC typically consists of a membrane sandwiched between two gas-diffusion electrodes, which are porous composites made of electrically conductive material. The assembly is pressed between two current collectors. The electrodes are hydrophobic so that gaseous reactants can be transported through the electrodes during cell operation. The outer faces of the electrodes are exposed to the reactant gases that enter and exit the gas chamber. Heat generated during the electrochemical reaction must be removed from the system, and proper water and heat management is essential for obtaining high power density at high energy efficiency. A heat pipe, replacing a coolant pump, heat exchanger, and thermal and other controls, can remove waste heat from the system.

Effective removal of the liquid water product is required to prevent flooding of the electrode, which would prevent gas from reaching the catalyst-membrane interface, where the reactions take place. The water is liquid because of the low operating temperature (<100 °C) of the PEMFC. Some PEMFC designs have used a wicking arrangement to remove the liquid water. At the cathode, or air electrode, this process is a countercurrent and competing process. The oxygen flows toward the interface and product water moves away. Two-phase (gas-liquid) countercurrent flow in microgravity is a critical phenomenon that can greatly affect the performance of a PEMFC. Flooding hampers the rate of mass transfer and results in poor cell performance, which is characterized by the cell's inability to maintain high current at a given cell voltage.

Full hydration of the cell membrane is required for the fuel cell to perform well and reliably. The membrane's requirement for water is a function of the conditions of operation, the amount of water required for hydration, and how and where the water should be added to maintain a fully saturated membrane. Water management schemes that allow for complete membrane hydration have been and are commonly used, and for a broad range of practical current densities there are no external water requirements since enough water is produced at the cathode to adequately hydrate the membrane. There are, however, very limited data for PEMFC performance under reduced or microgravity conditions. In summary, there are a large number of gravity-related issues that are not fully understood and that could affect PEMFC performance and safety (e.g., explosion caused by the sudden reaction of H_2 and O_2 in the cell). These issues include, but are not limited to, the following: (1) the effect of gravity on capillary flow in a porous membrane; (2) the effect of gravity on membrane hydraulic permeability; (3) membrane dehydration due to boiling in the porous structure, which could occur if the temperature exceeds ~100 °C due to some problem with the thermal control; and (4) transport of species and mass through porous electrodes in the noncontinuum regime, including the conjugated heat transfer through the structures.

NASA is considering whether a flight experiment is needed to qualify the PEMFC and to resolve the issue of two-phase flow in microgravity. According to a recent report,[2] the Jet Propulsion Laboratory (JPL), the Johnson Space Center (JSC), and the Glenn Research Center (GRC) are working to resolve some of these issues, but no specifics were provided.

Capillary-Driven, Two-Phase Devices

Thermal management is relevant not only to power-generation systems but also to life-support fluid and thermal systems during long-duration HEDS missions. Put simply, during space travel waste heat must be rejected to space. Fluids can be circulated through the spacecraft components, collected, and then eventually transferred to the radiator, where the heat is rejected to space. Alternatively, capillary-driven, two-phase devices (heat pipes, capillary pumped loops, loop heat pipes, rotating heat pipes, etc.) may be used as key subsystems in thermal control systems of space platforms (Faghri, 1995; Andrews et al., 1997; Vasiliev, 1998). Such devices are characterized by capillary-driven flow of the liquid in a wick structure or in axial grooves. The pure liquid working fluid flows to the heat input section, where it evaporates and carries away thermal energy as latent heat (see Figure III.A.5, for example). The working fluid changes phase from vapor to liquid in the heat sink (condenser) section as energy is rejected. The working fluid is then transported back to the evaporator by means of the capillary forces in a wick.

Heat pipes with different designs and operating in different temperature ranges are two-phase devices that have been recognized as key elements in the thermal management and control systems of space platforms (Faghri, 1995). A heat pipe is an evaporator-condenser system in which the liquid is returned to the evaporator by capillary action. In its simplest form, it is a hollow tube with a few layers of a porous material (e.g., wire screen) along the wall to serve as a wick, as shown in Figure III.A.5. Typical working fluids are sodium or lithium for high-temperature applications, and water, ammonia, or methanol for moderate-temperature applications. If one end of the heat pipe is heated and the other end is cooled, the liquid evaporates at the hot end and condenses at the cold end. As the liquid is depleted in the evaporator section, cavities form in the liquid surface as the liquid clings to the wick. In the condenser section, meanwhile, the wick becomes flooded. The surface tension acting on the concave liquid/vapor interface in the evaporator section causes the pressure to be higher in the vapor than in the liquid. This pressure differential causes the vapor to flow to the condenser section, where the vapor and liquid pressures are nearly equal. Heat removal from the condenser causes the vapor to condense, releasing the heat of vaporization. The condensate is then pumped back to the evaporator section by the capillary force generated at the liquid/vapor interfaces of the pores in the wick.

Since heat pipes rely on surface tension to return the condensate to the evaporator section, they can operate in

[2]Singh, B.S., Glenn Research Center. Multiphase flow and phase change in space power systems. Presentation to the Committee on Microgravity Research on October 14, 1997, Washington, D.C.

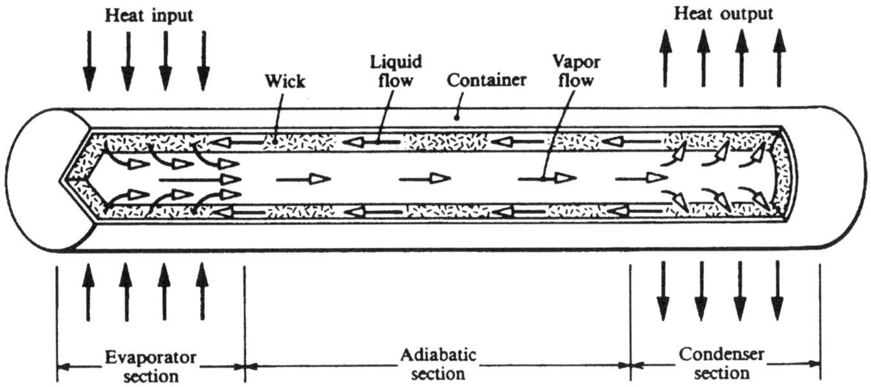

FIGURE III.A.5 Schematic of a simple heat pipe. Based on Faghri (1995).

a microgravity environment. A number of different heat pipes have been designed and used in U.S. and Soviet-Russian space missions. Heat pipe designs for low- (including cryogenic), intermediate-, and high- (using liquid metals) temperature working fluids are well in hand. According to a recent review (Vasiliev, 1998), more than 10 former Soviet space projects used different types of heat-pipe-based thermal control systems. In addition, the results of research and development on a heat-pipe-based radiator system to cool the Stirling cycle microcryogenic machine with a heat output of 110 W and an operating temperature range of 252 to 313 K have been reported (Vasiliev, 1998).

A capillary pumped loop (CPL) and a loop heat pipe (LHP) are two-phase heat transfer devices capable of transporting a large heat load over long distances with very small temperature differences across the system. Significantly, these devices require no external pumping (rather, they use the surface tension forces developed in a fine-pore wick to circulate the working fluid), and a large number of different designs have been proposed (Ku, 1997). Over the past two decades the CPL has been studied extensively with the aim of developing instrument thermal control for future spacecraft. Applications have included the Earth Observing System, the Mars Surveyor, the Hubble Space Telescope, and others (Ku, 1993, 1997). A CPL available for use in a low-gravity environment could represent a new, more effective way to transport thermal energy in space. It has several advantages over a standard heat pipe when used to transport thermal energy in space applications. A CPL system, for example, can provide heat rejection over a wider range of temperatures, and it avoids the limitations of the simultaneous countercurrent flows of liquid and vapor typically encountered in heat pipes (Herold and Kolos, 1997). Industry has developed an LHP system that uses deployable radiators to increase the heat rejection from inside the spacecraft to the space environment (Parker et al., 1999).

As noted above, capillary-driven, two-phase flow devices can operate reliably in a microgravity environment because the liquid flow is driven by surface tension. However, there are inherent limitations on the liquid flow rates. Moreover, a number of other physical phenomena and critical factors can affect the design, operation, and performance of such devices. Examples include obstruction of the liquid flow due to nucleate or film boiling in the wick as a result of overheating in the evaporator section of the heat pipe, freezing of the liquid due to operation of the device under off-design conditions, start-up (melting) from a frozen state, and critical heat flux limitations. In spite of the fact that CPLs and LHPs have been used successfully on several spacecraft, a number of phenomena that can affect the thermal performance of the devices are not fully understood. Examples include research issues such as incipient superheat, the effect of noncondensable gas generation, bubble dynamics, and two-phase behavior in the wick structure of the evaporator under microgravity conditions.

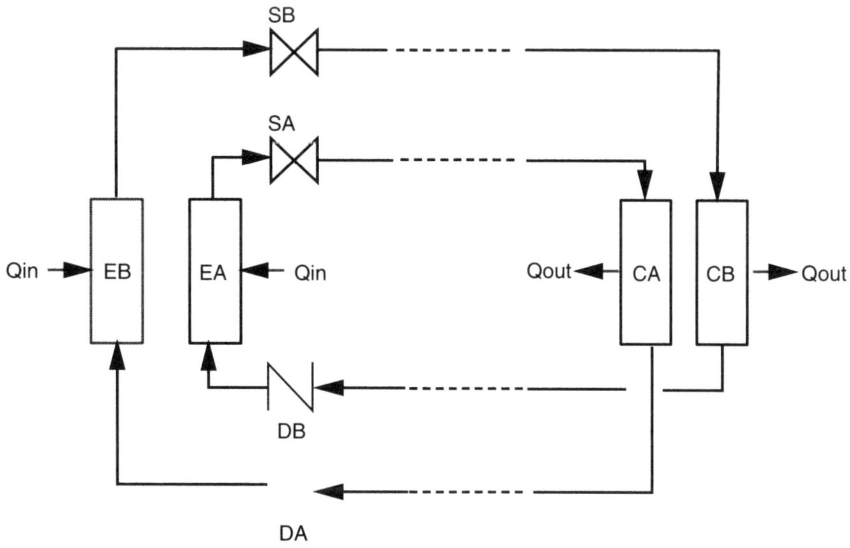

FIGURE III.A.6 Schematic of the vapor-pressure pumped loop concept. SOURCE: Lund et al. (1993). Reprinted with permission from Elsevier Science.

Vapor-Pressure Pumped Loops

A novel concept known as a vapor-pressure pumped loop (VPL) has been proposed (Lund et al., 1993) that uses differences in the vapor pressure between an evaporator and condensor to drive the convective heat transfer process. This device is shown schematically in Figure III.A.6. As can be seen, it is composed of multiple evaporators (two shown—EA and EB) and condensers (CA and CB). These are ducted together and the flow is controlled by solenoids (SA and SB) and check valves (DA and DB), which drive an oscillatory flow vapor and condensate around the loop.

This concept allows for much larger driving pressures and power densities than can be achieved with capillary-driven devices (e.g., heat pipes). Therefore, the VPL may prove to be an important type of multiphase system for use in a microgravity environment. Nevertheless, there will be inherent limitations in power density compared with other active phase-change power production and utilization systems (e.g., Rankine cycles). Moreover, since the VPL does not involve forced convection, there will be significant thermal limitations (i.e., critical heat flux).

Alkali Metal Thermal-to-Electric Conversion

With its inherently noise-free, high-efficiency operation, alkali metal thermal-to-electric conversion (AMTEC) has potential for use in space applications. Today's AMTEC cells are also compact and lightweight, and they have the potential for achieving 35 percent efficiency (Levy et al., 1997). Because of their potential advantages, these cells are being considered by JPL as power systems for Europa Orbiter, Pluto Express, and other space missions (Schock et al., 1997). These static converters can achieve a high fraction of Carnot efficiency at relatively low temperatures. The heart of the multitube AMTEC converter is a beta-alumina solid electrolyte (BASE) tube that is exposed to high-pressure sodium vapor on its inside and much lower pressure sodium on its outside. The tube wall, under a suitable pressure gradient, can conduct sodium ions but not neutral sodium atoms. In effect, the pressure gradient drives the electrical current through the external load resistance. The high pressure is produced by the high-temperature evaporator near the hot end of the cell, and the low pressure is produced by the low-temperature condenser at the cold end of the cell. The sodium condensate is then pumped back to the high-pressure anode by a wick-filled artery, similar to the wicked tubes in heat pipes. The evaporation and condensation processes are expected to depend on gravity, but AMTEC units have not yet operated under microgravity conditions.

The flow, thermal, and electrical components for analyses of an AMTEC unit are interconnected. The design of the AMTEC BASE tubes is complicated because the flow of sodium vapor in the tubes is not in the continuum or molecular-flow regimes but in the transitional regime. The AMTEC cell requires an evaporator and a condenser, which are connected by a wick for returning liquid sodium (condensate) to the bottom of the evaporator. Adequate temperature margin between the top of the BASE tube and the evaporator is required to prevent condensation in the tubes, which can lead to internal shorting of the multitube cell and degradation of performance. The processes of evaporation, condensation, and capillary-driven flow occurring in an AMTEC cell have not been studied under microgravity conditions, and the length of the wick for optimum cell power output has not yet been established. Liquid/vapor separation in microgravity is also of concern.

Summary of the Impact of Reduced Gravity on Selected Subsystems

The list of potential subsystems for power generation and storage is very large, and it is not practical in this limited account to be comprehensive. For the sake of conciseness, the more important subsystems have been identified and are presented in Table III.A.1. (It should be noted that AMTEC can use either a solar or a nuclear (radioisotope) energy source and is classified as an active electric power-generation system since thermal energy

TABLE III.A.1 Selected Subsystems for Passive and Active Power Generation and the Potential Impact of Microgravity on Their Operation

Representative Subsystem	Passive				Active			
	PV	TE	TI	TPV	BR	RA	ST	AMTEC
Batteries	L	L	L	L	—	—	—	—
Boiler	—	—	—	—	—	H	—	—
Capillary pumped loop	—	—	—	—	H	H	H	—
Compressor	—	—	—	—	L	—	L	—
Condenser	—	—	—	—	—	H	—	H
Converter	L	L	L	L	—	—	—	—
Controls	L	L	L	L	L	L	L	L
Evaporator	—	—	—	—	—	—	—	H
Heat exchanger	—	—	—	—	M	H	M	M
Heat pipes	—	—	—	—	H	H	H	—
Phase separator	—	—	—	—	—	H	—	—
Pipes	—	—	—	—	L	M	L	—
Pumps	—	—	—	—	—	M	—	—
Radiators	—	—	—	—	M	H	M	—
Regenerative heat exchangers	—	—	—	—	L	H	L	—
Regulators	L	L	L	L	L	L	L	L
Solar array	L	—	—	L	—	—	—	—
Solar collector	—	—	—	—	M	H	M	—
Storage	—	—	—	—	M	M	M	—
Turbine/alternator	—	—	—	—	L	L	L	—
Valves	—	—	—	—	L	M	L	—

NOTE: PV, photovoltaic; TE, thermoelectric; TI, thermoionic; TPV, thermophotovoltaic; BR, Brayton; RA, Rankine; ST, Stirling; and AMTEC, alkali metal thermal-to-electric conversion. The letters H, M, and L designate high, medium, and low (preliminary assessment) impact of reduced gravity on the operation of the subsystem. Where no letter is given, the subsystem is not applicable to the system listed.

is converted to electricity using a working fluid.) Table III.A.1 also indicates the estimated impact of reduced gravity on the operation of the subsystems as high, medium, or low (little or no impact).

References

American Institute of Aeronautics and Astronautics (AIAA), Aerospace Power Systems Technical Committee. 1995. Space nuclear power: Key to outer solar system exploration. AIAA Position Paper. Reston, Va.: AIAA.

Andrews, J., A. Akbarzadeh, and I. Sauciuc, eds. 1997. Heat Pipe Technology: Theory, Applications and Prospects. Amsterdam: Pergamon.

Bennett, G.L. 1998. Electric power technologies for spacecraft: Options and issues. AIAA Paper No. 98, p. 1022. Reston, Va.: American Institute of Aeronautics and Astronautics.

Bennett, G.L., R.J. Hemler, and A. Shock. 1995. Development and use of the Galileo and Ulysses power sources. Space Technol. 15:157-174.

Bennett, G.L., R.J. Hemler, and A. Schock. 1996. Space nuclear power: An overview. J. Propulsion Power 12:901-910.

Blomen, L.J.M.J., and M.N. Mugerwa, eds. 1993. Fuel Cell Systems. New York: Plenum Press.

Brandhorst, H.W., P.R.K. Chetty, M.J. Doherty, and G.L. Bennett. 1996. Technologies for spacecraft electric power systems. J. Propulsion Power 12:819-827.

de Monte, F., and G. Benevenuto. 1998. Reflections on free-piston stirling engines, Part 1: Cyclic steady operation. J. Propulsion Power 14:499-508; also Part 2: Stable operation. J. Propulsion Power 14:509-518.

Detwiler R., S. Surampudi, P. Stella, K. Clark, and P. Bankston. 1996. Designs and technologies for future planetary power systems. J. Propulsion Power 12:828-834.

Eckart, P. 1996. Spacecraft Life Support and Biospherics. Torrance, Calif.: Microcosm Press, and Dordrecht, Netherlands: Kluwer Academic Publishers.

Faghri, A. 1995. Heat Pipe Science and Technology. Washington, D.C.: Taylor and Francis.

Friedensen, V.P. 1998. Space nuclear power: Technology, policy, and risk considerations in human missions to Mars. Acta Astronautica 42:395-409.

Gilland, J., and J. George. 1992. Early track NEP system options for SEI missions. AIAA/SAR/ASME/ASEE 28th Joint Propulsion Conference and Exhibit, Nashville, July 6-8. AIAA Paper No. 92-3200. New York: American Institute of Aeronautics and Astronautics.

Herold, K.E., and K.R. Kolos. 1997. Bubbles aboard the shuttle. Mechanical Engineering (October):98-99.

Hirschenhofer, J.H. 1996. 1996 fuel cell status. Pp. 1084-1089 in Proceedings of the Thirty-First Intersociety Energy Conversion Engineering Conference, Vol. 2. New York: Institute of Electrical and Electronic Engineers.

Klaiber, T. 1996. Fuel cells for transport: Can the promise be fulfilled? Technology requirements and demands from customers. J. Power Sources 61:61-96.

Kroschwitz, J.I., and M. Bickford, eds. 1994. Fuel cells. Pp. 1098-1121 in Kirk-Othmer Encyclopedia of Chemical Technology, 4th Ed., Vol. 11. New York: John Wiley & Sons.

Ku, J. 1993. Overview of capillary pumped loop technology. Pp. 1-17 in Heat Pumps and Capillary Pumped Loops, HTD-Vol. 236. New York: American Society of Mechanical Engineers.

Ku, J. 1997. Recent advances in capillary pumped loop technology. AIAA Paper No. 97-3870. Reston, Va.: American Institute of Aeronautics and Astronautics.

Landis, G.A., S.A. Bailey, and M.F. Piszczor, Jr. 1996. Recent advances in solar cell technology. J. Propulsion Power 12:835-841.

Levy, G.C., T.K. Hunt, and R.K. Sievers. 1997. AMTEC: Current status and vision. Pp. 1152-1155 in Proceedings of the Thirty-Second Intersociety Energy Conversion Engineering Conference, Vol. 2. New York: American Institute of Chemical Engineers.

Lund, K.O., K.W. Baker, and M.M. Weislogel. 1993. The vapor pressure pumped loop concept for space systems heat transport. Aerospace Heat Exchanger Technology: Proceedings of the First International Conference on Aerospace Heat Exchanger Technology, Palo Alto, Calif. P.K. Shah and A. Hashemi, eds. New York: Elsevier Science Publishers.

Massardo, A.F., C.A. Tagliafico, M. Fossa, and A. Agazzini. 1997. Solar space power system optimization with ultralight radiator. J. Propulsion Power 13:560-561.

Mayer, A.J.W. 1992. Power sources for lunar bases. Pp. 763-773 in Proceedings of the Third International Conference: Engineering, Construction, and Operations in Space III, Vol. I. W.Z. Sadeh, S. Sture, and R.J. Miller, eds. New York: American Society of Civil Engineers.

National Critical Technologies Panel. 1993. Biennial Report. Washington, D.C.: U.S. Department of Commerce, National Technical Information Service.

National Research Council (NRC), Aeronautics and Space Engineering Board (ASEB). 1987. Space Technology to Meet Future Needs. Washington, D.C.: National Academy Press.

NRC, ASEB. 1990. Human Exploration of Space: A Review of NASA's 90-Day Study and Alternatives. Washington, D.C.: National Academy Press.

NRC, ASEB. 1998. Space Technology for the New Century. Washington, D.C.: National Academy Press.

Ohtani, Y., M. Fujiwara, and M. Watabe. 1998. Study of radiative heat transfer characteristics of droplet radiator. Pp. 441-442 in Proceedings of the 35th National Heat Transfer Symposium of Japan, Nagoya, Vol. II. Tokyo: Heat Transfer Society of Japan.

Parker, M.L., B.L. Drolen, and P.S. Ayyaswamy. 1999. Loop heat pipe performance—Is subcooling required? Proceedings of the 5th ASME/JSME Joint Thermal Engineering Conference, March 15-19, 1999, San Diego, California. Paper No. AJTE 99-6285. New York: American Society of Mechanical Engineers.

Schock, A., H. Norvian, and C. Or. 1997. Coupled thermal, electrical and fluid flow analyses of AMTEC converters, with illustrative application to OSC's cell design. Pp. 1156-1164 in Proceedings of the Thirty-Second Intersociety Energy Conversion Engineering Conference, Vol. 2. New York: American Institute of Chemical Engineers.

Shaltens, R.K., and L.S. Mason. 1996. Early results from solar dynamic space power system testing. J. Propulsion Power 12:852-858.

Sridhar, K.R., and B.T. Vaniman. 1997. Oxygen production on Mars using solid oxide electrolysis. Solid State Ionics 93:321-328.

Vasiliev, L.L. 1998. State-of-the-art on heat pipe technology in the former Soviet Union. Applied Thermal Engineering 18:507-551.

III.B SPACE PROPULSION

Introduction

Energy-conversion systems for the space propulsion required for HEDS missions have to be selected and developed. Propulsion requires an energy source and a corresponding power generation system (many of these are described in the preceding section) serving a propulsive jet that provides thrust by reaction. A wide range of propulsive capabilities will need to be provided, and therefore a large number of system possibilities must be considered, each comprising the many devices and subsystems needed to carry out essential functions within each system. Each potential system and its subsystems will need to operate reliably and efficiently at various, and perhaps variable, gravity levels. It is concern for the effects of gravity level that motivates this report and provides the focus for the descriptions and discussions that follow.

In this section the anticipated capability requirements are first listed and then potential systems are identified and discussed. Judgments and preferences are, however, avoided; while there is no more portentous decision for a HEDS mission than the selection of a propulsion system, it is not the purpose of this report to urge particular choices. No "baseline" selections are recommended. Rather, the aim is to include a range of systems sufficiently wide to bring out the significance of gravity level; any emphasis of one system over another reflects only its interest for microgravity research.

After the potential systems are described, major subsystems, or components, are identified by the functions they fulfill (e.g., a boiler). Then, those functions are associated with gravity-dependent phenomena, and, again as appropriate, specific subsystems are discussed in terms of requirements for microgravity research.

Required Space Propulsion Capabilities

Many quite different propulsive capabilities are needed for the various potential HEDS operations, including the ability to do the following:

- Achieve Earth orbit;
- Transfer and adjust orbit;
- Transit to the Moon and the planets of the solar system and return;
- Transit beyond the solar system and return;
- Achieve orbital capture at destination or on Earth return;
- Descend to surfaces of large bodies with significant gravity;
- Rendezvous with small bodies such as asteroids, which have very low gravity;
- Send and receive sampling probes; and
- Perform actions involving momentum or position change, such as the deployment or aiming of antennas.

Providing these propulsive capabilities will require that attention be paid to a variety of issues, including availability, security and dependability, and duration of the energy source; human safety, health, and effectiveness; and endurance, range, versatility, and reliability of the propulsion system. Commonality with other needs, especially those of spacecraft and station electric power, will also be an important concern, as will cost and the feasibility of timely development.

While judgments and choices concerning these issues are not part of this study, awareness of such general concerns will of course lead to specific concerns about microgravity effects. For example, human health may require microgravity countermeasures, and providing for operations in a range of gravity levels may seriously affect system reliability and cost.

Space Propulsion Systems

Various potential propulsion systems may be postulated to provide the capabilities mentioned above; some are now commonplace, while others are speculative in the extreme. They all convert or absorb energy in order to effect momentum change, usually by means of a propulsive jet. Before describing these systems and identifying salient low-gravity issues pertaining to them, the committee first summarizes them according to energy source.

As a space-propulsion system, the chemical combustion rocket can in principle provide all propulsion capabilities, but for distant missions its economic performance may not compare well with that of other possibilities. The nuclear thermal rocket (NTR) may be suitable for transit beyond Earth orbit, but probably only within the solar system. Quite different is nuclear electric propulsion (NEP), which uses ion and electromagnetic thrusters: it would be of particular interest for distant transits beyond Earth orbit, especially travel beyond the solar system. Solar thermal or solar electric (also using ion and electromagnetic thrusters) is of interest for all tasks within the inner solar system. The solar sail is also of interest for transits within the solar system. A laser thermal system may be of special interest for minipropulsion applications, such as sampling probes. For very long distances, the laser sail may be useful, especially for small or moderate loads. Finally, systems deriving energy from planetary fields and atmospheres may be mentioned. The tether is of potential interest for some orbital transfer tasks, while the aeroassist technique may facilitate orbital maneuvers or capture operations.

These systems are described and discussed below in varying degrees of detail, depending on their perceived significance for microgravity research. Besides the capability issues already mentioned, a number of specific technical themes appear, many of which, such as method of energy conversion and heat transfer, were thoroughly discussed in Section III.A. Other technical themes of interest include fluid handling and storage, cryogenics, mechanical machinery, structures, scale (micropropulsion), range of applicable gravity level, and complexity. Gravity level, be it microgravity or fractional gravity, will be an important consideration in all these technical topics, especially as it affects heat transfer and fluid handling.

Chemical Rocket

In a chemical rocket, the combustion of propellants produces hot gas at high pressure. Expansion of this gas through a nozzle converts its high thermal energy to the directed kinetic energy of a high-velocity jet exhaust; this kinetic energy provides the thrust needed for propulsion. Various types of chemical rockets may be of interest for providing propulsive capabilities for HEDS. These types differ in the way the propellants are provided for combustion; liquid, solid, or hybrid (solid/fluid) systems may each play a role in future HEDS missions.

Typical liquid-propellant rockets for space propulsion include Russia's kerosene-powered engines and the NASA space shuttle's main engine, in which cryogenically stored H_2 and O_2 are pumped to a combustion chamber, where they are injected and mixed and where they then react to produce water vapor at high temperature and pressure (Figure III.B.1). The thrust that a rocket produces is proportional to the flow rate of reaction product through the nozzle. The nozzle of the engine must be cooled by liquid H_2 on its way to the combustor. A measure of the effectiveness of propellant flow in producing thrust is the specific impulse, which is the ratio of thrust force to the rate of consumption of stored fuel. Commonly, the rate of consumption of fuel is expressed as Earth weight per unit time and the specific impulse is given in seconds. The specific impulse of a hydrogen-oxygen rocket can be on the order of 360 s (Hill and Peterson, 1965).

A system of the sort just described is quite simple and reliable; it has already taken humans to the Moon and has propelled unmanned spacecraft to the outer edge of the solar system. Chemical combustion rockets in

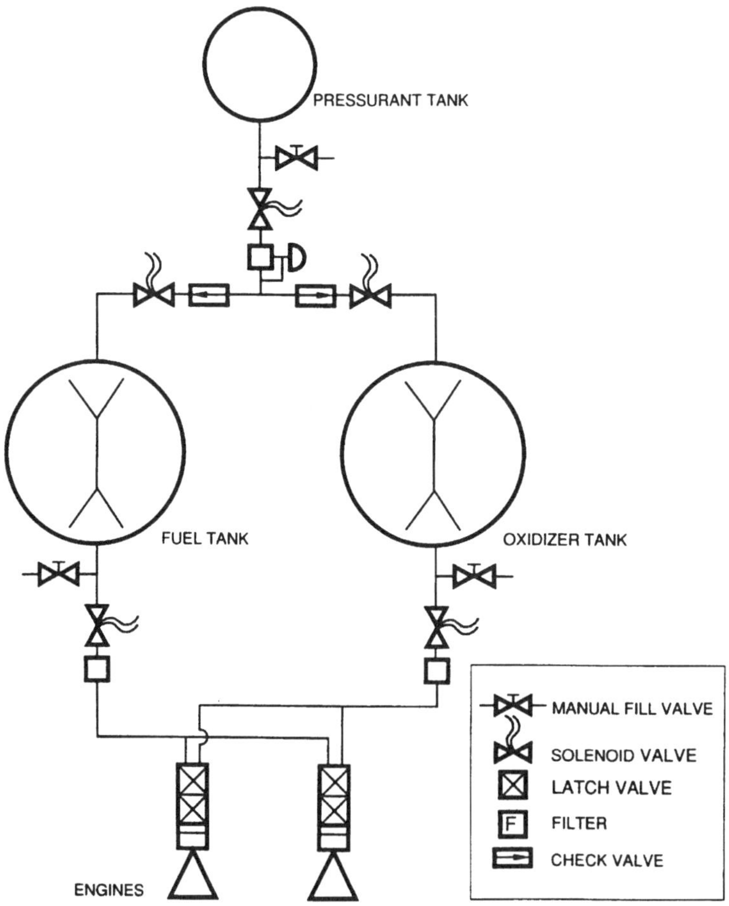

FIGURE III.B.1 Schematic of bipropellant chemical rocket. SOURCE: Cassady (1990). Reprinted with permission of the American Institute of Aeronautics and Astronautics.

principle are capable of any mission, at any scale, that one can imagine. They can produce high thrust and can be well controlled. However, the hydrogen-oxygen rocket produces this high thrust for a rather short period of time before the fuel is consumed, which limits mission flexibility, unless fuel can be replenished from in situ resources. In fact, its economic use probably does not extend beyond Mars (AIAA, 1995).

For this report, which focuses on issues of reduced gravity, a dominant technical feature of the hydrogen-oxygen system is the cryogenic storage apparatus; liquid gases must be refrigerated and kept cold in large tanks for long periods of time before use. Fuel replenishment, or transfer between tanks in space, is not a fully developed process. Fuel must also be acquired from the tanks and transferred, by high-pressure pumps, to the combustion chamber. The operation of the refrigeration system and of the various pipes, valves, seals, and bearings involved in handling the propellants can be expected to be sensitive to gravity, yet they must all operate reliably at various gravity levels.

In the rocket engine itself, including the combustor and propulsion nozzle, the fuel, propellant, and coolant are all subject to forced flow with high accelerations and therefore will not be affected by gravity level. That is, none of the engine's essential processes will depend on gravity. However, any start-up processes in space could be affected by gravity because initial velocities will be small.

Liquid or solid propellants that can safely be stored in stable form can be used in chemical rockets, thus avoiding the need for cryogenic storage. Many well-developed, storable liquid propellants (such as kerosene and ethanol) are now employed in space operations. However, these propellants are less energetic than the hydrogen-oxygen mixture and hence suffer the penalty of lower specific impulse. One concept for avoiding cryogenics while keeping the high-impulse benefit of high-energy fuel is to diffuse a light gas (such as hydrogen) into the interstices of a solid matrix, where the gas would be held and then "boiled off" as needed (Carrick and Harper, 1998). This option is in the research stage and is therefore of uncertain applicability for HEDS. The boiling-off process would presumably be affected by microgravity.

There is great interest now in very small (micro) low-cost rockets for special purposes such as sample acquisition and return (JPL, 1998). Because of their small scale, cryogenic storage would not be feasible for these devices. However, they could well be based on chemical combustion of reactants stored at high pressure, as in the pressure-fed rocket, or even in solid form. Such micropropulsion devices would often be used in low or microgravity situations but would not usually be sensitive to gravity changes, owing to their small dimensions.

The solid-propellant rocket merits further discussion. As suggested above, it is free of the fluid-handling issues that become problematic in reduced gravity, and it will surely be of interest for HEDS use, especially in micropropulsion applications. In the solid-propellant rocket, the fuel-oxidizer mixture is formed into a solid matrix (known as a grain) that burns exothermally after ignition (Sarner, 1968). Thus, while there are complex fuel-handling problems during manufacture, no such problems arise during operation. No microgravity issues appear during steady operation, although ignition in microgravity could be problematic. Nevertheless, a very severe operational limitation of a solid rocket is the lack of control over thrust after ignition; once started, it can be stopped only by burnout or destruction.

Because of the complexity of the manufacturing processes involved, it seems unlikely that solid-propellant rockets would be constructed extraterrestrially during HEDS activities. Their use would probably be limited to launches from Earth (as with the solid rocket boosters attached to the space shuttle during launch) or to small-thrust applications of rockets previously transported from Earth.

For HEDS applications, the so-called hybrid rocket may be more interesting than the pure solid type. In the hybrid rocket, a liquid or gaseous oxidizer is injected into a combustion chamber already loaded with solid fuel (Seifert, 1968). Unlike the solid-propellant rocket, a hybrid rocket can be throttled to control thrust by adjusting the rate of injection of the fluid oxidizer, thus providing operational flexibility. Of course, the hybrid incurs the complications of handling the fluid oxidizer, but these complications are much less severe than those of the liquid-propellant rocket, in which both fuel and oxidizer must be managed. Thus, any reduced-gravity issues for the hybrid rocket would concern only the handling of a liquid oxidizer. The hybrid rocket is easier to construct than the solid one, because the grain contains only fuel. As a result, fuel processing is less complex and perhaps could be managed in space or in a planetary colony. For example, carbonaceous fuel could conceivably be extracted from the Martian atmosphere and then formed into solid fuel matrices for hybrid rockets using fluid oxidant derived from in situ sources to serve local propulsion needs or perhaps even enable liftoff from the Martian surface.

Nuclear Thermal Rocket

In the NTR system, a reactor generates heat by nuclear fission. This heat is transferred, within the reactor, to a propellant gas that has been pumped at high pressure from a cryogenic storage tank. This thermal energy is then converted to kinetic energy, and hence thrust, in an exhaust nozzle (Dearian and Whitbeck, 1990; NASA, 1991; Rosen et al., 1993). The nozzle is cooled by the propellant on its way to the reactor (Figure III.B.2).

The NTR is likely to be a very large system if it is based on a massive, graphite-moderated nuclear reactor.[3] Because of safety concerns, such a rocket would probably be assembled and launched from Earth orbit. Thus, an NTR would generally be less flexible in design and operation than a chemical rocket, but it would be useful for

[3]It should be noted that very small NTRs have been proposed that have perhaps one-tenth the power of a conventional NTR.

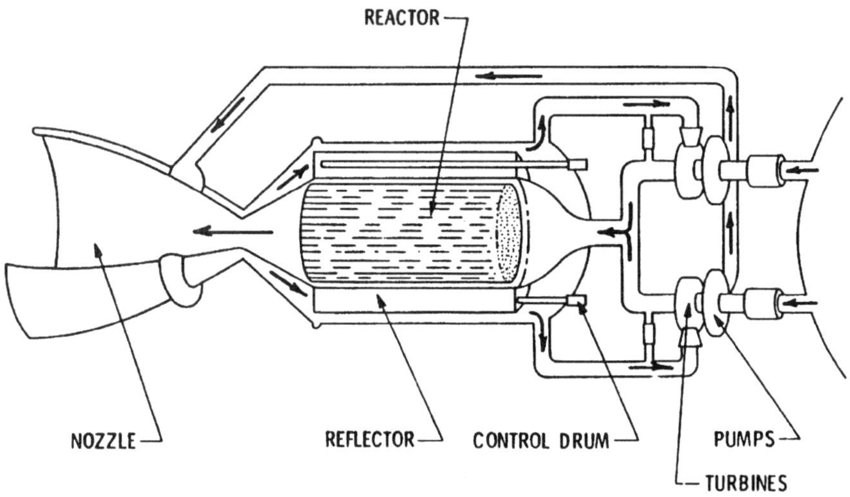

FIGURE III.B.2 Sketch of a nuclear thermal rocket. SOURCE: NASA (1991).

missions of transit between Earth, the Moon, and other destinations no more distant than Mars (Bennett and Miller, 1992). The nuclear electric propulsion option (discussed below) should prove preferable beyond Mars. Like the chemical rocket, the NTR would be a high-thrust, short-duration device.

The advantage of the NTR is that its specific impulse, based on propellant consumption, is about twice that of the chemical rocket, because pure H_2 can be used and there is no need for a relatively heavy oxidant. This advantage would be about 3:1 if based only on a comparison of H_2O and H_2 molecular weights, but it is diminished to about 2:1 by the rather lower allowable temperature to which the propellant can safely be heated in the reactor, as compared with a combustor.

In addition to the nuclear reactor and its associated control system, a cryogenic system will be needed to store and maintain hydrogen in a liquid state for the length of time needed for the mission. Propellant pumps and gas turbines to drive them, with their bearings and seals, will be required. Liquid hydrogen piping and control valves will be needed, both to provide the main propellant flow and for auxiliary purposes, such as to drive pump turbines and provide nozzle cooling. Provisions must be made to ensure the proper rates of heat transfer to the hydrogen propellant in the reactor and to the liquid H_2 coolant in the nozzle protection system.

As in the case of the combustion rocket, the engine itself, including the reactor and probably its heat-transfer processes, will have forced flow and high velocities and should be insensitive to gravity during operation at or near design thrust. However, start-up and shutdown in space will be required for nuclear propulsion systems (Dearian and Whitbeck, 1990), and fluid flow and heat transfer processes will generally be affected by gravity level during those operations, depending on reactor design.

The NTR is far beyond the conceptual stage; under the nuclear engine for rocket vehicle application (NERVA) program of 40 years ago, NASA and the Atomic Energy Commission together developed and ground-tested a full-scale NTR (Rosen et al., 1993; Bennett et al., 1994). Though it is not a "shelf" item, there is confidence that the conventional NERVA type of NTR is a practical option, probably achievable at a rather low cost for final development. However, the future use of NTR propulsion for HEDS may depend on advances in high-power-density reactors involving, for example, rotating-bed or fast-fission approaches. Indeed, fast reactors do not require neutron moderation (with, e.g., graphite) and thus are inherently much lighter than thermal neutron fission reactors.

The rotating-bed reactor might be particularly attractive for use in variable gravity. Centrifugal force would be expected to maintain fuel pellets in a stable bed through which the propellant is forced. High power density and relative insensitivity to gravity level could thereby be achieved, although issues of pellet-bed stability and of start-up would be of concern.

The gas-core nuclear rocket (GCNR) represents a variant of the NTR in which the necessary heat transfer between the fissile material and propellant would occur across a stable gas interface rather than a material wall. The gases in question might be uranium hexafluoride and hydrogen. If such a stable interface could be achieved and maintained, the material temperature limitations, mentioned above, on NTR specific impulse would be avoided, and full advantage could be taken of the low molecular weight of hydrogen.

The difficulty with this propulsion system is that no feasible method has been found to achieve the necessary stable gas interface, despite vigorous research emphasizing vortex flows, which peaked about 30 years ago (Schneider and Thom, 1971). The GCNR idea lingers, however, because of the hoped-for large potential increase of specific impulse.

Despite the impracticality of the foregoing application, the general problem of vortex containment of dissimilar fluids should be revisited with microgravity applications, such as phase separation, in mind. A NASA publication (Schneider and Thom, 1971) contains interesting studies on this topic.

Nuclear Electric Propulsion

The energy source for NEP would be nuclear fission, with the generated heat transferred from the reactor to a suitable working fluid rather than directly to a propellant gas, as in the NTR. The working fluid is then used to generate electric power through a thermodynamic cycle, and that electric power drives a plasma or ion thruster (Barnett, 1991). Such a system escapes the temperature limitation on specific impulse that characterizes the NTR, and specific impulse can be thousands of seconds. Electromagnetic thrusters are used very successfully today for small-scale intermittent applications (station keeping, for example).

Even when used for spacecraft propulsion, the thrust itself will still be quite small, as will the reactor, which would typically be one-tenth the size of the comparable NTR reactor. When used for long-distance transit propulsion, the NEP system must therefore operate continuously for periods comparable to the duration of the mission itself, in an environment of very low effective gravity. Clearly, NEP must meet a special burden of reliability for long missions.

Since the effective gravity during acceleration by NEP will be quite low (of the order of 0.1 g_0), microgravity issues are likely to arise, depending on the thermodynamic cycle used. The closed Brayton cycle using helium as a working fluid would probably be the simplest cycle and the least subject to microgravity concerns.[4] A heat exchanger in which heat is transferred from the reactor coolant (perhaps liquid lithium) to the gaseous helium working fluid would involve no phase change. Heated helium would pass through a turbine, which would drive the electric generator and, incidentally, the helium compressor. Then the helium would reject heat to space, probably through a secondary heat-transfer loop, which might involve heat pipes and phase change of a medium such as ammonia. Naturally the helium pump and turbine would require high-performance bearings and seals, as well as associated piping and controls.

The Rankine cycle is nonetheless commonly proposed for NEP because it promises higher thermal efficiency. It has many components, however, and its advantages depend on phase change, making it inherently more sensitive to gravity. In a typical Rankine cycle NEP (Figure III.B.3) the working fluid (potassium, for example) is evaporated in a boiler. Then, following power generation in a turbine-generator, the working fluid must be condensed in a heat exchanger, with heat rejected to space via a radiator system. Then the condensate is pumped to a high pressure and fed into the boiler. Special care must be taken to be sure that there is no liquid carry-over to the turbine and vapor carry-under from the condensor. For this purpose a phase separator, probably operating on the centrifuge principle, must be used. Fluid handling equipment such as piping, valves, pumps, turbines (with

[4]Nieberding, J., Lewis Research Center. Propulsion, power, and cryogenic fluid systems. Presentation at Workshop on Research for Space Exploration, May 8, 1997, Lewis Research Center, Cleveland, Ohio.

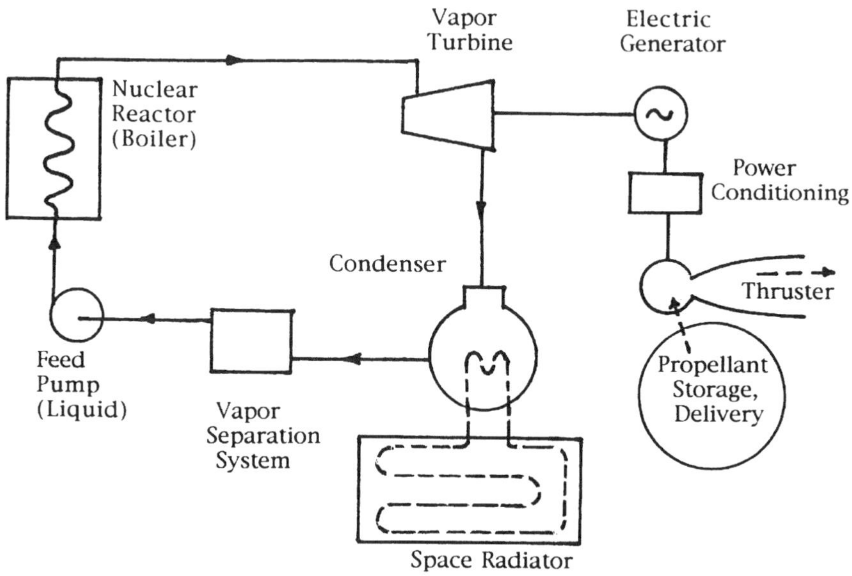

FIGURE III.B.3 Schematic showing major elements of a nuclear electric propulsion system.

associated bearing systems), seals, and controls tends to be especially complex for the Rankine cycle, and design for high reliability will be correspondingly important.

Like NTR, the NEP system would presumably be assembled and launched from Earth orbit. Start-up or shutdown in orbit poses complex issues of design and operation, especially for a Rankine cycle with a liquid-metal working fluid and a nuclear heat source. Thawing (in zero gravity) of the nuclear reactor is required and the entire system must, in a controlled way, be brought to an equilibrium appropriate to steady-state operation. Transient, intermittent, or variable operations generally must also be considered carefully, taking account of system dynamics and possibilities for system instabilities.

NEP would not depend as much on cryogenic storage as would chemical or NTR systems; however, the propellant material, be it liquid metal or noble gas, must be stored and maintained for especially long times.

A special feature of the NEP concept is the potential for dual use; electric power can be produced for purposes other than propulsion, i.e., for spacecraft or colony needs, as is more fully discussed above in Section III.A. These uses might include the powering of various microthrusters needed for spacecraft positioning or other mechanical actuation. It has also been proposed[5] that commonality between the propulsion and power systems might only exist with respect to the reactor, which would supply heat for a thermal rocket and/or, via a suitable thermodynamic cycle, electric power.

Another possible application for electric power generated by nuclear or solar means is to power an electromagnetic accelerator, or "mass driver" (Snow and Kolm, 1992), perhaps to convey materials between the Moon and Earth, for example.

Fusion propulsion ideas have been proposed and should continue to be studied with the aim of achieving specific impulses adequate for missions to the edge of or beyond the solar system. If plasma can be confined well enough to achieve "ignition," that energy could presumably be converted to a high plasma velocity in a rocket nozzle. Obviously, such ideas have no near-term relevance to HEDS, but in the far future they may be of crucial importance for space exploration. The same remarks apply to antimatter propulsion, and it should be noted that antimatter has been proposed as a trigger for fusion.

---

[5]Neiberding; see footnote 4.

Solar Thermal

Absorbed solar radiation can serve as an energy source for propulsion, although beyond Mars, the solar flux is too small to be useful. Very large collector arrays and concentrators could deliver heat directly to a propellant gas, such as hydrogen, to power a thermal rocket. Such an approach would be attractive for small thrusters. The heat collection process could require extensive fluid distribution systems, perhaps utilizing phase change, with difficult problems in structural dynamics and attitude control.

Absorbed solar radiation could also heat a working fluid in a thermodynamic cycle, such as Brayton or Rankine, to generate electric power for plasma propulsion (solar electric propulsion) as well as for the spacecraft or outpost needs discussed above in Section III.A. The issues described for NEP apply here as well, as would the structural and fluid-handling problems of large collector arrays peculiar to the solar option. Solar propulsion systems would inherently have high specific impulse but low thrust and therefore would be assembled and launched from orbit.

Solar Sail

A propulsion system deriving momentum change directly from solar radiation pressure and the solar wind would require a very large solar "sail" (Staehle, 1981). This sail would be subject to severe problems of structural dynamics and control in a microgravity environment but would of course be free of fluid-handling problems. The solar sail might be attractive for small-scale spacecraft in the inner solar system, where solar radiation is sufficiently high. The probability of meteorite impact would be of particular concern for such a system.

Laser Thermal

An interesting possibility, especially for small rockets designed to communicate between an orbiter and a planet or asteroid surface, would be use of the highly concentrated, continuous-wave radiation of a laser to heat a propellant gas, which would then power a rocket (Mead and Myrabo, 1998; Harris, 1999). The laser would be housed on an extraterrestrial site, using electric power assumed to be available there. Apart from the obvious tracking, control, and lost-line-of-sight problems of such a system, the propellant would pose heat-exchange and fluid-handling issues for the target vehicle.

Laser Sail

Studies by NASA and DOD have suggested the feasibility of "sail" propulsion powered by a fixed laser that illuminates a large collector on a spacecraft (Perry and Powell, 1998). Radiation pressure due to reflected radiation would provide thrust useful for small payloads and long distances, taking advantage of laser beam coherence. The greatest microgravity issues for such a scheme would presumably be those of positional stability and control of the very light collecting antenna.

Tether

A tether can be used to exchange momentum between bodies in space (Cutter and Carroll, 1992) and thus to provide a direct mechanical propulsive effect that can be especially useful in achieving orbital changes. Also, if a conducting cable, or tether, is suspended from a spacecraft orbiting a body having a substantial magnetic field (Earth and, especially, Jupiter are examples), the tether will experience a force provided current is caused or permitted to flow along the tether. The force may be a drag on the spacecraft or it may be propulsive, depending on the relative motion of the spacecraft and the magnetic field lines carried by the rotating body (Gallagher et al., 1998). The presence of an ambient plasma, or ionosphere, would be important in establishing the appropriate current flow. The electrodynamic tether concept may prove useful for orbital adjustments, including de-orbit for descent. Another potential use is, of course, the generation of electric power. Obviously, there would be a gravity gradient along such a tether, and the dynamic behavior of such a system would be problematic. Moreover, protecting the system against the effects of a meteorite impact would seem to require redundancy in the tether design (Hoyt and Forward, 1998).

Atmospheric Drag—Aeroassist

A spacecraft approaching a body to orbit around it, or perhaps descend to its surface, can achieve the necessary velocity reduction by descending into the outer edges of its atmosphere, assuming the body has an atmosphere of sufficient depth and density. The resulting drag would replace the reverse thrust that would otherwise be required from the propulsion system (French, 1981). This would save energy and, depending on the propulsion system, mission time as well. Such a system would obviously require exquisite control accuracy and dependability. Both crew and spacecraft systems would be subject to great changes of apparent gravity, from zero to many times Earth's gravity, during typical atmospheric drag, or aeroassist, maneuvers, posing important issues of human and technical system performance.

Major Subsystems, Their Purposes, and Their Sensitivities to Reduced Gravity

What follows is a list of the major subsystems and important devices, together with some discussion of their uses or purposes, that would make up the various propulsion systems described above. The list is selective in that only those devices are mentioned that seem to pose issues for microgravity science, directly or indirectly. For example, thruster nozzles, although important, are not listed, because their large pressure differences and high velocities will mask any effects of microgravity. Of course, nozzle coolant passages may contain cryogenic liquids subject to flashing, and they would therefore be subject to multiphase effects of reduced gravity.

Many subsystems of interest are of course common to a number of propulsion types and, indeed, to other functions such as power generation or construction and maintenance, as discussed in later sections of this report. Furthermore, some design elements, such as pipings and bearings, for which there are microgravity concerns are themselves common to different subsystems. These are mentioned later in this section, and then discussed more fully in Chapter V.

Nuclear Fission Reactor

A large nuclear reactor would provide heat for NTR, or a small reactor would serve an NEP system or provide for electric power generation. Coolant flow and heat transfer to a working fluid or propellant in such a reactor could be sensitive to gravity level, particularly during the transient operations of start-up or shutdown, as the coolant changes phase from a solid to a liquid. In addition, the required reactor control system would include robotic elements, bearings, seals, and actuators, whose performance may be affected by gravity level.

A hypothetical gas-core nuclear fission reactor would presumably need a special control system to maintain core containment. Such a reactor, if developed, could provide heat for NTR or NEP. It would be especially sensitive to gravity level with respect to the containment process (as yet undefined) essential to its steady operation.

Cryogenic Storage System

A cryogenic storage system for maintaining liquefied gases[6] comprises the following elements: storage tank, gas liquefaction system, refrigeration, insulation (passive cooling), liquid acquisition device, feed and fill systems, and pumps for pressure control and liquid transfer. The purpose of such a system would be to store and provide

---

[6]Neiberding; see footnote 4.

propellant and reactant gases for chemical rockets or any rocket in which a propellant gas is heated, perhaps by a nuclear reactor or by solar or laser radiation, or is accelerated electromagnetically as a plasma.

Such a fluid-handling system is inherently a multiphase system because it must maintain liquid and vapor in proper equilibrium, condense vapor, and, finally, deliver liquid via pumps, valves, and pipes to the rocket chamber. As such, a cryogenic system is highly sensitive to gravity level in all its aspects. For example, in microgravity, the storage tank poses problems of ullage location, temperature control, and filling procedure, while the valves and pumps are subject to flashing. Cryogenic storage problems are discussed in more detail in Section III.E.

Radiator System

Heat rejection to space requires a system that has a large surface area, a pipe and header system for coolant, and heat pipes to convey the heat from source to radiator. Space radiators and heat pipes are discussed above in Section III.A in connection with power generation. Here, it is emphasized that space radiators are necessary in order to reject waste heat from the energy conversion processes used for electric propulsion. The need for a heat-rejection subsystem is a feature of closed power cycles such as those used to generate electric power, for whatever purpose. In contrast, chemical or thermal propulsion (nuclear or solar, for example) are open cycles, which reject heat via the propulsive stream itself and therefore have no need for a radiator system and its attendant microgravity problems.

It should also be noted that the power requirement for electric propulsion of a large spacecraft implies a very large radiator area having the form of a large, light (and therefore flexible) structure. When gravity is absent, such a structure would be subject to dynamic mechanical behavior, which may be difficult to control.

Solar Collector

For propulsion systems depending on heat from the Sun, a large collector surface is required, probably containing a tube and header system of phase-changing fluid to absorb heat and deliver it to the spacecraft. Therefore, as in the case of the heat-rejection radiator, multiphase fluid mechanics in reduced gravity would presumably be important, as would the structural dynamics of an extended structure.

Boiler for a Rankine Cycle

If a Rankine cycle is adopted, a heat exchanger for evaporation and separation and collection devices for gas and liquid phases are needed. As pointed out in Section III.A, such a boiler would be an essential part of any method of electric power generation by means of the Rankine cycle, whether the energy source is chemical, nuclear, or solar, and whether the purpose is to make electric power for propulsion or for spacecraft or station power. The performance of a heat exchanger in which the working fluid is evaporated is expected to depend on gravity level. On Earth, the separation of vapor from liquid is often accomplished with the help of gravity; in microgravity, other effects, such as viscosity or acceleration, may be effective, depending on the design. The completeness of phase separation is vital for the physical integrity of the equipment and for the efficiency of the cycle; in reduced gravity, special effort must be made to achieve complete phase separation. On Earth, cyclone separators are often used, and since their workings are independent of gravity, they should be carefully considered for space applications.

Gas or Vapor Turbines

Gas or vapor turbines, including control valves, seals, and shaft bearings, will be needed, as on Earth, to drive electric generators for electric-propulsion schemes and perhaps to drive the various liquid and vapor pumps for the cycle working fluids, combustion reactants, coolants, and propellants that are used in any of the propulsion systems of interest. Owing to the high fluid velocities expected in such devices, no microgravity effects in the fluid streams are to be expected, unless cavitation or flashing of vaporizable liquids occurs. Such effects frequently occur in control valves, which suggests careful attention to shaft seals. Film bearings, particularly cryogenic bearings, may be sensitive to load changes associated with variable or reduced gravity.

Liquid Pumps

Pumps, which also require control valves, seals, and shaft bearings, perform liquid transfer duties in all propulsion systems. In particular, high-pressure pumps are needed to pressurize working-fluid condensate in Rankine-cycle electric propulsion, and pumps are needed to circulate working fluids generally, combustion reactants, coolants, and propellants for all propulsion systems. The same concerns about microgravity effects apply as for turbines, except that cavitation is not to be expected at the high-pressure end of pumps.

Compressor

A compressor is needed to raise the pressure of the (gas) working fluid in the Brayton-cycle option for electric propulsion, raise the vapor pressure in the refrigeration cycle for cryogenics, and propel gas-phase fluids as needed in other systems. In the gas stream, unless condensation occurs, no microgravity effects should arise. Microgravity would be an issue for valves, seals, and bearings, as it would be for pumps and turbines.

Condenser for a Rankine Cycle

A condenser for a Rankine cycle would contain a heat exchanger for use between working fluid and heat-rejection coolant as well as devices for separating and collecting fluid phases. In any electric propulsion system employing the Rankine cycle, the working fluid passes through a vapor turbine and then must revert to the liquid phase before being pressurized and returned to the boiler. This condensation process releases heat at a rate proportional to the power generated, and that heat must be delivered to space via the radiator system mentioned above in Section III.A. Therefore, a high-capacity heat exchanger must connect the working-fluid vapor and the radiator surfaces either directly or through an intermediate coolant such as ammonia.

On Earth, the condensed fluid is removed from the heat exchanger surfaces by gravity. In reduced gravity or microgravity, other mechanisms, such as surface tension in a heat pipe, centrifugal phase separators, or direct contact condensation, must be used. Significantly, these condenser designs may incur size and weight penalties because of microgravity. Moreover, the condensate pumps that return the condensate to the boiler must be carefully designed to make sure that there is sufficient net positive suction head in microgravity environments. To assure the integrity and efficiency of a Rankine cycle, it is very important that there be no liquid carry-over or carry-under. On Earth, gravity helps accomplish this task. In microgravity, the centrifuge effect (e.g., a rotary fluid-management device, or RFMD) would presumably be used. This kind of microgravity countermeasure is discussed in Chapter V.

Vaporizer for Propellant

In electric propulsion systems, the propellant must be provided to the nozzle system as a vapor. However, the propellant might be a noble gas stored as a liquid, cryogenically, or it might be a metal such as cesium. Vapor might be formed by adding heat or by reducing pressure (flash boiler), depending on the propellant (Cassady, 1990). In microgravity, vaporization methods dependent on buoyancy to collect the vapor would have to be avoided; therefore, vaporization by sudden reduction of flow pressure would presumably be preferred.

Switch Gear and Electric Power Conditioning

For electric propulsion, the power output of the generator must be modified and controlled to provide the proper voltage and current to the thruster system. Such power conditioning generates waste heat in large quantities, which must be delivered ultimately to a space radiator. Thus, heat-rejection methods needed for the thermodynamic power cycle itself must also deal with large losses in the electrical system. Further, the electrical apparatus would be kept at rather low operating temperatures, which would further burden the heat-rejection design.

Common Design Elements

The following paragraphs discuss specific elements or components of technology that seem to be important for propulsion systems and that are also important for almost all functions described in this chapter, especially those that involve fluid flow and heat transfer. These elements are affected by microgravity but often indirectly or secondarily, in ways easily overlooked but nevertheless important.

Heat exchangers between unmixed elements, fluid or solid, are needed for thermal management of spacecraft systems and as thermal connections between primary heat collection fluids, cycle working fluids, and propulsive gases. Heat exchangers in which phase change is not expected and which handle high flow rates, such as the cooling passages in a fission reactor, would not be expected to perform differently in microgravity. Any heat exchanger for which, on Earth, buoyancy might be expected to propel the fluid, due either to phase change or to thermosiphon phenomena (a two-phase riser would be an example of the latter), would be problematic in reduced gravity. An interesting multiphase heat-transfer device, the vapor-pressure pumped loop (Section III.A), may be helpful in microgravity.

Heat exchangers, especially those with a fine flow-passage structure, are subject to loss of performance by surface fouling, and this would be especially troublesome for space systems that are hard to clean or replace. In microgravity, atmospheres in spacecraft or space colonies might contain high concentrations of suspended particles or droplets that lead to fouling of heat exchanger surfaces. Thus, microgravity could indirectly but substantially affect heat-exchanger performance. The purity of the fluids passing through heat exchangers requires careful assessment.

Piping systems, including valves, would be used for fluid handling in all propulsion and power systems, and in a wide variety of other systems, for all fluids, including liquids, gases, mixed phases, slurries, and suspensions. The various microgravity phenomena that can occur in multiphase flows in straight pipes and channels are discussed in Chapter IV. Still other phenomena will come into play at pipe bends and fittings, where transverse fluid accelerations occur. Certain mechanical problems can also be foreseen. Because flow passages used in reduced gravity would presumably be designed to be of low mass, the piping is likely to lack the strength and rigidity one would expect for terrestrial installations. Piping designed for reduced gravity would therefore be expected to be highly vulnerable structurally, especially at bends and elbows, to surge and liquid-hammer effects and to flow-induced vibrations, especially if multiple phases are present. Abrupt valving may initiate or aggravate such behavior, for example because of cavitation. These issues are discussed again in Section V.A.

It has been mentioned that HEDS technology will require various machines for fluid-handling tasks, such as pumps, turbines, and motors, and these will typically need bearings. Fluid handling will certainly require valves for control, and bearings and valves both require seals to segregate liquids or gases from the surrounding spaces. Often, the operation of bearings and seals will not be directly affected by microgravity, but the magnitudes and nature of the loads for which the bearings are to be designed will be affected. This indirect influence of microgravity is discussed further in Section V.A.

For many HEDS missions, robots of various types will probably be used, for long periods, in order to protect the crew from the environmental hazards of space. As mechanical devices, these robots will experience wear and decay, which are processes affected by gravity level. Microgravity will also have indirect consequences (in terms of structural dynamics) for robot design and performance, as is discussed in Section V.A. Antennas for communication and solar collectors and space radiators for distant HEDS missions will be large and will require accurate control of orientation. Tanks for storage of liquids and gases will also be large. These components, to be used in low gravity, will also present structural dynamics problems, as is discussed in Chapter V.

TABLE III.B.1 Selected Subsystems Found in Propulsion Technologies and the Potential Impact of Microgravity on Their Operation

	Technology			
Subsystem	Chemical	Nuclear Thermal	Nuclear Electric (Rankine)	Nuclear Electric (Brayton)
Reactor (fluidized)	—	H	H	H
Evaporator (liquid or metal)	—	—	H	H
Cryogenic system	H	H	—	M
Condenser	—	—	H	—
Liquid/vapor separator	—	—	H	—
Radiator (two-phase)	—	—	H	H
Heat pipes	—	—	M	M
Liquid pumps	M	M	M	—
Vapor compressor	—	—	—	M
Vapor turbines	—	—	M	—
Vapor heat exchange	—	—	—	M
Solar collector	—	—	—	—
Sails (solar or laser)	—	—	—	—
Reactor controls	M	M	M	—
System controls	L	L	M	M
Seals	L	L	M	L
Valves	L	L	M	L
Film bearings	L	L	M	L
Piping	L	L	M	L
Robots	L	L	L	L

NOTE: The letters H, M, and L designate high, medium, and low (preliminary assessment) impact of reduced gravity on the operation of the subsystem. Where no letter is given, the subsystem is not applicable to the system listed.

General Concerns Regarding Propulsion and Power in Reduced Gravity

Certain general issues affecting power and propulsion component design and performance, seemingly important in microgravity, merit more thorough study in the HEDS context than they have received so far. Touched on in Section III.A on power and above in this section on propulsion, they deserve further emphasis here to round out the committee's discussion of the key topics of power and propulsion.

Variable Gravity

Most propulsion and power systems should be able to operate in a range of gravity or acceleration environments, from coasting to full thrust, or on a station on the surface of Mars; this is especially true of the electric power-generation modes of the bimodal NTR or NEP[7] systems. The exploration of asteroids, with its varied acceleration levels during flight and complex docking maneuvers, would test the ability of components to accommodate various gravity levels. While it is true that a device designed for Earth gravity may not work well in zero gravity, the converse is also true, that a device designed for zero gravity might fail to perform properly under Earth gravity.

[7]Neiberding; see footnote 4.

Solar Thermal	Solar Electric (Rankine)	Solar Sail	Laser Thermal	Laser Sail	Tether	Aeroassist
—	—	—	—	—	—	—
—	H	—	—	—	—	—
H	—	—	—	—	—	—
—	H	—	—	—	—	—
—	H	—	—	—	—	—
—	H	M	—	M	—	—
M	M	—	M	—	—	—
L	M	—	—	—	—	—
—	—	—	—	—	—	—
—	M	—	—	—	—	—
—	—	—	M	—	—	—
H	H	—	—	—	—	—
—	—	M	—	M	—	—
—	—	—	—	—	—	—
L	M	M	H	H	H	H
L	M	—	L	—	—	—
L	M	—	L	—	—	—
L	M	—	L	—	—	—
L	M	—	L	—	—	—
L	L	L	L	L	L	L

Transient Operation and Unsteady Processes

Intermittent or variable operation, including start-up in space, must be considered carefully. System dynamics and system instabilities are very important issues for power generation generally; the generation system and its load must be managed together. Some issues of this kind, especially nuclear start-up, have been carefully studied (Kirpich et al., 1990), but the full range of transient issues probably has not

Unsteady processes can be very important for system performance and can be favorable or unfavorable. Harmful effects include instabilities: in rotating systems, either fluid or solid, instabilities are common (Yih, 1965), liquid-hammer loads may occur in conduits (Streeter and Wylie, 1985), and structural fatigue can occur due to cyclic loading. Beneficial effects may include enhanced mixing or diffusion by wave processes. Both ultrasonic and very low frequency waves may be of interest. Generally, gravity level must be considered a factor in these various processes. Unsteady flow is the basis of the proposed vapor-pressure pumped loop.

Multiphase Flow

It appears that the power and propulsion systems needed for HEDS will require devices that depend on multiphase flow and heat transfer processes to achieve low mass, high efficiency, and low cost. However, as is discussed in Chapter IV, there are currently large uncertainties surrounding the behavior of multiphase flows in reduced gravity. In particular, the transverse distribution of phases within flow passages will be quite different from what it is in Earth's gravity. Moreover, surface-tension-induced forces (e.g., Marangoni forces) will be much more important in space than they are on Earth. If multiphase systems and processes are to be used in space, then reliable, physically based predictive tools will need to be developed and used by NASA for the design and analysis of candidate propulsion and power systems and subsystems.

Need for Artificial Gravity

Amid concerns for the effects of low or variable gravity, it is important to keep in mind that technical means exist for supplying artificial gravity, and these should be explored by NASA, the purpose being not to make things more familiar, but to counter the real technical penalties suffered when body force is not available. The technical means include, but are not limited to, rotation to provide centrifugal body force. Rotation might be imposed on the scale of the spacecraft itself or on the smaller scales of components such as swirl separators of liquids and vapors. The costs and benefits of a full range of microgravity countermeasures should be studied. This topic is discussed more fully in Chapter V.

Reliability

The more complex a system, the more opportunities it provides for failure and accident. This is not a matter of being careful in initial design; it is a matter purely of probabilities. Designs of 30-year-old aircraft are still being corrected, after thousands of mission cycles. In contrast, HEDS systems must be designed once and for all to be reliable, without repair, for years. For HEDS missions then, designers must surely learn to be fanatical on the subject of simplicity and reliability. Systems and components should be simple, with a minimum of moving parts, shafts, bearings, valves, or anything that may fail or wear out. This issue of reliability has affected the growth of commercial nuclear power; NASA should learn from that history and, especially for HEDS, put special emphasis on design for simplicity and reliability. Techniques of probabilistic risk assessment (PRA), discussed in Section V.C, should be useful for this purpose.

It should be emphasized that variability of gravity, or loading due to accelerations, also affects design and performance and, as such, can increase the possibility of failure and corresponding decreases in reliability.

Nuclear System Development

Nuclear fission can support electric power generation for propulsion and power and can also directly support thermoelectric, thermoionic, and advanced thermoelectric and thermoionic converters. Such systems are not affected by gravity and have the potential to provide higher power levels in the future, if development is pursued. However, NASA has stopped funding research and development on nuclear space power (Bennett et al., 1996; Bennett, 1998), and it appears that the United States will soon lose its ability to develop nuclear power and space propulsion systems.

Since it appears that nuclear fission power will be essential for the success of the long-range goals of HEDS, there is a need for NASA to maintain a steady effort in this field, with attention paid both to newer reactor types and to the many advanced components needed to ensure desired performance at various gravity levels.

Summary of the Effect of Reduced Gravity on Selected Subsystems

Summarized in Table III.B.1 are the various subsystems and components discussed in this section and the various propulsion systems where they are found. The impact of reduced gravity on the operation of these subsystems is estimated as high, medium, or low (little or no impact). It should be remembered, however, that the impact of the gravity level on these technologies will depend greatly on design context, which cannot be predicted.

References

American Institute of Aeronautics and Astronautics (AIAA), Aerospace Power Systems Technical Committee. 1995. Space nuclear power: Key to outer solar system exploration, AIAA position paper. Reston, Va.: AIAA.

Barnett, J.W. 1991. Nuclear electric propulsion technologies: Overview of the NASA/DOE/DOD nuclear electric propulsion workshop. Pp. 511-523 in Proceedings of the 8th Symposium on Space Nuclear Power Systems. College Park, Md.: American Institute of Physics.

Bennett, G.L. 1998. Electric power technologies for spacecraft: Options and issues. AIAA Paper No. 98, p. 1022. Reston, Va.: American Institute of Aeronautics and Astronautics.

Bennett, G.L., and T.J. Miller. 1992. NASA program planning on nuclear electric propulsion. AIAA paper 92-1557. New York: American Institute of Aeronautics and Astronautics.

Bennett, G.L., H.B. Finger, T.J., Miller, W.H., Robbins, and M. Klein. 1994. Prelude to the future: A brief history of nuclear thermal propulsion in the United States. In A Critical Review of Space Nuclear Power and Propulsion, 1984-1993. M.S. El-Genk, ed. New York: Springer-Verlag.

Bennett, G.L., R.J. Hemler, and A. Schock. 1996. Space nuclear power: An overview. J. Propulsion Power 12:901-910.

Carrick, P., and J. Harper. 1998. High energy propellants at the Air Force Research Laboratory. Pp. 169-175 in Proceedings of the 9th Advanced Space Propulsion Workshop. Pasadena, Calif.: Jet Propulsion Laboratory.

Cassady, R.J. 1990. Propulsion systems. Pp. 69-80 in Thermal-Hydraulics for Space Power, Propulsion and Thermal Management System Design. AIAA Progress in Astronautics and Aeronautics, Vol. 122. New York: American Institute of Aeronautics and Astronautics.

Cutler, A.H., and J.A. Carroll. 1992. Tethers. Pp. 136-144 in Space Resources—Energy, Power, and Transport. NASA SP-509, Vol. 2. Linthicum Heights, Md.: National Aeronautics and Space Administration Center for Aerospace Information.

Dearien, J.A., and J.F. Whitbeck. 1990. Advanced multimegawatt space nuclear power concepts. Pp. 41-67 in Thermal-Hydraulics for Space Power, Propulsion and Thermal Management System Design. AIAA Progress in Astronautics and Aeronautics, Vol. 122. New York: American Institute of Aeronautics and Astronautics.

French, J.R. 1981. An expedition to Mars employing shuttle-era systems, solar sails, and aerocapture. Pp. 245-250 in The Case for Mars (American Astronautical Society), April 29-May 2, Boulder. Science and Technology Series, Vol. 57. P.J. Boston, ed. San Diego: Univelt.

Gallagher, D.L., L. Johnson, F. Bagenal, and J. Moore. 1998. An overview of electrodynamic tether performance in the Jovian system. P. 427 in Proceedings of the 9th Advanced Space Propulsion Workshop. Pasadena, Calif.: Jet Propulsion Laboratory.

Harris, H.M. 1999. Light sails. Sci. Am. 280(2):90.

Hill, P.G., and C.R. Peterson. 1965. Mechanics and Thermodynamics of Propulsion. Reading, Mass.: Addison-Wesley.

Hoyt, R.P., and R.L. Forward. 1998. The terminator tether deorbit system. P. 467 in Proceedings of the 9th Advanced Space Propulsion Workshop. Pasadena, Calif.: Jet Propulsion Laboratory.

Jet Propulsion Laboratory (JPL). 1998. Micropropulsion session. Pp. 341-424 in Proceedings of the 9th Advanced Space Propulsion Workshop. Pasadena, Calif.: Jet Propulsion Laboratory.

Kirpich, A., A. Das, H. Choe, E. McNamara, and D. Switick. 1990. Startup thaw concept for the SP-100 space reactor power system. Pp. 143-169 in Thermal-Hydraulics for Space Power, Propulsion and Thermal Management System Design. AIAA Progress in Astronautics and Aeronautics, Vol. 122. New York: American Institute of Aeronautics and Astronautics.

Mead, F. and L. Myrabo. 1998. Overview of Air Force laser propulsion program and goals and laser propelled flight experiment. P. 427 in Proceedings of the 9th Advanced Space Propulsion Workshop. Pasadena, Calif: Jet Propulsion Laboratory.

National Aeronautics and Space Administration (NASA). 1991. Integrated Technology Plan for the Civil Space Program: Nuclear Propulsion. Washington, D.C.: NASA.

Perry, M.D., and H.T. Powell. 1998. Ultra high power lasers state of the art and beyond. P. 311 in Proceedings of the 9th Advanced Space Propulsion Workshop. Pasadena, Calif.: Jet Propulsion Laboratory.

Rosen, R., G. Reck, and G. Bennett. 1993. Application of nuclear propulsion to Mars missions. Space Technology 18(5):467-478.

Sarner, S.F. 1968. Solid propellant rockets. Pp. 308-331 in Jet, Rocket, Nuclear, Ion, and Electric Propulsion: Theory and Design. Applied Physics and Engineering Series, Vol. 7. W.H.T. Loh, ed. New York: Springer-Verlag.

Schneider, R.T., and K. Thom (eds.). 1971. Research on Uranium Plasmas and Their Technological Applications. Various articles. NASA SP-236. Washington, D.C.: National Aeronautics and Space Administration.

Seifert, H.S. 1968. Hybrid rocket theory and design. Pp. 332-355 in Jet, Rocket, Nuclear, Ion, and Electric Propulsion: Theory and Design. Applied Physics and Engineering Series, Vol. 7. W.H.T. Loh, ed. New York: Springer-Verlag.

Snow, W.R., and H.H. Kolm. 1992. Electromagnetic launch of lunar material. Pp. 117-135 in Space Resources—Energy, Power, and Transport. NASA SP-509, Vol. 2. Linthicum Heights, Md.: National Aeronautics and Space Administration Center for Aerospace Information.

Staehle, R.L. 1981. An expedition to Mars employing shuttle-era systems, solar sails, and aerocapture. Pp. 91-108 in The Case for Mars (American Astronautical Society), April 29-May 2, Boulder, Colo., Science and Technology Series, Vol. 57. P.J. Boston, ed. San Diego, Calif.: Univelt.

Streeter, V.L., and E.B. Wylie. 1985. Fluid Mechanics, 8th ed. New York: McGraw-Hill, pp. 521-543.

Yih, C.S. 1965. Dynamics of Nonhomogeneous Fluids. London: Macmillan.

III.C LIFE SUPPORT

Introduction

Arguably, the central and paramount HEDS function is to provide an environment consistent with the sustained existence of personnel outside of Earth's atmosphere at a comfort level that will enable high performance. The systems necessary to fulfill this function include those that protect against ionizing radiation; control temperature, pressure, humidity, and waste products to within prescribed limits; provide adequately balanced food, potable water, and hygienic water; and afford adequate physical activity. Because the psychological aspects of life support may be even more problematic than the physiological aspects, additional amenities consistent with creature comforts will be important. Systems considered important for life support are described qualitatively in an NRC report (NRC, 1997) and summarized in an excellent monograph (Eckart, 1996). The former report provided a useful framework for categorizing the various technologies cited in this section, while the latter provided invaluable design details for a wide range of technologies. Neither document identified scientific or technical issues associated with possible failure modes arising from reduced gravity. This section emphasizes these issues as it looks at selected technologies that are likely to be both important to life-support design and significantly affected by gravity levels.

Key to understanding life-support systems and their subsystems is the concept of homeostasis, or maintaining constant, optimal levels of the various physical, chemical, and biological systems necessary for life support. An important goal is to achieve as nearly as practical a closed ecological system requiring the input of a minimum of mass and energy and in which as many subsystems as possible utilize recycling. In order to close cycles, different forms of matter and of energy must be interconverted and stored, preserving the necessary balance of each within a habitable module. These are precisely the processes most heavily affected by the reduced influence of density differences on gravity-density coupling and interfacial phenomena (discussed in Section IV.A) in microgravity.

Systems required for life support fall into five principal areas of activity (AIAA, 1990):

1. Regulation within suitable limits of the temperature and relative composition, purity, and pressure of the ambient gas phase of the habitat and/or the equipment for extravehicular activity;
2. Management of the quantity and quality of drinking and hygienic water and the associated recovery and processing systems;
3. Collection, processing, and recovery procedures associated with biological waste and trash;
4. Food management, including production, preparation, and storage; and
5. Crew safety management, including radiation shielding and fire detection and suppression.

The first four activities consist of mass and energy conversions involving interrelated processes. For example, atmospheric parameters include the relative humidity, which is at the same time an issue for water homeostasis. Food production and human waste management are potentially closely coupled and interact with water and atmospheric management systems. The fifth area requires countermeasures to limit the biological effects of ionizing radiation and is discussed further in Section III.D. The adverse effects of microgravity on certain aspects of human physiology such as bone metabolism are covered in detail in a recent NRC report (NRC, 1998) and are not considered further here.

Conceptual guidelines dealing with interconnectedness for the purpose of design and development efforts are outlined in NRC (1997) and detailed in Eckart (1996). To understand these issues, regenerative and nonregenerative systems must first be distinguished. Regenerative systems explicitly recover and recycle in order to minimize the mass required for a mission and the ultimate waste that must be stored and jettisoned from it. They usually involve closed loops in which the flow of mass to and from the habitat is limited. It is useful to note briefly that wastewater recycling can reduce the relative supply mass by nearly 50 percent and that water, carbon dioxide, and oxygen recycling can reduce it by nearly 90 percent. Trace contaminant and particulate removal and temperature, pressure, and humidity control cannot be easily accomplished by regenerative systems. Other requirements, including oxygen generation and carbon dioxide reduction, are nearly always carried out by regenerative cycles.

TABLE III.C.1 Gaseous Component Management Systems

Carbon Dioxide Removal	Carbon Dioxide Reduction	Oxygen Generation
Molecular sieve	Bosch cycle	Static feed electrolysis
Electrochemical depolarization concentration	Sabatier cycle	Water vapor electrolysis
Solid amine-water desorption	Carbon formation reactor	Carbon dioxide electrolysis
Photochemical	Photocatalysis	In situ resources
LiOH	Direct electrolysis	Plants
Electroreactive carriers	Catalytic decomposition	Cryogenic storage
Plants	Ultraviolet photolysis	High-pressure storage
	Plants	

SOURCE: Adapted from Eckart (1996).

Generally, closed-loop systems are superior wherever resupply costs are high, whereas open-loop systems are simpler to implement because they require less technological development initially and have lower power requirements. The advantages of the former are offset somewhat by their more intensive dependence on phase changes and multiphase (fluid and solid mixtures) processing, which makes them harder to implement in microgravity. However, they will become increasingly necessary as enabling technologies as missions become more distant and of longer duration.

Life support is distinct from other HEDS functions owing to the introduction of biological, as opposed to physicochemical, strategies into systems that interchange the chemical compositions of different supply and waste streams. Such biologically based systems, or bioreactors, will probably be involved in both nutrient production and waste management.

Crucial to the successful development of suitable closed ecological life support systems (CELSS)[8] will be the development of interfaces between bioregenerative and physicochemical systems. Biological and hybrid systems exploit the metabolism of living organisms to effect some or all of the requisite transformations, while physicochemical systems rely exclusively on well-understood mechanical or chemical processes. In general, biological systems are bulkier, more difficult to maintain, consume less power, and respond much more slowly than do physicochemical systems. However, bioreactors have the unique potential to provide food. Moreover, many bioreactor subsystems rely heavily on capillarity and other interfacial phenomena whose behaviors in microgravity cannot yet be adequately modeled (see Section IV.B). Improved modeling capabilities would decrease the difficulty and expense of bioreactor development and testing.

Atmospheric Homeostasis

Systems in this category regulate and monitor atmospheric composition and supply (pressure), temperature and humidity, decontamination, and cabin ventilation. They must function with little or no human intervention for long time periods.

Air Revitalization

The most important functions in maintaining an appropriate atmosphere are those that regulate the partial pressures of oxygen and carbon dioxide (Table III.C.1). These two gases are interconverted by the dominant metabolic cycles of humans and plants. This reciprocity may ultimately be exploited to help regulate both gases by balancing the respective metabolic outputs. However, physicochemical control over both gases will remain

[8]Note that this acronym has appeared in the literature with slightly varying designations depending on the literature source and period.

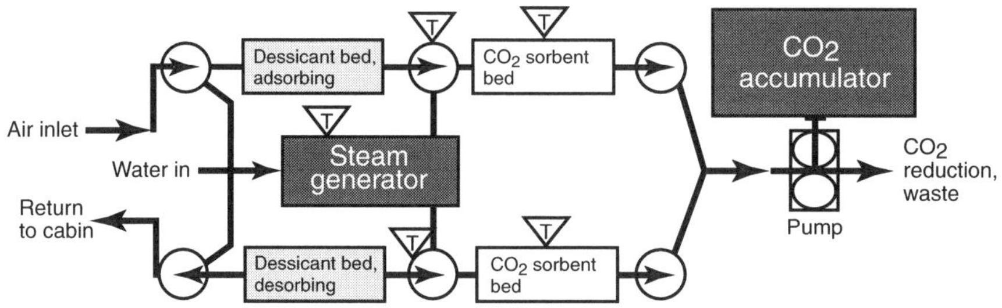

FIGURE III.C.1 Generic carbon dioxide (CO_2) collector. Gray sections represent the desiccant beds used in the technologically mature four-bed zeolite system. Sorbent beds represent either zeolite (four-bed system), carbon molecular sieves, or solid amine resin beds, which must be steam heated. Subsystems adversely affected by microgravity are shown in reverse contrast. SOURCE: Based on Eckart (1996).

central in the short term and may serve to fine-tune even advanced bioregenerative systems. The physicochemical cycles used to interconvert carbon dioxide and oxygen are independent, however, and systems with a high technical readiness level (TRL) therefore deal with three separate processes: carbon dioxide removal, carbon dioxide reduction, and oxygen generation.

Carbon Dioxide Removal and Concentration

Carbon dioxide is a waste product produced by human respiration at the rate of ~1 kg/man-day and is potentially toxic. Treatment involves concentration followed by chemical reduction to a reduced form of carbon, such as methane, and molecular oxygen. Concentration or "removal" can be effected using a variety of regenerable and nonregenerable systems with a range of technological readiness, weights, ambient operating conditions, and power requirements. The carbon dioxide removal system with the highest TRL is a four-bed molecular sieve incorporating synthetic zeolites or metal ion aluminosilicates to collect carbon dioxide (Eckart, 1996). These collecting materials cannot tolerate excess moisture, so a preabsorbing sieve is necessary to dry the air before carbon dioxide absorption. Moreover, in the desorption cycle the sieve is heated while it is exposed to space vacuum, so the carbon mass is lost. Thus, although the system is "regenerable," it cannot be incorporated into a closed system without the development of additional vacuum technology to recover and concentrate the carbon dioxide. Moreover, use of this system has shown that it periodically builds up residual carbon dioxide, which must be eliminated by a bakeout at 478 K.

Regenerative carbon dioxide collectors compatible with closed systems, illustrated schematically in Figure III.C.1, are likely to be affected by microgravity. One near-term technology is the solid amine-water desorption (SAWD) system, which uses chemical reaction with solid amines in the absorption phase, similar to aqueous ion exchange. Subsequent desorption with steam requires a boiler, which is highly vulnerable to the gravity level and adds to the loading of heat rejection systems. Newer, carbon-based molecular sieves have been developed that are insensitive to water vapor and that offer more efficient trapping and can be cycled through desorption at closer to ambient pressure and temperature. These materials can be incorporated into two-bed sieve systems that do not require the desiccant beds and are roughly twice as efficient as four-bed systems. Two-bed systems can be integrated more readily into closed-loop regenerative systems. However, no prototypes have as yet been designed. The remaining systems listed in Table III.C.2 currently have very low technological readiness but are of interest because of their potential economies of power and/or weight. Of these, ultrafiltration using osmotic membranes (membrane removal), electroactive carriers (ion-exchange dialysis), and, possibly, combinations of these could potentially provide for the most direct processing. Moreover, all closed-cycle regenerative systems require a carbon dioxide condenser/separator, which would be very sensitive to microgravity.

TABLE III.C.2 Physicochemical Systems for the Concentration of Carbon Dioxide

Nonregenerative Systems				Regenerative Systems			
System	TRL[a]	Efficiency	Products	System	TRL[a]	Efficiency (%)	Products
LiOH	8	—	$LiCO_3$,	Four-bed molecular sieve	8	66	
Sodasorb			H_2O	Two-bed molecular sieve	2-3	90	
Superoxides				Solid amine-water desorption	6		
				Electrochemical depolarization	6		H_2O, heat, DC power
				Air polarization	2		
				Membrane removal	2		
				Electroactive carriers ion-exchange electrodialysis	1		

[a]TRL, technical readiness level, the maturity of a system ranging from level 1 (a basic principle observed and reported) to level 8 (a design qualified for spaceflight).

Reduction of Carbon Dioxide

The Bosch and Sabatier reactors, particularly the Sabatier reactor, are crucial for generating oxygen for various propulsion systems, and both are described further in Section III.E. Both produce water vapor, which must be condensed and separated from the vapor phase. On Earth this separation would be driven by gravity. Subsystems involved in steam generation involve two-phase fluid management, so they would operate much differently in a microgravity environment.

Oxygen Generation—Water Electrolysis

A regenerative CELSS will almost certainly require a water electrolysis system. Electrolysis involves the delivery of electrons via an electric current to water, yielding hydrogen, and the removal of an equivalent number of electrons from the hydroxide ion, yielding molecular oxygen. The process is inherently two-phase, as the dissolved gases must be recovered and concentrated. It is thus also highly sensitive to gravity level.

Oxygen Supply and Regeneration

The chemical storage of oxygen is the least affected by operation in microgravity, but it is not suited to regenerative cycles. Cryogenic or supercritical oxygen storage has the significant advantages of high density and low pressure, and liquefied gases could serve as an important heat sink for integration into larger thermal control cycles. Development of these technologies is therefore crucial for HEDS. However, they are highly sensitive to gravity, the former requiring a condenser and the latter a compressor.

Temperature and Humidity Control

Control of ambient temperature and humidity within tolerable limits involves the physical processes of heat exchange, multiphase flow, phase changes, and phase separation, all of which are strongly affected by the gravity level, as described in Chapter IV. The system responsible for all of these processes is a condensing heat exchanger (CHX). Cooling water is used to bring the incoming air to a temperature below the dewpoint of water, inducing formation of a condensate, which must then be separated from the airstream. The phase separation generally takes place via a "slurper," which is a flat, thin volume bounded by plates, one of which is in contact with the airstream and condensate film. Holes in this plate serve to entrain the condensate into the volume between the plates (Kuhn, 1988).

TABLE III.C.3 Physicochemical Systems for Recovering Water from Urine and Hygienic Waste

Process	TRL[a]	Efficiency (%)
Distillation		
Vapor-compression distillation	5	70
Thermoelectric integrated membrane evaporator	4-5	91
Vapor-phase catalytic ammonia removal	3	95
Filtration		
Reverse osmosis		
Multifiltration		
Electrodialysis		

[a]TRL, technical readiness level, the maturity of a system ranging from level 1 (a basic principle observed and reported) to level 8 (a design qualified for spaceflight).

Trace Biological, Chemical, and Particulate Contaminant Control

The isolated nature of a spacecraft environment introduces the critical need to detect and eliminate airborne hazards arising from outgassing of insulation and plastics, from human metabolism, and from accidental spills and degradative processes. Contaminant detection by gas chromatography, mass spectrometry, and/or Fourier transform infrared spectroscopy would seem to pose no obvious problems in microgravity.

Water Homeostasis

Recovery

Water can be recovered from aqueous waste by either fractional distillation or membrane filtration (Table III.C.3) (Eckart, 1996). Urine and waste wash water present significantly different problems for water recovery. Urine contains urea, sodium chloride, and various organic salts and acids at total concentrations of up to 5 percent. It supports bacterial growth. In contrast, the contaminants in waste wash water are considerably more diverse, ranging from dead skin, hair, and dirt to fats, soaps, detergents, and other organic compounds found in sweat. These differences in composition have led to the perception that distillation is the preferable treatment for urine, while filtration is to be preferred for other waste water sources. The water condensates dealt with by humidity management (from respiration, perspiration, and transpiration sources) are technically, although not explicitly, in the former category.

Distillation requires paired phase changes and separations and is strongly affected by gravity. The four distillation schemes in Table III.C.3 incorporate different solutions to the problems posed by a microgravity environment. Vapor-compression distillation (VCD) uses a compressor to raise the saturation temperature of the process water vapor, while the actual condensation is engineered to take place in direct contact with the evaporator, exploiting the latent heat of condensation for the evaporation process (Eckart, 1996; Schmidt, 1989). Vapor-phase catalytic ammonia removal (VAPCAR) combines distillation at 523 to 723 K with catalytic decomposition of contaminating volatiles to molecular nitrogen, hydrogen, and oxygen by a series of two platinum/ruthenium catalytic reactors. The high ambient temperature ensures that the water produced is not only chemically very pure but also free of microbial contamination, requiring only adjustment of the pH to render it potable.

The low TRL ratings for these subsystems reflect, in large part, the technical difficulty of using multiphase systems in microgravity without further substantial research and development. VCD and VAPCAR technologies use a compressor and separator, respectively, that must operate in microgravity. The thermoelectric integrated membrane evaporator uses reduced pressure outside a hollow membrane fiber system to evaporate water from urine. The extensive pretreatment required to fix free ammonia and prevent microbial growth currently uses oxone, a hazardous component that must be used as a liquid in microgravity, under which conditions it becomes unstable. Reverse osmosis, electrodialysis, and filtration may become alternatives to two-phase systems if they can be brought to maturity.

Solid Waste Management

Whereas liquid waste poses physical separation problems, recycling requires that solid waste be chemically transformed, generally by oxidation. To date, solid waste treatment has been limited to drying the various organic solids, with or without recycling of the extracted water. Three unique problems impede further development:

1. Isolating and managing the organisms and toxins present in biological wastes;
2. Processing mixtures of solid materials whose composition varies significantly; these mixtures may be characterized as ill-defined, multiphase systems; and
3. Coupling solid waste processing with plant growth chambers and other biologically based waste treatment systems, as well as with physicochemical gas/liquid/solid mixers and separators.

Both incineration and supercritical wet oxidation are being studied for recycling solid wastes (Eckart, 1996; Bilardo and Likens, 1991). Both systems show promise for future spacecraft designs. Incineration systems process mixtures of human feces, urine, and other solid biological wastes previously dried (by evaporation) to water content levels of less than 50 percent. Subsequently the feed material is combined with air or oxygen in a reactor operated near ambient pressure and heated to temperatures above 800 K to effect combustion. Incomplete combustion, produced by supplying the reactor with substoichiometric oxygen, can produce partial pyrolysis and a more stable exhaust gas stream of nearly pure water and carbon dioxide and with larger particulate sizes and fewer toxic emissions than are produced by providing excess air (or oxygen). Partial incineration of nitrogenous waste also produces ammonia, rather than nitrate and nitrite ions, and the ammonia can be converted to nitrogen and water in an afterburner.

Supercritical wet oxidation utilizes the unique solvent capabilities of supercritical water above 647 K and 22.1 MPa. Above these conditions, and at significant levels of oxygen, normally insoluble organic solids can be absorbed in water, enabling rapid oxidation of undesirable organic waste products almost completely, in a single supercritical phase. Because supercritical water does not have separate liquid and vapor phases, many of the microgravity phase-separation problems are avoided. However, the high temperatures and pressures of the reactor, as well as the corrosiveness of the supercritical water itself and the toxicity of the product gases, raise safety concerns.

As noted, the solid waste recycling systems have not been operated in space and are not currently justified for Earth-orbiting systems like the International Space Station. Because of the inherent variability and the low immediate value of the recycled materials that can be recovered, waste recovery systems should probably not be developed as stand-alone systems but eventually should be integrated into bioregenerative systems that produce food and oxygen from plants while recycling solid waste at some level. Accordingly, several types of biologically based solid waste recovery systems are being considered at this time. To develop those systems for operation on spacecraft, advances are required in the following areas:

- Systems that chop or grind solid wastes effectively, collecting the product;
- A process for blending solid wastes with water to produce a controlled liquid feed compatible with pumping;
- Systems that dissolve gases in liquids;
- Systems that remove dissolved gases from liquids;
- Liquid water/nutrient management systems that do not flood plant chambers; and
- Systems that can maintain immobilized cell colonies efficiently in microgravity.

TABLE III.C.4 Major Subsystems Found in the Various Life-Support Systems and the Potential Impact of Microgravity on Their Operation

	System				
Subsystem	Molecular Sieve	Solid Amine-Water Desorbtion	Electrochemical Depolarization Process	Bosch	Sabatier Process
Adsorber (gel)	L	—	—	—	—
Adsorber (solid matrix)	M	L	—	—	—
Boiler	—	H	—	—	—
Catalyst bed	—	L	—	L	L
Centrifuge	—	—	—	—	—
Chopper/grinder	—	—	—	—	—
Compressor	—	—	—	L	L
Condenser/separator	—	H	—	H	—
Electrochemical membrane	—	—	L	—	—
Fan/blower	L	—	L	—	L
Filter	L	L	—	L	L
Heat exchanger	L	L	L	L	L
Heat pipes	—	—	—	—	—
Heater	L	—	—	L	L
Mixer	—	—	—	—	—
Oven	—	—	—	—	—
Pipes and valves	L	L	L	L	L
Pump	L	L	L	L	L
Scrubber	—	—	—	—	—
Slurper	—	—	—	—	—
Sparger	—	—	—	—	—
Storage tank	L	H	H	H	L
Thermoelectric refrigeration	—	—	—	—	—

NOTE: The letters H, M, and L designate high, medium, and low (preliminary assessment) impact of reduced gravity on the operation of the subsystem. Where no letter is given, the subsystem is not applicable to the system listed. TIMES, thermoelectric integrated membrane evaporator; VAPCAR, vapor-phase catalytic ammonia removal.

Food Production

In contrast to some areas discussed above, food production from recycled wastes is expected to make a relatively minor contribution to life support in the near future (Eckart, 1996). However, longer term advances will depend increasingly on improvements in this area. Heading the list of systems required for regenerative food production is the soil-bed reactor (SBR), in which plants are grown for their respiratory production of oxygen, conversion of carbon dioxide into carbohydrate, and conversion of carbohydrate and ammonia into protein for human consumption. Unsettled questions about how microgravity can affect plant physiology are discussed in detail in a recent report (NRC, 1998) and will not be pursued here. The remaining technical issues presented by the SBR involve the simple question of how, in the absence of drainage, to supply water and nutrients to the plants without drowning them. The interfaces between roots, solid support, and liquid nutrients change in microgravity because of the increased dependence of liquid transport, as well as the solubility and diffusion of gases in liquid and porous media, on surface tension. Adequate computational models of transport, diffusion, and wetting and their mutual interactions under microgravity are critical for the ground-based research into SBR design.

In addition to the issues of fluid management, there are clearly complex biological cycles relating food

Cabin Atmosphere Loop	TIMES	VAPCAR	Incinerator	Bioreactor	Supercritical Wet Oxidation
—	—	—	—	—	—
L	—	—	—	L	—
H	H	H	—	—	—
L	—	L	L	—	L
—	—	—	—	L	—
—	—	—	H	H	H
—	—	—	L	—	L
H	H	H	—	H	—
—	—	—	—	—	—
—	L	—	L	L	—
L	—	L	L	M	L
L	L	L	L	L	L
M	—	—	—	—	L
—	—	L	L	L	L
—	—	—	—	M	M
—	—	—	L	—	—
L	L	L	L	L	L
L	L	—	L	L	L
—	—	—	M	M	—
M	—	—	—	—	—
—	—	—	—	M	—
H	H	H	L	H	H
L	—	—	—	L	—

production to waste management in a closed, regenerative system. However, these are outside the scope of this study and are not discussed here.

Summary of the Impact of Reduced Gravity on Selected Subsystems

Microgravity has pervasive impacts on the subsystems found in each of the major systems involved in life support (Table III.C.4):

- Oxygen and carbon dioxide generation and recovery procedures use a variety of differential liquefaction processes, together with separations of liquid and gas phases.
- Solid waste processing involves complex and ill-defined multiphase systems requiring the separation of dispersed solids from the continuous liquid phase.
- Temperature and humidity control depends on a variety of heat exchangers, condensers, and separators.
- Water recovery involves controlled distillation, which in turn requires successive phase changes and separations.

• Food production involves special problems associated with the altered wetting and capillary behavior in microgravity, as well as with the management of fluids and multiple phases.

Summarized in Table III.C.4 are the various subsystems and components discussed in this section and the various propulsion systems where they are found. For each subsystem that appears in a given system, the impact of reduced gravity on its operation is estimated as either high, medium, or low (little or none). It should be remembered that the impact of gravity level on these technologies will depend greatly on the design context.

References

American Institute of Aeronautics and Astronautics (AIAA). 1990. Final Report to the Office of Aeronautics, Exploration and Technology, National Aeronautics and Space Administration, on Assessment of Technologies for the Space Exploration Initiative. Washington, D.C.: American Institute of Aeronautics and Astronautics, December 31.

Bilardo, V., and W. Likens, eds. 1991. In-house Life Support Technology Review Book. Document No. 90-SAS-R-003. NASA Ames Research Center, Advanced Life Support Division.

Eckart, P. 1996. Spaceflight Life Support and Biospherics. Torrance, Calif.: Microcosm Press, and Dordrecht, Netherlands: Kluwer Academic Publishers.

Kuhn, P. 1988. Condensing heat exchangers for European spacecraft ECLSS. Pp. 193-197 in Space Thermal Control and Life Support Systems. European Space Agency (ESA) SP-288. Noordwijk, Netherlands: European Space Agency.

National Research Council (NRC), Aeronautics and Space Engineering Board. 1997. Advanced Technology for Human Support in Space. Washington, D.C.: National Academy Press.

NRC, Space Studies Board. 1998. A Strategy for Research in Space Biology and Medicine in the New Century. Washington, D.C.: National Academy Press.

Schmidt, R.N. 1989. Water recovery by vapor compression distillation. Paper 891444 in Papers Presented at the 19th Intersociety Conference on Environmental Systems. Washington, D.C.: Society of Automotive Engineers.

III.D HAZARD CONTROL

Introduction

This section is focused on hazards to humans, defined as detrimental phenomena that may directly impair the ability of humans to function or ultimately lead to loss of life. Traveling to distant locations in space, in particular the Moon and Mars, living there, and returning to Earth is, by the nature of the activities involved, dangerous. The relative unfamiliarity and remoteness of HEDS environments increases the risks to a point that careful attention to hazard control must become an essential and specific function. Hazards include long-term exposure of personnel to electromagnetic and particle radiation, meteorite damage to spacecraft and habitats, fire, and chemical and biological contamination of the environment. These hazards, some of which are also faced on Earth, must be eliminated or at least controlled to reduce the risk of loss of life and life support function, and the approaches and technologies used must be functional in different gravitational environments, from microgravity to the partial gravity of the Moon and Mars to the gravity of Earth.

Hazards often are addressed in a piecemeal fashion that would be undesirable for HEDS missions. To address hazard control optimally, total-system concepts are needed that use the methods of probability risk assessment, explained in Chapter V. For example, strategic use of particulate and carbon dioxide detectors, electrical monitoring, automated fire-suppression response, and manual protocols are needed for effective overall fire-protection systems for HEDS. Nearly all such systems on Earth are strongly affected by buoyancy phenomena; indeed, they often rely on buoyancy effects, such as the rising of hot gases, for their operation. Special ventilation control systems and procedures need study for each separate HEDS environment to enable design of the best system. The large gas-density changes in fires raise many reduced-gravity habitat issues that are poorly understood today (Ross, 1996; Friedman, 1998).

The usual technologies proposed for elimination and/or control of hazards include shielding and chemical radioprotection drugs taken internally to reduce the biological effects of radiation exposure, debris shielding to guard against meteorite damage to spacecraft, and chemical and particle cleaning to scrub the environment in the case of contamination. For fire safety, there is a need to select, insofar as is possible, materials that are nonflammable. Because complete elimination of flammable materials is essentially impossible (in fact, minimizing the quantity of flammable materials becomes more difficult as the length of a mission increases), it is necessary to employ systems that have the ability to detect faults. For instance, electrical systems, which might lead to smoldering or flaming combustion, might require fire detection and suppression systems, as well as systems for postfire particle and gas cleanup.

Likely technologies for fire safety detection, suppression, and postevent cleanup include fire signature detectors, which sense smoke, gaseous fire precursors or combustion products, flame radiation, etc.; suppression systems, e.g., handheld and automated extinguishers using chemical scavenging, oxygen-depletion blanketing, or cooling agents; and absorbtion or chemical conversion systems for particle filtration and toxic-gas removal. These systems are influenced by gravity level not only because their operation can be affected by gravity but also because the behavior of the fire and its products is influenced by gravity.

Protection against hazards is addressed below in several categories: fire protection, spill and cleanup, radiation shielding, and protection from chemical and biological contaminants.

Fire Protection

Electrical System Fault Diagnostic and Response System

A possible source of ignition, leading to smoldering and/or flaming combustion, is the spacecraft or habitat electrical system. This hazard is a consequence of the presence of current next to electrical insulation, which generally is made of flammable material. In view of this situation, the installation of diagnostic systems to survey the health of distribution systems (e.g., to compare current levels with set points and ranges) is a reasonable approach to protecting against a substantial ignition source. Detection would be coupled to feedback to take action should a problem be detected (e.g., isolation and the suppression of ventilation), because slight ventilation generally invigorates a smoldering or flaming event, particularly in microgravity. Such diagnostic systems would not be expected to be particularly affected by gravity level unless the fault detection technique incorporates the detection of flammable offgas resulting from overheating.

Smoke Detectors

Smoke detectors used in space applications include ionization detectors (space shuttle) and photoelectric detectors (planned for the International Space Station). In an ionization detector, particulates interrupt an ionization current; in a photoelectric detector, the particulates alter the transmission of light through a fixed path length. Regardless of the detection scheme, the smoke must be transported to the detector, a process driven by natural convection on Earth. In the reduced-gravity environments of spacecraft and the Moon and Mars, smoke transport to smoke detectors must be provided for via forced environment gas flows, because smoke production and transport in reduced-gravity fires are not the same as in normal-gravity fires. In particular, microgravity fires in a quiescent environment produce little smoke, while gentle ventilation results in substantial smoke, much of which can agglomerate into large particles.

Fire Extinguishers

Handheld Halon 1301 (bromotrifluoromethane) extinguishers are in use on the space shuttle. However, production of this compound is now prohibited because of its deleterious effect on the ozone layer. Additionally, Halon presents major cleanup problems following its use. As a result, International Space Station plans call for carbon dioxide extinguishers. While these may be generally effective, a better understanding of the techniques for their application in reduced gravity needs to be developed. With such knowledge, the Martian atmosphere itself becomes an attractive extinguishant. Fire-detection systems, including smoke and other signature detectors, can include automatic suppression systems to supplement handheld extinguishers that must be actuated by human beings. Gravity levels significantly affect the transport of suppressants to the base of the fire, where they are most effective, so gravity-related phenomena should be studied in this connection. These phenomena involve the effects of varying buoyancy on the motion of gases, liquids, and solid clouds of particles in the vicinity of burning materials. In microgravity, net flow from fires tends to be outward in all directions, and suppressants need to be applied in a manner that will overcome the outward flow effects and penetrate to the fire.

Postfire Cleanup

Following a fire, atmospheric cleanup is necessary, and this is particularly important in spacecraft. Filtering of fire-generated particulates and removal of products that are toxic to either personnel or the structure is necessary. While it would be possible, on a lunar mission, to return the vehicle to Earth for cleanup, such a scheme on a Martian mission would be essentially impossible. In HEDS operations the material cleaned up must, therefore, also be recovered and stored or disposed of safely during the mission. Again, because formation of smoke and other combustion products is affected by gravity, a need exists to determine the makeup of the materials to be removed during postfire cleanup. Currently this is not a priority in space flight planning. Instead, efforts are aimed at avoiding fire and refurbishing space vehicles after their return to Earth.

Spill Cleanup

Fires are not the only events that require attention to cleanup during interplanetary missions. Carefully planned housekeeping procedures will be needed to maintain a healthy environment for humans on extensive voyages. Special attention will need to be paid to automatic sealing and filtering requirements as well as to the design of handheld tools that can aid astronauts in cleaning unwanted spills.

Spills of solid and liquid materials behave differently and therefore need to be considered separately. Both types of materials will behave differently in reduced gravity from the way they behave on Earth. One may expect that there will be less of a tendency for spilled materials to accumulate on one face of an enclosure. Greater dispersal on all faces should be anticipated. In addition, there may be more of a tendency for spills to accumulate and remain in the atmosphere. While this can exacerbate health hazards it can also facilitate control by automatic filtration and sealing.

Surface-tension phenomena are expected to play a greater role in spill behavior and cleanup, especially under microgravity conditions, than they normally would on Earth. Microfiber technology, for example, might therefore find greater application. These areas appear not to be well studied and involve reduced-gravity issues in need of further research.

Radiation Shielding

Gamma and particle radiation constitute a serious but reducible threat to long-term survival in space environments (Wilson et al., 1997). There are three general types of radiation to which personnel are exposed during missions. First, Earth is surrounded by the two Van Allen radiation belts, in which Earth's magnetic field has concentrated charged particles from the solar wind and solar flares. Second, the Sun is a continuing source of solar wind (mostly a plasma of protons and electrons) and solar flares (plasma containing a richer variety of atomic weights). Solar-flare radiation fluctuates in amplitude by more than four orders of magnitude; the occurrence of these flares is unpredictable, but their probability density is correlated with the 11-year sunspot cycle. Solar flares, which pose the most danger to personnel, require sophisticated systems to monitor their activity and, possibly, to deploy protective shielding rapidly. The third type of radiation is galactic cosmic radiation (GCR). This radiation includes particles of high atomic weight, which are termed HZE (high atomic number and energy) radiation. The biological impact of HZE radiation is difficult to assess and has very large uncertainties (NRC, 1996). In the absence of solar-flare activity, however, GCR is considered to constitute about 90 percent of the exposure to personnel, and HZE particles represent 1 percent of this exposure, which could be responsible for an appreciable fraction of the long-term biological effects.

Considerable data have been accumulated, primarily from Soviet long-term flights, on the nature and extent of radiation hazards in low Earth orbit. Generally speaking, such an experience results in a dose equivalent to 150 mSv in a 115-day Mir flight, much greater than the 2.0 mSv normally encountered annually in terrestrial environments. Data from the U.S. lunar exploration program suggest that a lunar round trip involves exposures an order of magnitude greater (50 mSv) than background. Current estimates indicate that in the absence of serious solar-flare activity, a mission to Mars equipped with the shielding used on lunar missions would involve an increase in dose equivalents of yet another order of magnitude at a minimum. Such an amount would be sufficient to induce radiation sickness in about 10 percent of exposed personnel and is within one order of magnitude of the level (10 Sv) at which no survivors would be expected from acute exposures. These simple statistics underscore the importance of developing alternative systems for protecting crew from radiation hazards.

Such statistics do not assess the longer-term chronic problems associated with cellular damage. Of these, the most evident is the risk of cancer. This risk is compounded by the fact that spaceflight results in alterations of immune responses that may result in changes in resistance to neoplasia. Thus, an important area for further research concerns the interactions between microgravity, the immune system, and exposure to radiation (NRC, 2000). The additional risks to organ systems, the central nervous system, and especially germ-line cells could be comparably severe and must be taken into account in any planning for human penetration into extraterrestrial spaces.

This brief consideration of the estimated risks from the three sources of radiation clearly demonstrates that radiation poses an unacceptable risk to HEDS activities unless effective countermeasures are implemented. The countermeasures for limiting exposure to radiation fall into four categories: passive bulk shielding, electromagnetic shielding, electrostatic shielding, and chemical radioprotection. Much of the current emphasis in this area is on passive bulk shielding, but other systems, while more speculative, may deserve attention because they have the potential to offer alternative and perhaps superior protection. Combination systems, particularly those that can make use of rapid deployment when the precursor phase of solar flare activity is detected, would seem to have a high priority for further study.

Passive Bulk Shielding

The atmosphere on Earth provides considerable passive shielding. It has been estimated that 5 meters of Martian regolith will be required (Hepp et al., 1994) to provide radiation shielding on Mars comparable to that provided by the terrestrial atmosphere. Despite the common perception, due to their use in terrestrial applications, that materials like lead provide optimal shielding, lighter elements and their compounds are more effective passive shields for the HZE particles than are heavier elements. For instance, polyethylene sheet, with a surface density of 0.19 g/cm^2, provides far better shielding efficiency per unit mass than do lead or other metals (Eckart, 1996). Water is nearly as good. An important design consideration for both spacecraft and extraterrestrial habitats will be the integration of construction and other materials so as to optimize the use of the total mass for shielding purposes. Distributing resources such as drinking water over large surface areas can contribute materially to radiation protection.

Electromagnetic Shielding

Magnetic fields are very effective at deflecting energetic light-nuclei radiation but cannot deflect heavier particles with high kinetic energies. This property could make magnetic fields attractive for protecting against solar flare radiation but would limit their usefulness against galactic cosmic radiation (GCR). However, electromagnetic shielding offers the additional possibility of dynamic control, which could be integrated with dosimetric monitoring of, for example, solar-flare activity. Dynamically controlled electromagnetic shielding subsystems might therefore be a very appropriate component in an integrated radiation protection system, because they could be used when the assault from light particles in solar flares is high. Issues that would need to be examined in a theoretical study of such a system include the field strength and power requirements and the effects of the field on onboard electrical systems and humans.

Electrostatic Shielding

Shielding provided by electrostatic fields apparently was extensively studied by the Soviet space program, but details are unavailable (Helmke, 1990). This strategy is analogous to electromagnetic shielding; in this case, Coulomb forces are employed to deflect or retard incoming charged particles. Since GCR particles have such high kinetic energies, one can speculate that novel ways might be found to exploit this energy while at the same time diverting it from crew members. Of course, such a capability would require that the energy of the particles be captured by a shield material, such as plastic, and concentrated in particular regions of the spacecraft, which could itself result in a potentially dangerous electric field. Nevertheless, physical and chemical interconversion of various forms of radiation would seem to be one of the more interesting areas for novel research.

Chemical Radioprotection

Considerable effort has been devoted to the development of chemical scavengers of radiation. These compounds react with the by-products of ionizing radiation within the body, thereby limiting the effective dose exposure (Helmke, 1990). Their effectiveness has been found to be significant. However, it seems likely that more remains to be done to optimize their performance. Radiochemistry and radiobiology therefore constitute fruitful areas of future research aimed at successively more sophisticated applications of chemical radioprotection. The compounds most often cited, including aminopropyl-aminoethylthiophosphoric acid (APAETF), are to be taken internally (Eckart, 1996). Other such compounds may be developed in the future. There is an appreciable amount of information available on this subject, but there also are many unknowns. For instance, serious side effects have been seen with chronic use of some radioprotective compounds (NRC, 1996). Much more needs to be learned if chemical radioprotection is to be applied with confidence in HEDS missions.

There are immune-system concerns (one of the unknowns referred to above) in using chemicals to scavenge radiation. Otherwise, apart from the construction of shields, there are few if any reduced-gravity issues associated with radiation countermeasures. However, weight is a critical consideration when selecting a shielding material for spacecraft.

Protection from Chemical and Biological Contamination

The dispersal of contaminants to nonhazardous levels is facilitated on Earth by the atmosphere and oceans, which serve as large reservoirs for the dilution and removal of toxic materials. In HEDS environments, which involve essentially closed systems that rely on recirculation for life support, the long-term buildup of contaminants is potentially a much more severe problem. Chemical or biological agents produced at very low rates can, over time, concentrate to hazardous levels. Special systems are likely to be needed in HEDS habitats to deal with such threats.

There are two problems associated with these two types of contamination: detection and control. Knowledge of the types of contaminants that can be of concern is needed to design suitable detection systems. Much knowledge of this kind is obtainable from existing experience in the space program and more should result from operation of the International Space Station. It will, however, also be necessary to give special consideration to each specific HEDS environment. For example, contaminants unique to the lunar or Martian habitats could become of major concern. Anticipation of these possibilities and improvement of more general all-purpose detectors could help to address these problems.

TABLE III.D.1 Subsystems Found in Hazard Protection Systems and the Potential Impact of Microgravity on Their Operation

	System				
Subsystem	Fire Protection	Spill Cleanup	Meteorite Protection	Radiation Shielding	Contamination Protection
Electrical fault detector	L	—	—	L	L
Smoke detector	H	—	—	—	H
Fire extinguisher	M	—	—	—	—
Cleanup filter	L	L	—	—	L
Manual cleanup device	M	M	—	—	M
Meteorite shield	—	—	L	—	—
Pressure loss detector	L	—	L	—	—
Bulk shield	—	—	L	L	—
Electromagnetic shield	—	—	—	L	—
Electrostatic shield	—	—	—	L	—
Chemical radioprotector	—	—	—	L	—
Contaminant detector	H	M	M	—	M
Contaminant remover	L	—	—	—	L

NOTE: The letters H, M, and L designate high, medium, and low (preliminary assessment) impact of reduced gravity on the operation of the subsystem. Where no letter is given, the subsystem is not applicable to the system listed.

Contaminants can be controlled by use of systems for their removal, and such systems will need to be designed and installed for HEDS activities. Some such systems were discussed in Section III.C. Additional systems, unique to specific environments, should be addressed. There may or may not be reduced-gravity issues associated with such systems, depending on the methods employed (filtration, two-phase flow, etc.). Gravity effects will have to be considered on a case-by-case basis. Consideration can be given also to more-general removal procedures that do not require the detection or identification of specific contaminants.

Summary of the Effect of Reduced Gravity on Hazard Protection Systems

Table III.D.1 lists the subsystems that make up the hazard protection systems and indicates which ones may be affected by gravity levels, as discussed in the preceding sections. The degree of commonality of subsystems among the systems is also evident. As can be seen, the expected impact of gravity level on the phenomena taking place in the subsystems is appreciable in a substantial number of cases.

References

Eckart, P. 1996. Spaceflight Life Support and Biospherics. Torrance, Calif.: Microcosm Press, and Dordrecht, Netherlands: Kluwer Academic Publishers.

Friedman, R. 1998. Fire safety in extraterrestrial environments. Pp. 210-217 in Space 98: Proceedings of the Sixth International Conference and Exposition on Engineering, Construction, and Operations in Space, April 26-30, Albuquerque. R.G. Galloway and S. Lokaj, eds. Reston, Va.: American Society of Civil Engineers.

Helmke, C. 1990. Synopsis of Soviet manned spaceflight radiation protection program. USAF Foreign Technology Bulletin, FTD-2660P-127/105-90.

Hepp, A.F., G.A. Landis, and C.P. Kubiak. 1994. A chemical approach to carbon dioxide utilization on Mars. Pp. 799-818 in Resources of Near-Earth Space. J. Lewis, M.S. Matthews, and M.L. Guerrieri, eds. Tucson and London: University of Arizona Press.

National Research Council (NRC), Space Studies Board. 1996. Radiation Hazards to Crews of Interplanetary Missions: Biological Issues and Research Strategies. Washington, D.C.: National Academy Press.

NRC, Space Studies Board. 2000. Review of NASA's Biomedical Research Program. Washington, D.C.: National Academy Press, in press.

Ross, H.D. 1996. Combustion processes and applications in reduced gravity. Pp. 527-532 in Space V: Proceedings of the Fifth International Conference on Space '96, Albuquerque, June 1-6, Vol. 1. S.W. Johnson, ed. New York: American Society of Civil Engineers.
Wilson, J.W., F.A. Cucinotta, J.L. Shinn, M.H. Kim, and F.F. Badavi. 1997. Shielding strategies for human space exploration. Shielding Strategies for Human Space Exploration: A Workshop, Chapter 1. J.W. Wilson, J. Miller, and A. Konradi, eds. Washington, D.C.: National Aeronautics and Space Administration.

III.E MATERIALS PRODUCTION AND STORAGE

Introduction

Two distinct ranges of gravity level will be encountered most frequently in space exploration. One is generally referred to as microgravity ($g < 0.01\ g_0$), and the other is denoted as fractional gravity ($0.01\ g_0 < g < 1 g_0$). Microgravity, for example, is experienced on board spacecraft, on the International Space Station, and on the satellites of Mars, Phobos, and Deimos (which could be a source of raw materials for a colony on Mars). In operations on the Moon and Mars, one encounters fractional gravity: 0.169 g_0 on the Moon and 0.38 g_0 on Mars. While these levels would cause some problems in directly applying many terrestrial processes to life sustenance on these bodies, in many cases the harsh local environments provide far greater challenges than those presented by the reduction in gravity. The lunar surface is characterized by a hard vacuum with a silty, fine-grained sand regolith in a cold ambient with long days and nights (each is about 14 Earth days). The atmosphere of Mars is composed almost entirely of carbon dioxide at a pressure of 7 mbar, and although its surface temperature may reach 25 °C at the equator in mid-summer, it is generally much colder, with large temperature swings. In addition, dust storms occur about 100 days out of almost every year, with the dust consisting of ~5 μm particles.

The processing of local resources to obtain various materials will be crucial to space exploration. This in situ resource utilization (ISRU) would include the beneficiation of regolith and the extraction of essential materials such as water, oxygen, and fuels by chemical processing. The production of hydrogen and oxygen by electrolysis of water is a critical enabling technology for long-duration life support systems and for a wide range of ISRU applications on a majority of planetary bodies found in the solar system. Water electrolysis has a critical dependency on gravity since the generated gaseous products require liquid-vapor separation, a process that fails at zero gravity. In addition, electrolysis can be used to produce oxygen from carbon dioxide using a solid electrolyte and to extract oxygen from lunar regolith where it is combined with metallic elements, primarily as silicates. While the gases produced (oxygen, hydrogen) can be liquefied for convenient storage, at $g = 0$ severe problems are encountered in the storage and transfer of cryogenic fluids.

Although water for the electrolysis process is potentially available from indigenous permafrost, other promising chemical processes that produce water as a by-product are reviewed briefly in this section. These include extraction of carbon dioxide from the Mars ambient, which could be reacted with hydrogen to form methane and water (Sabatier process) or to form carbon dioxide and water (reverse water gas shift). Except for the critical dependence of the electrolysis step on gravity level, these processes would be little affected in gravity. Other processes that are more speculative but potentially useful (and relatively independent of gravity) are pyrolysis, radio-frequency processing, and volatilization/condensation. The last-mentioned process could be an important step in obtaining water from lunar permafrost or ice frozen under a hard vacuum, where the vapor released by heating would have to be collected by an appropriate cold trap.

Because of the weight-driven costs associated with transporting material from Earth's surface into space, substantial savings can be realized by using materials that are already out of Earth's gravity well. In some cases, the ability to exploit in situ resources for the production of consumables such as rocket propellant, breathable air, and water could be the enabling technology in terms of keeping launch mass and mission costs within realistic limits, and current NASA planning for human missions to Mars relies heavily on utilization of local resources (NASA, 1997). Furthermore, as HEDS missions expand outward from Earth, the case for exploiting useful resources from other low-gravity bodies, such as specific asteroids (including dead comets) and planetary satellites, that are convenient to an extraterrestrial base can become a very logical element in the overall program evolution, provided that the microgravity environment has been mastered. Regardless of the extraterrestrial source, the key leveraging element is the ability to substitute mass acquired from outside of any deep gravity well (relative to the specific base) for mass that otherwise would come from Earth. Fortunately, there is a symmetry between (1) the opportunity to substitute relatively high-mass, low-value in situ units, such as radiation shields, construction elements, and consumables, for Earth launch mass and (2) the need to acquire low-mass, high-value units, such as digital devices and sophisticated instruments and machines, from Earth. The substitution of power-generation equipment for return propellant mass represents the fundamental difference between a touch-and-go exploration strategy and the establishment of outposts that can eventually support unlimited human stays. The development of the technology required to exploit extraterrestrial resources in the reduced-gravity environments, where the greatest leveraging exists, will ultimately enable the expansion of human presence in space. Indeed, material-processing stations distributed throughout the solar system will enable resupply and thus extend our ability to explore the far reaches of space.

The surface environments associated with resource-rich extraterrestrial bodies offer many challenges. The Moon and Mars both have reduced-gravity environments, widely varying local surface temperatures, and surface pressures ranging from a hard vacuum on the Moon to (approximately) water's triple-point pressure at Mars. Their surfaces have relatively severe dust problems, associated on the Moon with the abrasive character of very fine lunar dust and, on Mars, with the planetwide dust storms that are observed with some regularity.

The two best-characterized extraterrestrial resources at this time are the Martian atmosphere and the lunar regolith. The processing of the Martian atmosphere into oxygen (and its subsequent storage) is understood sufficiently well at this time to serve as the most probable first step toward exploiting extraterrestrial resources. However, if humans are going to be protected and sustained, it will be necessary to construct human shelters almost immediately upon arrival at Mars. As noted in a previous section, neither Mars nor the Moon has magnetic fields or atmospheres that are sufficient to protect humans from the harmful radiation that is produced intermittently by solar flares and storms. Because the Martian atmosphere cannot be processed readily into materials that can be used for radiation shields or shelters, there is a strong need to develop site preparation and mining capabilities. These needed capabilities will range from the ability to place soil material over spacecraft modules (to serve as radiation shields) to site preparation for the deployment of surface power generating and distribution equipment. Those operations would logically use material mined from the regolith.

For lunar operations, regolith must be collected, transported and processed for the basic site preparation, radiation shelter construction, and hardware deployment steps associated with any type of human outpost construction. It would then be logical to develop the ability to manufacture basic construction units, such as bricks, beams, and blocks, and to consider the processing of lunar material into oxygen and metals. The possible extraction of helium-3 from lunar fines has also been the subject of a great deal of research. The reduced gravity, vacuum conditions, and temperature extremes encountered on the Moon affect all of these processes. A similar set of problems will be encountered on other planetary bodies, which have not yet been characterized as accurately as the Moon.

Before such a wide range of surface operations can be carried out, a better understanding of the behavior of soil in reduced gravity environments will be needed. The geotechnical aspects of extraterrestrial soils not only control their load-bearing capacity but also affect their ability to support the forces that will be required for anchoring and/or excavating regolith. These granular materials are known to exhibit surprising levels of compaction on the Moon and to produce a kind of a hard crusty surface layer (called duracrust) on Mars. Furthermore, their behavior affects the geophysics of the planetary body.

Successful exploration of extraterrestrial regions, a primary HEDS goal, requires the ability to produce useful mass outside of Earth's gravity well. It is not logical to bring along all of the consumables needed for propulsion and life support on round-trip missions with characteristic times measured in years. Site preparation, material processing, habitat construction, mining, and infrastructure development in a robust HEDS program must evolve rapidly away from dependence on terrestrial systems that have been transported from Earth.

Mining

Terrestrial mining equipment is characterized by a heavy rugged design, and its use is both labor- and energy-intensive. This terrestrial equipment could not be directly incorporated into space activities even if the transportation costs were acceptable. Such factors as reduced traction at lower gravity, the high level of abrasive dust in the ambient, and operation at low temperatures in vacuum, where common lubricants fail by evaporation, preclude the direct use of conventional mining equipment. A mining operation requires planning and integrating an entire range of procedures and techniques such as beneficiation, material handling, secondary recoveries (e.g., of ores and fuels), and storage, as well as support programs such as habitat construction and the development of transportation modes. In addition, there is a need for new components and tools such as the vibratory bulldozer and penetrator studied by Szabo et al. (1994) and Nathan et al. (1992). These show a very significant increase in efficiency, typically a greater than 50 percent decrease in the force needed to advance the bulldozer. Similar studies in static augering (Klosky, 1997) showed, however, that no decrease in torque could be obtained by similar vibration of the auger.

The mining of bedrock, which adds several difficult steps to the ordinary mining process, has been discussed by DeLa'O et al. (1990). First, the bedrock must be penetrated, and then it must be broken into pieces that can be hauled to a site for comminution or pulverization. Because of numerous factors discussed in detail by Chamberlain et al. (1993)—including heavy, bulky equipment designs unsuited to the Moon's low gravity, vacuum, and temperature fluctuations—terrestrial blasting and mechanical mining techniques would not usually succeed in the lunar environment. In addition, just a few centimeters below the surface, the soil is extremely dense, even denser than can be obtained with heavy conventional compaction equipment on Earth (Carrier and Mitchell, 1990). Open pit mining on Earth uses about 0.14 kg of conventional explosives per ton of ore (Wempfen, 1973). The transport of large quantities of such hazardous material aboard spacecraft is not practical. Moreover, it is not clear that the terrestrial experience of such blasting is applicable at reduced gravity or that the distribution of explosives to achieve a desired consequence is known. For example, the explosive could generate shock waves that would cause extensive damage to previously constructed facilities in the habitat. In addition, safe haven considerations for the operator during explosions have not been evaluated.

In spite of these problems with bedrock mining, there are several advantages in pursuing this goal. For example, as the lunar base matures and the facilities become more sophisticated, underground mining into the bedrock may be desirable for some applications such as providing a comfortable, secure, pressurized habitat where one can move and work freely. Also, as discussed elsewhere, energy and processing efficiencies may be obtained by mining bedrock to recover oxygen and material for building block formation. There has been promising work on the use of electromagnetic energy from, for example, carbon dioxide lasers, microwave, or solar sources to produce thermal stresses or local melting that could effectively fracture the bedrock (Lindroth and Podnieks, 1998).

During the pulverization and comminution processes, safety considerations require monitoring and control of all yield gases and small particulates in the ambient. Even if the same terrestrial process for pulverization could be applied at reduced gravity, the process on Earth is both inefficient and energy-intensive (NRC, 1981). It is possible that all early habitat and mining operations will be limited to processing lunar regolith. Similarly, little is known of the Martian subsoil, and early ISRU activities will involve processing the regolith. One of the potential substances to be mined is water ice, which would be a valuable and essential resource; the extraction problems for this case are discussed in the following pages.

In conclusion, the reduced gravity issues relevant to mining are the design of mining equipment for excavating, bulldozing, transport, etc. at reduced gravity (or low traction conditions) and ejecta management during these operations. The processing and refining of regolith involve mainly problems of material handling, transport, and chemical separations and are discussed later in this section.

Volatilization/Condensation

Water, oxygen, and bound carbon and hydrogen molecules are available in such abundance on Earth that it is difficult to think of them as resources worthy of mining operations. However, hydrogen, oxygen, and carbon molecules are the primary building blocks for virtually all consumable materials needed for HEDS missions. Acquiring large quantities of those molecules outside Earth's gravitational field is a crucial step in the early deployment of HEDS systems.

Water ice is known to be present on most of the planets. Hydrogen and helium have been implanted by the solar wind in the surface layers of unshielded planetary bodies such as Mercury and the Moon. The helium-3 implanted in the surface layer of the Moon is sufficiently valuable for use in fusion reactors, that its acquisition and transport to Earth could become a viable industry. Most of the other volatile molecules are available throughout the solar system, either as ices or as adsorbed molecules in the surface materials of planetary bodies. Hence, a key HEDS mining goal will be to extract and collect volatile molecules. The obvious approach is to volatilize these molecules in situ and collect them either by capturing them in a container or by condensing them on a cold surface for subsequent collection. These processes are influenced strongly by gravitation and by the particular ambient pressure and temperature environment.

Because of the contrast between extracting adsorbed volatiles from small particles such as lunar fines and extracting water from extraterrestrial ice, this discussion will examine those two cases as a basis for identifying the microgravity research issues.

Lunar Volatiles

Adsorbed volatiles are found in the surface layer of lunar fines, with the highest concentrations on the Sun-facing side. In addition, the recent Clementine and Lunar Prospector missions have provided conclusive data showing that significant quantities of bound hydrogen (probably water ice) are deposited in the Moon's permanently shadowed polar craters. While more definitive data on the distribution and availability of lunar water ice is needed, it is likely that water ice will be a key resource on the Moon. However, since water extraction at a lunar site requires processing hardware that is similar to the water extraction hardware that would be required on a Galilean satellite or on many other low-gravity surfaces, that discussion is deferred to the next section.

Fegley and Swindle (1993) have reported on the relative abundance of lunar solar-wind-implanted molecules and on their thermal release when heated. Helium and arsenic, with abundances of 14 ppm and 1 ppm, respectively (the abundance of helium-3 was 4.2×10^{-9}), were released almost completely from lunar fine samples when heated to 1200 ºC. Similar release efficiencies were obtained for chemically reactive hydrogen (50 ppm), carbon (125 ppm), and nitrogen (100 ppm) over the same thermal range. It should be noted that Apollo samples showed that the volatiles were bound primarily to particles 20 µm in diameter or smaller, and the Apollo missions indicated that the soil layer that was rich in volatiles was confined nominally to the first 30 cm.

The capture and release of these volatiles is a significant technical challenge. Owing to the low concentrations of volatiles (in the parts per million range), large quantities of lunar soil must be processed in order to collect useful amounts of volatile material. Even if processing takes place only during daylight hours, the energy required to heat lunar fines to 1200 ºC is significant, so it would be desirable to segregate the lunar material containing significant volatiles prior to heating. Because the lunar surface pressure is on the order of 10^{-9} torr, it will not be practical to heat the lunar fines in an unconfined space, trying to collect the volatile molecules as they are liberated. Batch processing of segregated lunar fines in sealed containers would be desirable, particularly if energy could be recovered after the volatiles are released. Heat recovery from low-conductivity granular material in lunar gravity is not understood sufficiently to permit engineering designs.

The value of helium-3 warrants special consideration if it can be separated from helium-4 (after the helium has been separated from the other volatiles). If volatile recovery systems can be operated in the permanently shaded polar crater regions, where undisturbed surface temperatures can approach 4 K, helium separation and helium-3 recovery would be greatly enhanced. However, at those temperatures, major efforts will be required to prevent locally generated thermal pollution from destroying the environment and the associated resources. The subtle influences of lunar gravity on these heat and mass transfer processes must be understood fully before any sort of processing base can be designed.

Extraction of Water Ice on Low-Gravity Surfaces

Water ice is present in comet cores, on the Galilean satellites, in Saturn's rings, at Mars's north pole, and on some asteroids. Water is probably present as ice in the permanently shaded polar craters of the Moon, in Mars's permafrost, and perhaps even in the permanently shaded craters on Mercury. Water recovery is a major technical challenge on low- and very-low-gravity ice-covered surfaces, especially since their ambient environments typically are extremely cold and at very low pressures. At ambient pressures below 4.6 torr (triple point pressure), water cannot be converted to a liquid unless it is compressed. Therefore water extraction and recovery from ices on the Moon and at virtually all of the other extraterrestrial locations will present unusual difficulties. Aside from the obvious phase separation problems in these very-low-gravity environments, a variety of other problems associated with ISRU operations are present. For example, if an ISRU unit for processing water into oxygen is to be placed on almost any one of the icy surfaces identified here, the simple process of landing on that surface could be a major technical challenge. The problem of attaching a spacecraft to these low-gravity surfaces is discussed elsewhere, but presumably the hardware for an oxygen processing plant would be operated at temperatures well above Earth ambient (300 K). Between power-generation requirements for material extraction operations and the associated processing that must occur on the body's surface, heat rejection rates will be high and it will be very difficult to isolate the unit thermally from the surface environment. Contact surfaces warmer than the ambient temperature will promote sublimation of water on ice-covered objects. The very low ambient pressures, which are typical on these bodies, cause the released water vapor (cold steam) to sustain very high volumetric flow rates at very modest rates of heat rejection. Sonic water release speeds will be sustained whenever the contact pressure exceeds roughly twice the ambient pressure, and contact temperatures as low as 300 K (Earth ambient) can generate sufficient heating rates to levitate spacecraft, while ejecting rocks and other debris that could damage or destroy the unit. Unless system masses are very high, it will be necessary to attach these systems to the planetary surface mechanically and it may be necessary even then to refrigerate contact surfaces to avoid steam-driven debris bombardment problems due to heat rejection into these cold icy surfaces. Lunar and Martian gravity are sufficient to permit process designs that avoid these thermally driven levitation problems, but it will not be possible to maintain passive surface contact on an asteroid or an outer planet satellite such as Europa (Ash et al., 1981), because of low gravitation. Anchoring systems and/or alternative system designs will be needed to permit safe operation.

Studies are needed to determine whether water extraction equipment can be sealed reliably enough to an irregular, low-gravity surface to maintain pressure environments sufficient to liquefy (but not refreeze) water. If batch processing techniques must be used, major questions concerning how to break loose and/or gather icy objects for water extraction processing in extremely low-gravity environments must be answered. Because of the very low gravitational levels that characterize most of these surfaces, ejection of unwanted objects during any sort of surface material handling operation must be expected, and systems must be designed to minimize the resulting hazard. Cold trap designs for water capture should also be examined in these microgravity environments.

Material Handling and Transport

Material handling requires the design of lightweight, temperature- and dust-insensitive equipment such as bulldozers, bucket scoops, cranes, winches, conveyer belts, and trucks that can operate at reduced gravity over a large ambient temperature excursion and in a very abrasive, fine, silty soil that could extend to depths of between 3 and 10 meters. As discussed by Chamberlain et al. (1993), the key factors in the design of material-handling equipment include simplicity, flexibility, robustness, light weight, low energy consumption, potential for automation and teleoperation, tribology considerations (e.g., lubrication of bearings and seals), and ability to operate at low and wildly fluctuating temperatures, in a vacuum and/or at low pressure, and often in a very abrasive and dusty atmosphere. An additional consideration is the potential for bursts of electromagnetic radiation and particle bombardment that may be hazardous to humans and to equipment components such as electronic circuits.

Excavation equipment selection has been discussed by Matsumoto et al. (1990), by Okumura et al. (1998), and in an early study by Dalton and Hohmann (1972), but without an obvious choice of equipment. Given the wide range of environmental constraints that must be accommodated, new designs for material handling equipment, such as the vibratory bulldozer and penetrator (Szabo et al., 1994; Nathan et al., 1992) discussed previously, will have to be considered. To minimize the expenditure of energy, it is desirable to carefully select the material to be processed; processing requires the ability to collect and transfer material to a primary power node. Material handling can be facilitated by incorporating local resources into the operational planning. For example, it has been suggested that after a lunar base has been established, a railroad constructed from indigenous materials by teleoperated robots could be developed to provide surface transportation for expansion of human activities to other areas of the Moon (Schrunk et al., 1998). Although an ambitious undertaking, a lunar railroad would allow traveling on the silty, dusty lunar regolith without penetrating into the abrasive soil, and it could provide faster transport than robots or other individual surface vehicles. If a truck-type hauling vehicle were used to transport regolith to a processing area, traction and support in the soft regolith terrain could be provided by large belts connecting the drive wheels, as in an armored military tank.

The handling and transport of irregularly sized granular material at reduced gravity require further studies, because frictional contact between particle surfaces depends on the normal forces between particles, and hence on gravity. The phenomenological behavior of granular materials as it affects the construction of habitats and roads and the utilization of resources is discussed in Section IV.G.

Concentration and Beneficiation of Feedstock

Many methods of beneficiation of ores are used on Earth, including screening, filtering, flotation, differential sedimentation or settling, leeching, and electrostatic/magnetic separation. Of these, electrostatic and magnetic separation are judged to have potential for the beneficiation of lunar regolith. A particularly important characteristic is that they do not involve the use of scarce water resources.

Electrostatic and Magnetic Beneficiation

Magnetic beneficiation methods are based on differences in the magnetic susceptibility of the components of an ore. If one component has a higher magnetic susceptibility than the other components, application of a magnetic field to a stream of ore particles may be used to deflect and concentrate this component. The susceptibility of lunar ilmenite is approximately 7.6×10^{-5} cgs mass units, which is too low to allow for efficient magnetic beneficiation of the lunar soil for ilmenite. By contrast, ferromagnetic agglutinates (e.g., particles containing iron) of the lunar soil may be separated magnetically from the soil; this is desirable prior to the electrostatic treatment of the soil for ilmenite concentration.

Electrostatic beneficiation methods are based on the differential charge acquired by the components of an ore via contact charging, or by induction when grounded (on a slide or drum) and subject to an applied electric field, or by exposure to an ionizing electrode. An electric field then separates the particles during free fall, depending on their acquired charge. Electrostatic beneficiation is used to separate rutile and ilmenite from zircon and monazite in plants in the United States and Australia that process heavy beach sand (Fraas, 1962; Kelly and Spottiswood, 1982).

The input to an electrostatic or a magnetic beneficiation system is a stream of particles, which first may require the comminution of the raw ore (e.g., lunar regolith). The system consists of particle handling components such as hoppers, feed belts, and accumulation bins; in an electrostatic system an electrostatic generator and deflecting electrodes are required, whereas in a magnetic system deflecting magnets are required.

Reduced gravity levels will generally increase the difficulties of handling a particulate stream, including the generally diminished flow out of hoppers. A specific possible negative effect of reduced gravity levels on the beneficiation process would be a reduction in the force acting to separate the particulate stream from the charging drum or belt into the free-fall zone, where deflection and enrichment occur. Under these conditions, some additional measures to ensure separation might be used, such as a directed blower. A benefit resulting from the reduced gravity level would be an increase in the residence time and hence in the deflection of particles in a free-fall zone of fixed design. Otherwise, no serious issues due to the gravity level are expected, although there may be serious issues related to controlling the electric charge of particles in a hard vacuum.

Atmosphere Acquisition (and Compression) Systems

This discussion focuses on the problem of acquiring and compressing Martian atmospheric gases (95 percent carbon dioxide) for processing in a reactor to produce propellant and/or oxygen for life support. Carbon dioxide is a resource that is available everywhere on the Martian surface, and the processing steps are simple. Since the ambient atmospheric pressure on Mars is approximately 5 torr, compression by a factor of approximately 100 or more is required to reduce the volume of gas to be processed to a reasonable level. In principle, this could be achieved with mechanical compressors using two or three stages. Assuming that the first-stage dust filter is sufficiently rugged, this approach has the advantage of allowing dust filtration to be accomplished between stages using mechanical filters with reasonable pressure drops. The disadvantage would be the system's large energy requirement and the wear on its moving parts, especially that due to ambient dust.

An alternative is a sorption pump (Rapp, 1998; Ash et al., 1978). Such a pump consists of a bed of absorbing material such as a zeolite that is exposed to the Martian atmosphere at night, resulting in the absorption of gas from the atmosphere into the zeolite. During the day the zeolite bed is cut off from the atmosphere and heated while interfaced with the reactor. The gas absorbed during the night is expelled at the higher temperature and pressure to the reactor. The diurnal cycle of Mars is about 24 h. The data reviewed by Rapp (1998) indicate a compression ratio of 136:1 between 6 torr and 815 torr, with a temperature swing from 200 K to 450 K. The net release of gas is estimated to be 0.11 gram of carbon dioxide for 1 gram of zeolite.

The main problems to be addressed in operating these systems are (1) preventing dust accumulation in the bed and other components and (2) ensuring efficient heat exchange during the heating and cooling parts of the cycle. Dust is a major and ubiquitous problem on Mars, where the surface environment exhibits very different electrical and chemical behavior than Earth (see Kolecki and Hillard, 1992, for example). Passive filtration with mechanical filters seems very unlikely to be successful for the sorption compressor under consideration because of the low pressure drops available during the collection part of the cycle. Dust removal technology is discussed in Landis et al. (1997) with regard to solar cells; a favored proposed method is electrostatic dust precipitation, which may be an option for the atmosphere collection system described here.

The design of an efficient heat exchange system for cooling at night and heating in the daytime that uses vacuum jackets, thermal isolators, and thermal switches is discussed in Rapp (1998). It should be emphasized that because the absorption of gas during the night is exothermic, active cooling is necessary to prevent the bed from heating up. Other components include heat exchangers, temperature-controlled valves, and pipes and valves.

At the Martian gravity level of 0.36 g_0, none of the processes discussed above are expected to be greatly affected compared with operation at 1 g_0. Some changes may occur, however, in the heat exchange equipment that involves multiphase flow, as well as in the operation of packed zeolite beds in an adsorption pump and in the management of dust.

Filtration

Removal of particulate from fluids is for life support, protection of fluid-processing hardware, and material separation and processing. Filters typically incorporate a porous solid matrix of some type through which fluid flows and on which particles of a specific size range are captured. The simplest filtration mechanism is called straining and works literally like a strainer, trapping particles larger than the size of the strainer holes. Straining can only be considered in space applications where very low particle loadings (particles per unit volume) are encountered or where the particles themselves are being collected for a specific use. Otherwise, since the filter becomes clogged by trapped particles, that type of system cannot be used for unattended long-duration operation. In this sense, removal and capture of the strained particles in a space environment is a problem.

Terrestrial filter systems use a variety of other mechanisms to capture and remove particles from fluid streams (Hestroni, 1982; see, specifically, Cooper, 1982). Many of the filter mechanisms require that particles be captured on impact with a filter fiber or filter pore. Aerodynamic or inertial capture is effected when particle mass (and shape) produce flight trajectories that cannot follow the sharp turns required of the flowing fluid during transit through the filter matrix, resulting in particle capture on filter surfaces. Diffusion or Brownian motion capture occurs in filters with very low fluid velocities relative to characteristic filter passage lengths (and pore diameters), where particle transit times are so long that the particles contact and attach to the filter material, as a result of random motion, before egress. These types of filters fail or become ineffective when too many flow passages are blocked owing to particle buildup or because the filter material loses its ability to trap particles that reach its surface. Sloughing can also occur for these systems when the flowing fluid exerts aerodynamic forces on captured particles that are sufficient to pull them away from the filter surface. These mechanisms are virtually unaffected by gravity.

Electromagnetic forces can also be used in filter systems. Electrostatic fiber capture occurs when an electric field is maintained by the filter system such that the electric field lines converge on the fibers of the filter, resulting in electrophoretic capture of particles as the nonconducting, particle-laden fluid flows through the matrix. Electrostatic precipitation utilizes corona- or glow-discharge electric fields, generally across discrete electrodes, to charge particles as they are convected through the field. The charged particles are captured subsequently, either by attachment to one of the electrodes or by accumulation in a controlled region of the flow. Coulomb-type electric capture results when the particles to be collected carry a charge that induces an opposite charge on filter surfaces (fibers), thereby producing an attractive force that is strong enough to pull the charged particles out of the flowing fluid and attach them to the filter. Generally, the electromagnetic forces exerted by these systems greatly exceed gravitational forces, but the removal and collection of the particles captured and collected by these systems is affected by gravity.

Both the particle capture method and the filter design depend strongly on the particle size distribution and loading. Typical terrestrial particle size ranges are presented in Figure III.E.1, where it is noted that particle sizes encountered on Earth, and hence particle sizes that could be encountered within a spacecraft environment, span eight orders of magnitude. Associated with the various particle sizes are the specific particle capture methods that apply to the various size ranges, as represented in Figure III.E.2. Particles smaller than 0.1 µm tend to remain suspended in Earth's atmosphere. That particle size will likely behave similarly in Mars's atmosphere (and gravity). It is noted that particle sizes between 0.005 and 2 µm can act as nucleation sites in the formation of droplets and crystals (Hudson and Squires, 1973), so that great care may be required in certain types of scientific experiments if control of nucleation is necessary. For biological systems, filtration of particles of sizes between 0.01 and 0.2 µm is required to remove viruses, whereas bacteria are filtered at sizes slightly larger than viruses but small than 10 µm in diameter.

Filtration of particles from liquids is for the most part insensitive to gravity, and cloth and screen filters can be used to strain particles as small as 10 µm diameter from liquids. Exceptions are sedimentation separation, which obviously is controlled by gravity, and the foam and bubble fractionation technique for removing submicron-size particles, which is also influenced strongly by gravity. Since many of the small particle filtration systems in terrestrial laboratories utilize centrifuges and ultracentrifuges (for particle sizes in the 0.01 to 100 µm range), gravity is not a factor. Particles smaller than 0.1 µm can be removed from liquids using diffusion processes such as reverse osmosis and dialysis or by ion exchange or electrodialysis (Freeman, 1982). In the size range of these processes, particle masses are so small relative to surface area that terrestrial gravitational influences can be neglected all together.

Dust management in extraterrestrial environments deserves special consideration. Not only are settling processes controlled by interactions between atmospheric gases and local gravitation, but the process of producing new dust is also a concern. The creation of dust by particle impacts with a surface is called saltation, and that process is controlled by gravitation. Saltation has been studied for Mars atmospheric processes because of concerns related to planetwide dust storms (Greeley et al., 1980). Furthermore, as was observed in videos of the Apollo Moon Rovers, short-lived dust clouds are created by moving wheels, even in near-perfect vacuum conditions. Dust removal from lunar surfaces is a major design issue, whereas dust filtration from artificial lunar environments and filtration of the Martian atmosphere present more subtle research issues. Even though the Martian atmosphere is dusty and planetwide dust storms lasting on the order of 100 days occur nearly once every

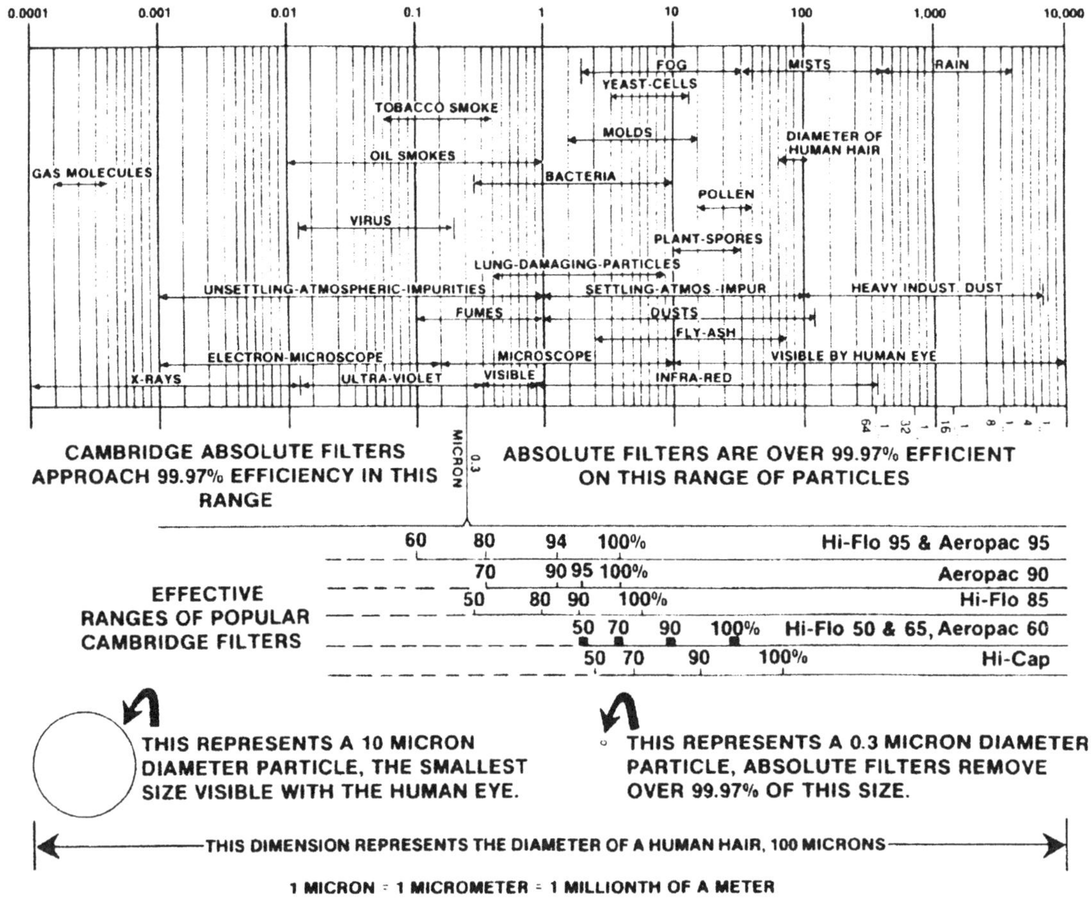

FIGURE III.E.1 Sizes and characteristics of atmospheric contaminants. SOURCE: Cambridge Filter, Syracuse, New York.

Mars year, Martian dust particles are thought to be smaller than 5 microns and claylike, with particle loadings on the order of one particle per cubic centimeter. The Martian atmosphere would not be considered to be dust-laden by terrestrial standards, but because of its low atmospheric density and the possibility that dust particles are either very abrasive or contain adsorbed volatile molecules (which could poison or otherwise damage chemical processing equipment) dust filtration is still necessary. Away from Earth, it is not necessary to maintain environmental pressures of 1 bar for systems that do not need direct human interaction. In particular, manipulation of Mars's atmosphere can be accomplished in ambient pressures of about 5 torr. Filtration of particles from uncompressed environments of this type requires designed filters with virtually no pressure drop, because even small pressure drops translate to very large increases in the volume of gas that must be moved through processing equipment. At very low operating pressures like Mars ambient, mechanical equipment must have larger dimensions to accommodate the high volume flow rate needed to achieve the necessary mass flow rate. This in turn increases energy consumption. Although the flow behavior is not influenced strongly by gravitation, the removal of dust from low-pressure-drop filter systems is dependent on the gravitational environment.

The research issues associated with filtration in a microgravity environment depend on the particular type of filter system and the range of particle sizes to be filtered. In general, the influence of gravity on particle agglomeration with respect to filter flow rates and surface interaction processes must also be understood. How

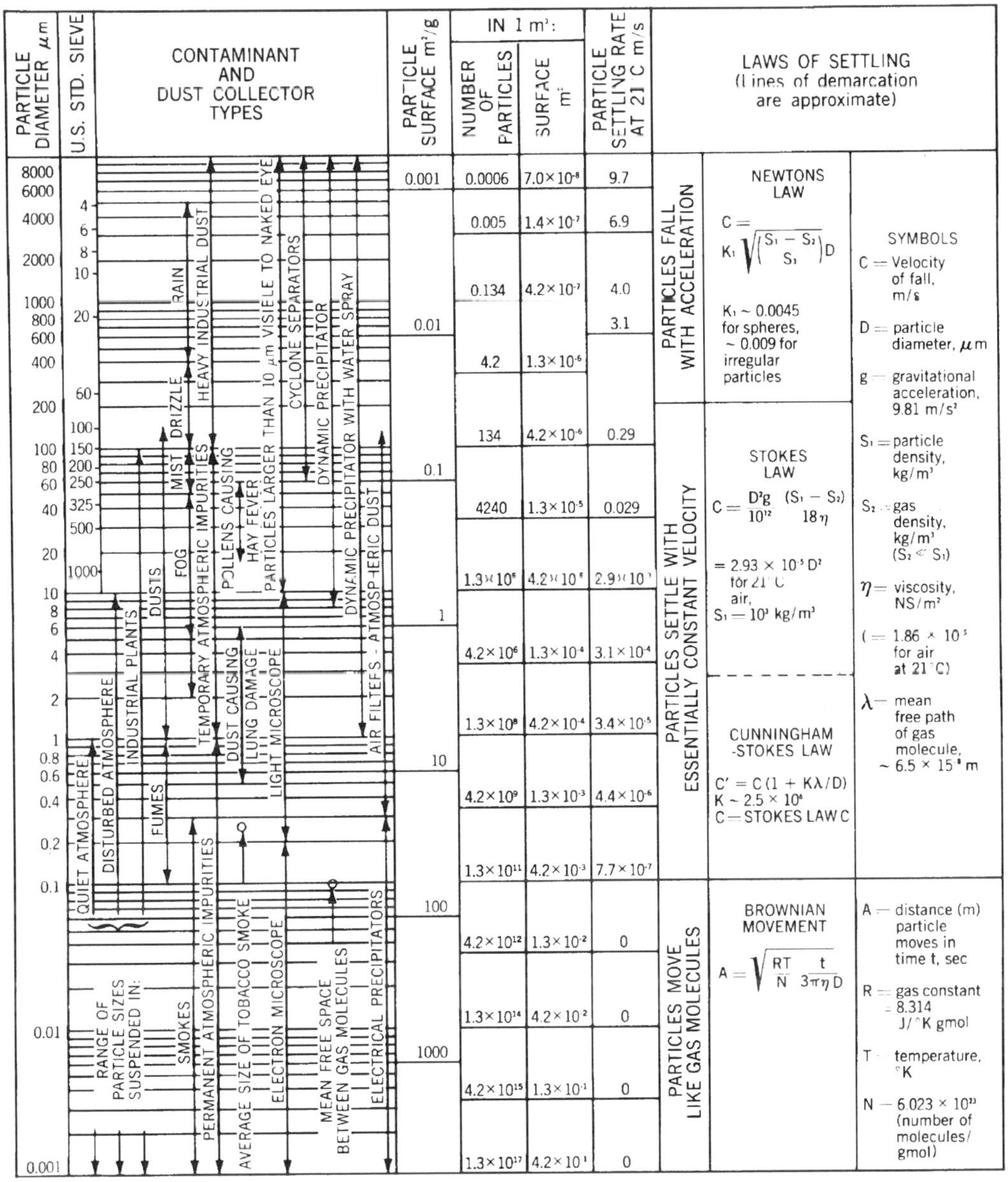

FIGURE III.E.2 Sizes and characteristics of atmospheric solids.

NOTE: The values for particle surface, settling rate, and number and area per cubic meter of air are based on the following: particle specific gravity = 1 (density = 1000 kg/m^3); mass concentration = 70 μg/m^3, typical of urban concentrations, particles are smooth spheres, all of equal size; and gas is air, with a density of 1.29 kg/m^3, temperature is 21 ºC, pressure is 1 atm, and viscosity is 1.86×10^{-5} kg/m/s.

SOURCE: Reprinted with permission from AAF International.

different gravity levels influence sloughing must be understood. Development of filter cleaning systems that can be operated in microgravity for long-duration missions is an important issue. Microgravity issues associated with filter recycling include (1) threshold gravity levels required for particle removal from filter surfaces, (2) the effectiveness of scraping and shaking processes for cleaning filters in microgravity environments, and (3) how to capture and remove the dust accumulations resulting from filter cleaning operations. Furthermore, the effect of the intermittent and random acceleration vector orientations that characterize typical spacecraft operations must be understood in terms of how filter systems will be affected during long-duration missions.

Fluid-Based Chemical Processing

Electrochemical Processing

Water Electrolysis

Production of oxygen from recycled water for life support during long-duration spaceflights is a key element in enabling manned missions to Mars and beyond. Production of oxygen for rocket propellant and for life support, using water extracted from a variety of ice-containing planetary bodies, will probably be a key step enabling HEDS missions to be extended throughout the solar system. Water electrolysis can be used to produce not only oxygen but also hydrogen for use as rocket fuel and/or as a feed gas for making a variety of hydrocarbons. The wide range of extraterrestrial water sources, coupled with the need for oxygen and hydrogen molecules in space operations, makes the development of ultrareliable, autonomous water electrolysis systems a pacing technology for the HEDS enterprise.

Water electrolysis is energy-intensive because electric energy is used directly to break the atomic bonds between water's hydrogen and oxygen atoms (5.3 kW-h per kg). The electric power required to sustain the desired hydrogen and oxygen production rates is high enough to cause problems with electrode corrosion and electrode life that have not yet been overcome. Reduced gravity will alter the gas-liquid interface behavior associated with the production of gaseous oxygen and hydrogen in lunar and Martian electrolysis systems, but those effects should be scalable. However, the design and operation of water electrolysis cells on board spacecraft or on very-low-gravity surfaces, such as on ice-containing asteroids or on inactive comet cores, will be determined by the microgravity environment and will require fundamental knowledge of multiphase processes in these microgravity environments to achieve necessary levels of system reliability and autonomy (see Humphries et al., 1991). Because the electrochemical production of oxygen (and hydrogen) from water in microgravity environments is of critical importance to a range of HEDS missions, a hardware demonstration program on the International Space Station for the purpose of establishing the necessary reliability levels and for developing fully autonomous systems would be fully justified.

Even though water electrolysis units have been operated intermittently on Mir since 1989, producing more than 700 kg of oxygen—sufficient for approximately 1,000 man-days of life support—the systems have not demonstrated enough reliability to be used as Mir's primary oxygen supply (Belaventsev et al., 1991). Furthermore, because of its explosive potential, generated hydrogen must be removed completely from the system. Belaventsev et al. (1991) reported that hydrogen mixtures had relatively low ignition energies for concentrations (in Mir gravitational environments) ranging from 4.1 to 96 percent.

Over the past 20 years, NASA engineers have studied static feed water electrolysis (SFWE) and solid polymer water electrolysis (SPWE) system designs; both were considered for life support use on the International Space Station, but neither system was selected. More recently, a water vapor electrolysis system has been studied that is based on operating principles similar to the SFWE but that operates directly on humid cabin air rather than by utilizing a separate water supply (Wydeven, 1988).

The SFWE system is represented schematically in Figure III.E.3 (Wood, 1992; Figure 8 in Wydeven, 1988). The SFWE reactor uses static pressure to supply liquid water to a water feed matrix, made up of thin asbestos sheets saturated with a hygroscopic aqueous potassium hydroxide solution. The cathode and anode cell matrix elements are configured similarly. The hydroxide ions provided by the potassium hydroxide in the cell matrix act

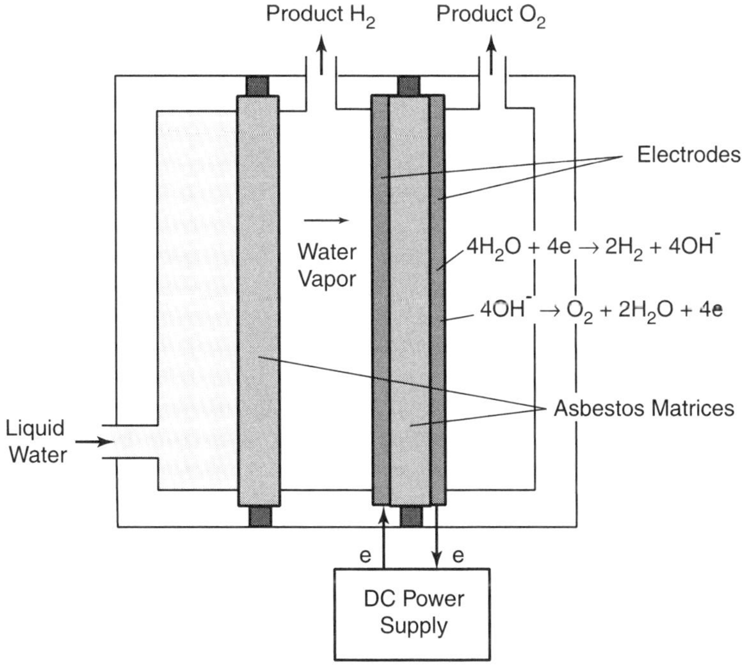

FIGURE III.E.3 Static feed water electrolysis cell. SOURCE: Wood (1992). Courtesy of NASA.

catalytically on the anode and cathode reactions shown in Figure III.E.3. As electrolysis occurs, the potassium hydroxide concentration is increased across the cell elements, creating a water-vapor pressure gradient and drawing additional water from the feed matrix into the cell matrix. Because the SFWE electrodes operate directly on water vapor, the cell compartments thus avoid two-phase flows, transferring the two-phase management problem back to the (liquid) water feed unit.

The SPWE reactor utilizes perfluorinated sulfonic acid polymer membranes, which act as electrodes when wetted. The system is shown schematically in Figure III.E.4. The SPWE half reactions at the cathode and anode differ from the SFWE reactions since protons, rather than hydroxide ions, diffuse through the membrane and act catalytically in the anode and cathode reactions shown in Figure III.E.4. Deionized water remains in contact with the cathode, generating sufficient heat to require cooling. The anode remains exposed only to the vapor phase, which results in the production of almost pure oxygen. Hydrogen must be removed from the water vapor in the cathode chamber. Both types of water electrolysis are influenced strongly by gravity level.

So-called water vapor electrolysis operates directly on cabin air and serves both to dehumidify the air and to produce hydrogen and oxygen. Moist air enters the electrolysis compartment, where oxygen is produced and released at the anode; the generated hydrogen ions diffuse to the cathode, where they recombine with electrons to produce molecular hydrogen, which is vented. Use of moist air as a water source eliminates the liquid-vapor separation problems associated with the SFWE water supply.

Phase separation problems abound in these systems. First, water vapor humidifies the generated hydrogen and oxygen gases. Because the electrolysis cells are operated typically at temperatures above 300 K, humidity levels can become quite high. The water vapor in the generated gases must be removed if the gases are to be stored cryogenically, in order to avoid ice formation in critical valve and flow-management elements. Second, because of its explosive potential, hydrogen removal is critical in all systems. It is very difficult to prevent hydrogen from leaking across membranes or diffusing through other containment barriers. In microgravity environments, the absence of significant buoyant forces makes it very difficult to monitor, collect, and/or remove unwanted hydrogen from gaseous volumes in the separation units. Third, dissolved gases will be present in the water that is being

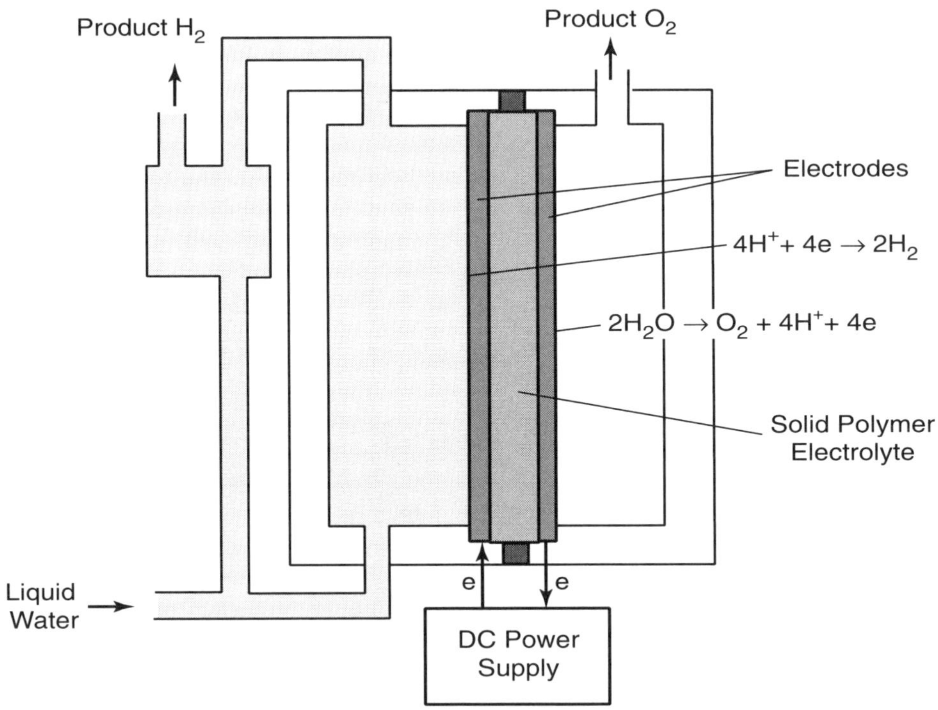

FIGURE III.E.4 Solid polymer water electrolysis cell. SOURCE: Wydeven (1988). Courtesy of NASA.

electrolyzed and it will be necessary to prevent gas buildup, even if the gases are inert, in order to prevent the electrochemical cells from becoming "polarized."

The separation and control of dissolved gases in liquids in microgravity environments are key elements in the development of water electrolysis systems in support of HEDS. Since vapor-phase electrolysis units rely on pressurized liquid water feeds for supply, they also require liquid-vapor phase management studies in microgravity environments in order to achieve the large production rates, over long periods of time, required for HEDS missions. It will also be necessary to develop instrumentation for autonomous operation and control. Many of the autonomous control and operation problems that must be resolved before water electrolysis can actually become a key building block technology in the overall HEDS program can be addressed through long-duration testing of autonomous water electrolysis units on the International Space Station. The significance of electrochemical production of hydrogen and oxygen is potentially so great that a long-term operational program can be justified.

Gas Phase Electrochemical Extraction

A variety of solid and molten electrolytes have been used as fuel cells, many of them in space. These devices, in addition to their ability to produce and store electrical energy (Sridhar and Foerstner, 1998), could also serve as important chemical processing units for the HEDS program. Fuel cells can use electrical power to extract hydrogen, oxygen, carbonate, or hydroxyl ions from mixtures of other atoms and molecules. Thus they can be modified for use as electrochemical processors, but the microgravity issues for fuel cell applications are essentially the same. Hence, it is more efficient to describe the operation of solid-electrolyte oxygen-extraction systems, which are currently scheduled to be demonstrated on the Mars 2001 mission, in terms of possible reduced gravity issues, noting that these devices can be used as fuel cells.

Solid Electrolyte Electrolysis: Extraction of Oxygen from Carbon Dioxide

Space-based extraction of ionic molecules from gaseous extraterrestrial materials allows avoiding the need for phase separations and the concomitant gravitational influences on chemical separation processes. The electrolyte barriers employed in these separation systems are, however, only capable of conducting specific ions—usually at rather high temperatures. The systems being considered here act like reverse fuel cells, consuming electrical energy to remove and then pump a specific ion across a solid membrane for collection. Because of their ability to extract oxygen directly from the Martian atmosphere for near-term space missions, solid electrolytes that conduct oxygen ions but offer very high electronic resistance are presently the most thoroughly studied systems for extraterrestrial resource processing. This approach was first suggested by Stancati et al. (1979) and evaluated experimentally by Richter (1981). More recently, Ramohalli and Sridhar (1991) and Sridhar and Vaniman (1995) have reported on advanced systems similar to the system that would have been operated on the Mars 2001 mission (now cancelled). Recent reviews of solid-electrolyte separation systems for Mars applications have been reported by Hepp et al. (1994) and Rapp (1998). Most of those systems utilize zirconia (ZrO_2), stabilized with yttria (Y_2O_3), but a variety of other dopants have also been used. Stabilized zirconia maintains a doped cubic form of the material so that it can carry oxygen ions when subjected either to a voltage or an oxygen concentration gradient. Solid electrolytes other than zirconia have been considered for ionic oxygen conduction as well, but stabilized zirconia currently appears to be the best choice as it exhibits satisfactory ionic conduction at temperatures above 850 °C and excellent ionic conductivity at 1000 °C.

Reduced gravity will influence these systems in subtle ways. Systems needs relating to using solid electrolytes for the extraction of oxygen from carbon dioxide include the following:

- Systems designed so that they can survive vibration loads, prevent leaks of critical importance, and incorporate adequate instrumentation and control. Because of the high-temperature operation and probable thermal cycling, mechanical behavior will be affected by gravity;
- Systems designed so as to minimize mass and volume;
- Systems designed so that they can tolerate thermal shock and thermal fatigue associated with temperature variations ranging from Mars ambient through steady-state cell operation. The Martian thermal environment controls the thermal shock conditions, and to some extent thermal shock is influenced by gravity;
- Development of electrodes that promote oxygen production on the feedstock side of the membrane, maintain good electrical contact with the electrolyte, and provide controlled voltage and current distributions across the electrolyte with minimum Joule heat losses. Gravitational concerns must be addressed;
- Development of hardware elements that minimize differential thermal expansion between components and that have seal systems able to withstand repeated thermal cycles without losing their integrity;
- Development of multiple cell designs that can incorporate flow path redundancies and distributed instrumentation and control and that assure proper operation of all of the active cell units;
- Development of oxygen separation units that can withstand exposure to sulfur compounds that may be present in Martian dust;
- Development of system designs that are self-cleaning and that have some ability to modify or repair themselves in situ;
- Development of systems that operate at maximum conversion efficiency using the least amounts of electrical and thermal energy; and
- Development of systems that can be used intermittently and that are capable of providing electrical power when operated in reverse (see Sridhar and Foerstner, 1998).

The principal solid electrolyte processing elements are the following: feedgas filter system, carbon dioxide concentrator and/or compressor, regenerative heat exchanger, feedgas heater, electrochemical cell array (oxygen extraction and compression), exhaust gas return (through heat exchanger), radiator, oxygen cryocooler, and oxygen storage system.

Molten Metal Electrolysis

Lunar Magma Electrolysis

Production of oxygen and metals from lunar rock and regolith is a more ambitious form of electrochemical processing. Beck (1992) has described a lunar oxygen plant design employing a pair of electrodes immersed in a pool of molten silicate produced from molten lunar rock and regolith. Using direct current, oxygen would be produced at the anode and various metals at the cathode. Systems of this type are difficult to operate reliably on Earth, and fundamental understanding of the physical chemistry pertaining to the entire system is needed, particularly when the basic unit operations are altered by reduced gravity.

The rate at which molten mass flows to the electrodes is a very important performance and design parameter. If mass transport is slow, say by diffusion, then electrolysis at a given cell voltage will be slow and larger cell voltages will be needed to increase reaction rates. Buoyancy-driven convection is a very desirable alternative mechanism for mass transport, and viscosity is an important operating parameter. In addition, sedimentation and buoyancy are important for product separation. These are highly gravity-dependent.

The effects of gas bubbles, the porosity of the electrodes, and simultaneous electrolyte reactions are complicating factors. Potential and flux variations affect the distribution of current on various electrode shapes via ionic conduction, which is controlled by the boundary conditions imposed by kinetics and mass transport. Reduced gravity is undoubtedly an important controlling variable in the design of molten metal electrochemical processes. For example, a comparison between terrestrial and lunar operation of a copper refining cell predicts that 57 percent more electrode area would be required for lunar operation (Beck, 1992).

Beck has also described concepts for systems to produce iron from basalt magma, silicon and aluminum from anorthite, and iron from iron-rich regolith. For the case of iron production from basalt magma, the metal will sink to the bottom of the melt, where it will be trapped. In the fused anorthite electrolysis, aluminum silicon alloy floats to the top, where it will be trapped. Both are continuous processes. The production of iron via regolith electrolysis is a batch process. At startup of the cycle for a given batch, the in-place regolith is melted (e.g., by a solar concentrator) before the lid of the vessel and the anode are put in place. After electrolysis starts, more regolith is added to the vessel. The top of the cathode grows by settling of molten iron, and the anode is moved up to accommodate the growth. After some time, the electrolysis is terminated, material is removed, and the cell is reloaded for the next cycle.

Production of Aluminum from Orbital Debris

Recovery of aluminum from orbital debris is the only known HEDS application of molten metal electrolysis that can be considered seriously at the present time: because it is possible to collect feedstock of a prescribed composition, the chemical processing requirements can be known accurately enough to pursue system designs. Processing could take place in Earth orbit with the retrieval of specific types of debris, such as nonfunctioning satellites or specific classes of spent upper stages. The hundreds of tons of aluminum orbiting Earth as spent boosters are a potentially attractive source of aluminum for solar-powered satellites and other orbital applications. In addition, the recycling of this material would reduce the hazards it poses for orbiting satellites. The recycling of orbiting aluminum would require solving problems of materials management in microgravity, including those associated with the separation and segregation of raw materials and the transport of nonuniform solid and liquid materials, including multiphase fluids containing dissolved gases. Many of the research issues associated with developing technology for recovering aluminum are similar to those encountered with technologies for welding in microgravity, which is discussed elsewhere in this chapter.

Radio-Frequency Processing of Materials

Radio-frequency (RF) processing of materials is a rapidly expanding segment of the semiconductor industry. By proper design of electrode geometries, processing temperatures, and pressures and by optimization of appropriate RF frequencies and voltages, it is possible to process gaseous molecules and particle suspensions (in gases) selectively and efficiently for a variety of terrestrial applications. RF processing of the Martian atmosphere, discussed below, has been identified only recently as a potential means of selectively dissociating the carbon dioxide into carbon monoxide and oxygen at low processing temperatures and at Mars ambient pressures (Vuskovic et al., 1997, 1998). The RF based glow discharge is preferred to direct current discharges (Wu et al., 1996) because it requires far less power to sustain the plasma. RF processing of molecules is an approach that can be used selectively to avoid high-temperature processes. Furthermore, because of the electrophoretic forces that can be produced and controlled by these systems, it is possible to use RF-generated forces to separate and or manipulate solid particles, particularly in gravitational fields.

RF Processing of the Martian Atmosphere

RF processing of the Martian atmosphere has many similarities to the well-developed research field associated with maintaining proper concentrations of carbon dioxide in CO_2 lasers (Byron and Apter, 1992), except that the CO_2 dissociation in RF-excited gas mixtures needs to be maximized rather than minimized. Hence, it is not surprising that the lower pressure of the Martian atmosphere (as compared with the pressure in laser cavities) is favorable to CO_2 dissociation because it reduces the potential for three-body recombination processes and also because the plasma-generated electron densities are higher, thus promoting increased interactions with CO_2 molecules. Although the volume fraction of carbon dioxide in the Martian atmosphere is much higher than in CO_2 laser cavities, the small concentrations of argon and nitrogen (in the Martian atmosphere) do serve as buffer gases, which sustain the discharge and provide electrons for CO_2 dissociation.

RF processing systems can be designed to effect almost complete dissociation of Martian atmospheric carbon dioxide to carbon monoxide and oxygen. Systems can be designed that utilize natural convection as the prime mover for transporting the Martian atmosphere through the RF field and through an oxygen extraction unit. If such systems are properly designed, Mars dust can be collected and removed during the RF processing step and the (dust filtered) exhaust gas from the oxygen extraction unit can be collected and used as a carbon monoxide fuel. The average gas temperature rise associated with this type of RF processing is only about 100 °C (or less), which is too small to create large convective flows under Martian conditions, and so the reduced-gravity environment becomes a major design consideration, affecting dust particle collection and removal and the flow of the feed gas through the processing unit.

Since the exhaust stream produced in the RF processing/oxygen extraction system cannot be captured and stored by utilizing natural convection, an oxidizer/fuel production system requires blowers and/or compressors to collect and compress the processed Mars exhaust (fuel). Blowers and compressors are especially vulnerable to dust damage by abrasion and from the clogging of small passages. Hence, removal of dust from the Martian atmosphere becomes an even more important requirement, and the RF approach relies completely on detailed knowledge of the influence of reduced gravity on particle behavior.

The RF plasma resides in a gas mixture that is no longer in thermodynamic equilibrium since the processing is a continuous flow process and the entering Martian atmosphere molecules are relatively cold. The flow behavior of this nonuniform mixture and the associated oxygen transport to the collector surfaces can be influenced by the altered buoyancy effects in Mars's gravity.

Separation or Filtration of Solid Particles (Dust) from RF-Processed Gas Streams

The RF plasma will interact directly with any particles that happen to be suspended in a gas stream, because the RF field will increase particle surface charges by several orders of magnitude. However, because the plasma is sustained by a very high frequency, alternating voltage field, particles are not collected on electrode surfaces, as happens with conventional electrostatic precipitators. Rather, they become trapped in specific volumes of the RF field. The electrostatic, ion drag, thermophoretic, and aerodynamic forces acting on dust particles act in different directions, with different relative magnitudes, depending on particle composition, size, shape, and density. Thus, when the RF-derived forces are combined with Mars gravity forces, the net force acting on each dust particle propels it to a particular spatial equilibrium location, depending on its physical characteristics and the feed gas flow rate. In fact, because of relative differences between the different forces acting on different size particles, the particles will be segregated by size and, to some extent, by their properties. In the case of an inductive RF-plasma system, the smallest particles are segregated in the vicinity of the centerline of the RF field, whereas larger particles become trapped in annular zones bounding the electrode sheath. These dynamic traps will become saturated with dust particles at some point, and it will be necessary to stop the plasma long enough to permit the particle clouds to be removed either by gravity or by mechanical means. Based on current limited knowledge of Martian atmospheric dust loads, it is likely that dust removal would need to occur once each Mars sol (24.66 h).

Oxygen Production

Oxygen requirements for life support and for most liquid chemical rocket propellants are so large for long-duration human missions that the ability to produce oxygen mass away from Earth's surface could enable near-term HEDS missions to be commissioned for planetary objects virtually across the solar system. Furthermore, when oxygen is produced from simple molecules such as from carbon dioxide and from water, the chemical processing steps are simple. Advanced life support systems and ISRU processing systems differ primarily in the scale of the hardware (Sridhar and Miller, 1994) and in the fact that life support systems will have to operate primarily in microgravity (during space travel) rather than in the fractional gravity associated with most ISRU processing sites. Hence, a discussion of direct chemical processing emphasizing oxygen production is justified.

Sabatier Reactors

Recovery of oxygen from carbon dioxide and water in microgravity environments has been an active area of research since the beginning of human spaceflight. The Sabatier process (Sabatier and Senderens, 1902) for transforming carbon dioxide into oxygen and methane has been one of the most thoroughly studied approaches for developing advanced life support systems. That process reacts hydrogen with carbon dioxide:

$$CO_2 + 4\,H_2 \rightarrow CH_4 + 2\,H_2O. \qquad \text{(III.1)}$$

The water is captured and pumped to an electrolysis unit, while the gaseous methane is collected and either vented or stored. Hydrogen and oxygen are generated by electrolysis of the water, and the generated gases are filtered dry. The desired oxygen is stored for life support, while the hydrogen is returned as feedstock for further processing with the carbon dioxide.

Hamilton Standard Corporation and NASA have been evaluating systems of this type more recently (see, for example, Cusick, 1974, and Martin et al., 1983, or Sullivan et al., 1995) for advanced space station life support and for extraterrestrial resource processing. A variety of resource processing options are possible because Sabatier reactions other than reaction [1] can be activated by using different catalysts and processing temperatures.

At temperatures near 1100 K, Seglin (1975) reported that it is possible to employ transition element catalysts (iron, cobalt, nickel, ruthenium, palladium, osmium, indium, and platinum) or silver to promote the following chemical reactions:

$$CO + 3\,H_2 \rightarrow CH_4 + H_2O \text{ and} \qquad \text{(III.2)}$$

$$CO + H_2O \rightarrow H_2 + CO_2. \qquad \text{(III.3)}$$

In addition, reactions such as the complete reduction of carbon monoxide to solid carbon and subsequent formation of carbon dioxide, as well as the hydrogenation of solid carbon to form methane, are possible but must be avoided because of difficulties in handling solid carbon, particularly in a microgravity environment. Reactions [1] and [2], both producing methane and water as products, are the primary reactions for ISRU systems. Although water electrolysis is considered to be an off-the-shelf technology, reliable operation of a liquid water electrolysis unit in a microgravity environment has not yet been demonstrated.

Zubrin et al. (1991) proposed and demonstrated an ISRU preprototype system for Mars that utilized Sabatier reaction (1) in a tube reactor filled with catalyst to produce methane and water in a highly exothermic process. The methane product was collected and stored as cryogenic liquid rocket fuel that could be used for a Mars return mission, while the water product was deionized and electrolyzed into hydrogen and oxygen. Subsequently the hydrogen was recycled back into the Sabatier reactor and the oxygen was collected and stored as a liquid cryogen to be used either for life support or for return propellant. Zubrin et al. (1991) proposed transporting hydrogen from Earth, thus avoiding the need for an in situ water requirement at Mars, which can only be avoided otherwise by transporting methane from Earth. More recently, Mueller (1998) reported an inability to identify hydrogen storage designs that could provide quantities of hydrogen, during the period between Earth launch and landing on the surface of Mars, sufficient to enable the round trip mission. However, Rapp (1998) reports that Zubrin et al. (1991) have recommended that only half of the required oxygen (and all of the methane) be produced via Sabatier, with the remaining oxygen needs being met using either a solid electrolyte carbon dioxide processor or a reactor employing reverse water gas shift, thus reducing the hydrogen mass required from Earth.

The major processing elements for a Mars-based system are a hydrogen storage and supply system or a water extraction system (from polar ice, permafrost, or other sources, including the atmosphere), an atmospheric filter, a carbon dioxide adsorption/desorption compressor or a Mars atmosphere compressor, a catalyst bed reactor and thermal management system, regenerative heat exchangers, a radiator and water collector, a methane-water separator, a methane dryer, a water collection and storage unit, a water electrolysis unit, hydrogen and oxygen dryers, methane and oxygen precooler(s), oxygen storage, methane storage, and cryogenic refrigerator(s).

The following critical questions remain to be answered by further research:

- At reduced gravity, will the gas bubbles at the electrolysis electrodes separate into collectable volumes of gas as they do at $g = 1\ g_0$?
- Is the maximum 7 mbar pressure difference between Mars ambient and the adsorber sufficient to transport the atmospheric carbon dioxide into the sorption pump?
- Will the conventional zeolite adsorption material require periodic bakeout, can it recycle indefinitely without changing its properties, and will it be subject to poisoning by constituents of the Martian atmosphere? How does the reduced gravity influence the matrix when subjected to the Mars temperature-pressure cycle?

Reverse Water Gas Shift

Oxygen can be produced from carbon dioxide via the reverse water gas shift (RWGS), which consists of the following reaction pair:

$$2\ CO_2 + 2\ H_2 \rightarrow 2\ CO + 2\ H_2O \text{ and}$$

$$2\ H_2O + \text{electricity} \rightarrow 2\ H_2 + O_2.$$

These reactions are the reverse of the industrial reaction that combines carbon monoxide and water to produce hydrogen. Since the reverse water gas shift reaction utilizes water electrolysis in connection with a carbon dioxide (and hydrogen) feedstock, it is very similar to the Sabatier electrolysis process. The primary differences are that they use different catalysts and operate at different temperatures. Below 400 °C, the Sabatier reactions, producing methane and water, dominate. Above 650 °C, the water gas shift reactions, producing carbon monoxide and water, dominate. Because RWGS produces carbon monoxide as a product in carbon dioxide reduction, it offers the potential to exploit a variety of other hydrogenation reactions (Fischer-Tropsch reactions) for ISRU. Therefore, the reverse water gas shift approach offers the possibility of producing such chemicals as methanol and dimethyl ether (CH_3OCH_3), which in turn opens up the possibility of producing other hydrocarbons and plastic (Zubrin et al., 1997). The system consists of a catalyst bed reactor, with a copper on alumina catalyst, that combines hydrogen and carbon dioxide in an exothermic reaction at temperatures on the order of 400 °C. For oxygen production, the reactor exhaust is cooled and the water is condensed and removed. When terrestrial hydrogen is used, it is necessary to dry the carbon monoxide in order to minimize hydrogen loss. The collected water is transported subsequently to an electrolysis unit.

Microgravity issues associated with operating reverse water gas shift systems include the condensation and separation of liquid water from a gas stream and the operation of liquid water electrolysis systems. In addition, the opportunity to utilize carbon monoxide, methane, and hydrogen in a Fischer-Tropsch synthesis gas reactor to produce more complex hydrocarbons opens up a variety of issues associated with the extraction and processing of specific chemicals for more advanced material production systems in a reduced-gravity environment, which will surely alter the unit operations.

A Mars-based system using RWGS for oxygen production can incorporate a hydrogen storage and supply system or a system for extracting water (from polar ice, permafrost, or other sources, including the atmosphere), an atmospheric filter, a carbon dioxide adsorption/desorption compressor or a Mars atmosphere compressor, a catalyst bed reactor and thermal management system, regenerative heat exchangers, a radiator and water collector, a water condenser, a carbon monoxide dryer and exhaust, a water electrolysis unit, an oxygen dryer, radiator(s), an oxygen liquefaction system, an oxygen storage system, and cryogenic refrigerator(s).

Ilmenite Reduction as a Source of Oxygen

One of the more promising processes for the production of oxygen on the Moon is the use of hydrogen to reduce ilmenite (Zhao and Shadman, 1993; Gibson et al., 1990; Taylor et al., 1993), an oxide of iron and titanium ($FeTiO_3$) present in the lunar regolith. The concentration of ilmenite in the lunar regolith varies but is thought to be the highest in the mare (basaltic) regolith, where it is believed to reach 5 to 10 percent by volume. Surface mining the 2 to 5 meter thick mare regolith is considered to be the most promising source of ilmenite. Mining and beneficiation methods are discussed by Vaniman and Heiken (1990), Sharp et al. (1990), and above in this report. Electrostatic separation techniques have been proposed by Agosto (1985). The remaining discussion here is focused on the reduction process to obtain oxygen.

The hydrogen reduction process is described by the following reaction,

$$FeTiO_3 + H_2 \rightarrow Fe + TiO_2 + H_2O,$$

in which hydrogen reduces the ilmenite to yield iron, titania, and water. The water is then electrolyzed to obtain the desired oxygen, and the hydrogen is recycled. Metallic iron is a by-product; further reduction of the titania is difficult and not usually considered. The reaction is slightly endothermic, absorbing 9.7 kcal/gram mole at 900 °C, with a partial pressure equilibrium ratio P_{H_2O}/P_{H_2} of about 0.1. The overall yield of oxygen from the ilmenite is about 10 percent by weight, which means that about 90 percent of the reduced feedstock must be removed as solids on a continuing basis.

Figure III.E.5 shows a schematic of a plant developed by the Carbotek Company under contract with NASA. The process utilizes a fluidized-bed reactor and a solid-state electrolysis cell that electrolyzes water in the vapor state. A stream of hydrogen is forced through a fluidized bed of ilmenite particulate at about 900 °C to achieve a reasonable reaction rate; higher temperatures tend to sinter the particulate and reduce the porosity. In the Carbotek process, the exiting gas is passed through a solid-state electrolysis cell that decomposes the water vapor into oxygen, which is collected, and hydrogen, which is recycled through the bed.

If all hydrogen reacted according to the ilmenite reduction equation and was recovered subsequently during the electrolysis step, it could be reused indefinitely without the need for additional hydrogen. In practice, however, there are small losses, principally from the unreacted hydrogen entrained in the spent feedstock. Most of the unreacted hydrogen waste could be recovered in a batch processing operation by vacuum pumping sealed discharge hoppers, but a need for makeup hydrogen would remain.

The main components of the plant are shown in Figure III.E.5. The ilmenite, stored in a hopper, is conveyed

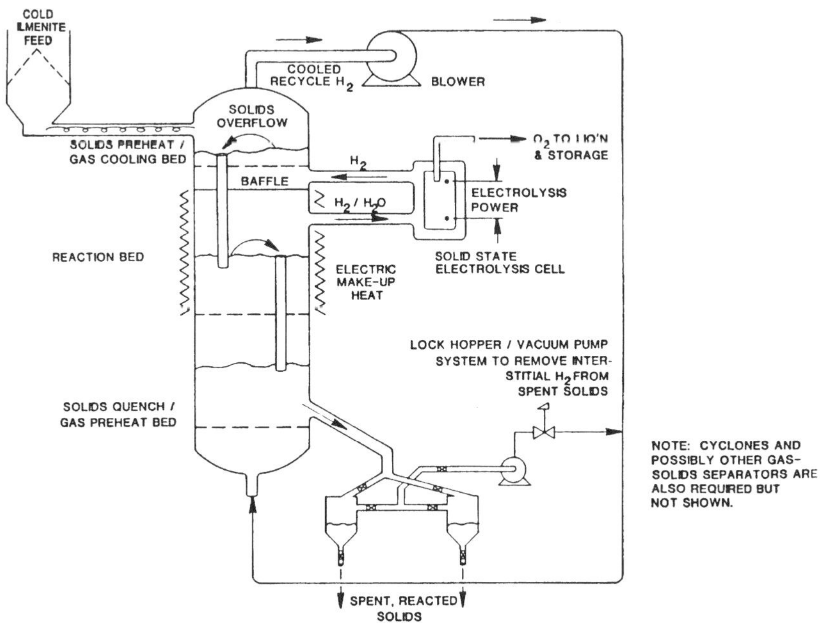

FIGURE III.E.5 Continuous, fluid-bed ilmenite reduction/O_2 production. SOURCE: Gibson and Knudsen (1990). Reprinted with permission from the American Society of Civil Engineers.

by a belt to the fluidized bed, where it reacts with the hydrogen coming up through the bed. The product gas containing water exits through the top of the fluidized bed and enters the solid-state electrolytic cell, where the oxygen is bled off to cryogenic storage and the hydrogen is recycled through the bed. The spent ore is discharged alternately into one of two hoppers, which may be locked and pumped out to recover adsorbed hydrogen.

The components most affected by reduced gravity would be the fluidized-bed reactor (Gibson et al., 1990; Ness et al., 1990) and cyclone separators, if present. Fluidized beds are proposed not only for the ilmenite reduction reaction but also are considered desirable for the removal of contaminants such as hydrogen sulfide in the gas stream. A model two-dimensional, fluidized-bed test article has been operated in a KC-135 NASA research airplane experiment at 1/6 g_0 (Gibson et al., 1990) and found to operate predictably. Additional tests related to design optimization would be desirable for the reactor and cyclone separators.

There appear to be no serious microgravity issues associated with the solid electrolyte process, but a number of design problems with gas-phase electrolysis cells (principally zirconia) remain (Rapp, 1998). Liquid-state electrolysis, on the other hand, is well developed and extensively used on submarines in Earth gravity. It is not clear, however, how well these or similar units would operate at reduced gravity. Thus, which form of electrolysis can ultimately be used is an open question.

Pyrolysis

Pyrolysis of lunar regolith has been proposed as another approach for large-scale oxygen production on the moon (Senior, 1993). It is possible to produce oxygen by vaporizing metal (or semiconductor) oxides. By heating lunar regolith sufficiently, some of its metal oxides reduce to other oxides, liberating oxygen in the process. That liberation of oxygen in a reducing environment is similar to terrestrial pyrolysis processes, except that liberated oxygen rather than pyrolyzed solid is the desired product on the Moon. This approach is very energy-intensive but it avoids reliance on nonabundant molecules containing hydrogen or carbon. After the regolith has been pyrolyzed the processed material is cooled in a condensation step to remove the waste, the metals, and/or the reduced oxides.

Subsequently, the liberated oxygen is recovered. One example of a regolith pyrolysis reaction is the reduction of silicon dioxide:

$$SiO_2 \text{ (solid)} \rightarrow SiO \text{ (metastable solid)} + {}^1/_2\, O_2.$$

Another reaction that can be sustained is the thermal reduction of ferrous oxide, where

$$FeO \text{ (liquid)} \rightarrow FeO \text{ (gas)}$$

$$FeO \text{ (gas)} \rightarrow Fe \text{ (gas)} + {}^1/_2\, O_2 \text{ (gas)}.$$

Two variations of this process have been proposed for obtaining oxygen. One is called vapor separation, or thermal pyrolysis. The other is called selective ionization, or plasma pyrolysis. In the former, material is heated to about 2000 K. The metal species, or reduced oxides, are condensed from the hot gases, leaving gaseous oxygen. In the plasma pyrolysis process, the vapor is heated to very high temperatures (approaching 10,000 K). The plasma thus generated is passed through an electrostatic field in which the ionized metals are separated from the neutral oxygen. Little work has been done on the plasma process, and extensive study would be required to determine its viability.

Vapor pyrolysis oxygen production rates depend on heat and mass transport in the liquid, evaporation and oxygen dissociation, mass transport in the gas, and condensation of metal-containing species. Thermodynamic and transport properties of the molten feed material are critical to process design. Experimental and theoretical work would be needed on both the vaporization and condensation processes.

Molten material that is being reduced will have to be contained and, since the temperatures are likely to be very high, selection of construction materials will be an important problem. Avoidance of tap plugging in the removal of waste must also be considered.

Because of the abundance of solar energy on the Moon and the absence of an atmosphere, it is possible to sustain very high temperatures by solar means. Hence vapor pyrolysis could be competitive with other oxygen production systems that do not rely on water as their feedstock. The unit operations associated with this process will be altered by reduced lunar gravity, as will the manipulation and transport of lunar materials.

Cryogenic Storage

Reduced-gravity cryogenic fluid management issues fall into two distinct areas depending on whether a fractional gravity or microgravity environment is being considered.

It is anticipated that the technology used to efficiently store cryogenic fluids terrestrially should transfer with few difficulties to the gravity on the Moon and Mars, where the generally much lower ambient temperature will reduce the problem of heat flux. However, in space, where the effective gravity is small, the location of the liquid in a tank is determined by the competing effects of gravity and the liquid's surface tension in order to minimize the sum of the gravitational and surface energies (Dodge, 1990), and the storage problems are much more complex.

Reynolds and Satterlee (1966) showed that, in the absence of gravity, the liquid will form one or more interfaces of constant spherical curvature, with interface radii determined by the container geometry and size, the volume of liquid, and the contact angle at which the interfaces meet at the container wall (this angle is near zero degrees for most liquids and container materials of aerospace interest). They also show that the most stable configuration of liquid minimizes the total capillary energy, so that the liquid will collect in a single volume and the gas will form a single bubble that is attached to the walls at a definite location. The introduction of gravity with perturbing accelerations such as *g*-jitter or brief engine firings for attitude control would negate these predictions of stable distributions of contained liquid.

Depending on the details of the fluid's properties and previous force cycles, the fluid could be smoothly spread over the container walls or dispersed in the container as a collection of liquid globs or perhaps in many other configurations. The position of the liquid in a container is an important factor in the process of transferring the liquid out of the container into another container, e.g., in the very essential process of resupplying the International Space Station with propellants and liquids from a supply ship. Since, in general, some of the liquid could be dispersed as drops or globs in random parts of the container, the gauging of liquid quantity, both contained and transferred, presents new, important problems at $g = 0$.

Surface tension of the stored liquid is an important force at $g = 0$, and the associated capillary forces have been used to design many liquid acquisition devices that use fine mesh screens or similar porous materials to control fluid motion in a container (Dodge, 1990). When wet with the stored liquid, they can withstand a pressure differential from the gas side to the liquid side. The maximum possible differential is a function of the mesh opening dimension and the surface tension of the liquid. The pressure required to force fluid through the mesh openings is just the pressure exerted in a gas bubble and is proportional to the surface tension and inversely proportional to the diameter. Thus, until the fluid exceeds this pressure, it is separated from the vapor. This design principle permits the important operation of venting vapor to reduce the system pressure while saving the stored liquid.

Again, taking advantage of surface tension in the wetting of the liquid to the container walls, the fluid location relative to an outlet can be controlled effectively by using appropriate vanes (Dodge, 1990), which have been shown to provide stable, nonturbulent flow during container filling at $g = 0$.[9]

Excellent single-phase fluid control can be obtained with metal bellows tanks or collapsible flexible bags with a single outlet port, where the tank shrinks as fluid is drawn, so that the tank remains completely filled with fluid (no vapor) and where flow is regulated by pressure exerted on the exterior of the tank (Dodge and Kana, 1989).

In the absence of gravity there are no convection currents in the stored liquid, so that when heat is transferred into the tank through the container wall, a significant portion of the liquid can become superheated, even at very low heat fluxes. A superheated liquid represents a metastable condition that is ultimately susceptible to explosive boiling or flashing, causing pressure spikes of varying magnitude in a closed system. The system design must take into consideration the pressure rise rate of stored fluids and be capable of containing these pressure surges.

Ring baffles around the container walls can minimize sloshing since the baffles interfere with the free up-and-down motion of the slosh waves. Although at Earth gravity, g_0, the height of the contained fluid is minimized, at zero gravity the surface tension minimizes the area of free surface. During sloshing at g_0, fluid displacement is resisted, but at $g = 0$ surface tension resists the creation of more surface area.

The specific microgravity issues are related to the design of equipment for storing, transferring, and controlling large quantities of cryogenic liquids, or more specifically, the problems encountered in transferring liquid from supply ships to the International Space Station storage tanks. At $g = 0$, the fluid location in a storage tank is not a priori determined, so that movement of fluid from one container to another using conventional transfer tubes may not be effective. Since some fraction of the cryogenic fluid vaporizes during transfer, it must be ensured that no liquid escapes when a vent is opened to relieve pressure. To successfully accomplish such a transfer, it is necessary to develop methods to measure the quantity of cryogenic fluid stored in the receiver tanks and monitor this during the transfer.

NASA has long recognized the problem of liquid containment and management in microgravity (Reynolds and Satterlee, 1966) and had planned a systematic, 5-year study of liquid dynamics at $g = 0$ in a project designated the Cryogenic On-Orbit Liquid Depot-Storage, Acquisition, and Transfer satellite, or COLD-SAT for short (Dodge and Kana, 1989), but the project was never completed.

Summary of the Effect of Reduced Gravity on Selected Subsystems

Summarized in Table III.E.1 are the various subsystems and components discussed so far in this section and the various materials processing and storage systems in which they are found. For each of the subsystems in a

[9]Hasan, M.M. Cryogenic fluid management for ISS and HEDS missions. Presentation to the Committee on Microgravity Research, October 14, 1997, Washington, D.C.

TABLE III.E.1 Subsystems Found in Materials Production and Storage Systems and the Potential Impact of Reduced Gravity on Their Operation

	System			
Subsystem	Atmospheric Acquisition	Direct Chemical Extraction	Electromagnetic Beneficiation	Filtration
Blower pump	—	L	—	L
Compressor	—	—	—	—
Condenser	—	—	—	—
Desiccator	—	—	—	—
Dust filters	M	—	—	M
Electric current source	—	—	—	—
Electrostatic generator	—	—	L	L
Electrostatic/magnetic separator	—	—	L	—
Fluidized-bed reactor	—	H	—	—
Furnace or heater	M	—	—	—
Gas-ionizing electrode	—	—	L	—
Gravity collection bins	—	—	H	—
Heat exchanger (counterflow)	L	—	—	—
High-temperature crucible	—	—	—	—
Hopper	—	—	H	—
Liquid/matrix electrolytic cell	—	—	—	—
Particle feed systems (conveyors)	—	—	M	—
Pipes (multiphase)	H	H	—	H
Radiator	L	—	—	—
Refrigerated tank	—	—	—	—
Rotating drum charging unit	—	—	L	—
Solid electrodes	—	—	—	—
Solid-state electrolytic cell	—	—	—	—
Vacuum pump	L	—	—	L
Valves	L	L	—	L
Zeolite adsorption bed	L	—	—	—

NOTE: The letters H, M, and L designate high, medium, and low (preliminary assessment) impact of reduced gravity on the operation of the subsystem. Where no letter is given, the subsystem is not applicable to the system listed.

given system, the impact of reduced gravity on the operation of these subsystems is estimated as high, medium, or low (little or no impact). It should always be kept in mind, however, that the impact of gravity level on these technologies will depend on the design.

Additional Processes of Interest

The processes discussed below are considerably more speculative in nature than those covered in the preceding sections, so the impact of gravity on their operation is more difficult to assess. Nevertheless, their potential significance to HEDS missions warranted their inclusion in this chapter.

Ejecta Capture from Asteroids and Comets

The ability to establish propellant fueling depots in the vicinity of Earth but outside deep gravity wells can dramatically reduce the mass that must be launched from Earth and, accordingly, the costs associated with large-scale exploration missions throughout the solar system. Water ice is highly probable in dead, short-period comets

Ilmenite Reduction	Molten Electrolyte Electrolysis	Pyrolysis	Reverse Water Gas Shift	Sabatier Process	Solid Electrolyte Electrolysis	Water Electrolysis
L	—	—	L	L	—	—
—	—	—	L	L	L	—
—	—	—	H	H	—	—
M	—	—	M	M	M	M
—	—	—	M	M	M	—
—	L	—	—	—	L	L
—	—	—	—	—	—	—
—	—	—	—	—	—	—
H	—	H	H	H	—	—
M	M	M	—	—	M	—
—	—	—	—	—	—	—
—	—	—	—	—	—	—
—	—	—	L	L	L	—
—	M	M	—	—	—	—
H		H	—	—	—	—
—	—	—	—	—	—	M
M	—	M	—	—	—	—
H	—	—	H	H	H	H
—	—	L	L	L	L	—
—	—	—	M	M	M	M
—	—	—	—	—	—	—
—	L	—	—	—	—	—
L	—	—	—	—	L	—
—	—	—	—	—	—	—
L	—	L	L	L	—	L
—	—	—	—	—	—	—

and in carbonaceous chondrite asteroids that are in the vicinity of Earth (Lewis et al., 1993; Lewis and Hutson, 1993). If asteroids or short-period comets contain water ice, they can be exploited as sources of hydrogen and oxygen for propellant or for other resources, such as nickel or platinum. However, these objects are likely to have masses that are insufficient to produce significant gravitational attraction forces.

To exploit asteroids and comets, it will be necessary to come into physical contact with them. Furthermore, it will be necessary to attach spacecraft elements to them and to process materials in their low gravitational fields. Strategies for processing these objects will probably require the deliberate ejection or removal of large chunks of material and their subsequent capture for resource utilization. In the case of water, it is not possible to extract water ice and separate it from other materials without subjecting it to a phase change. Gradual collection of released (sublimated) water molecules via some sort of cold trap maintained on a spacecraft collector surface is possible, but more robust and efficient high-volume collection systems can be utilized if chunks of dirty ice are broken away from the asteroid or comet and collected in a container that can be sealed and pressurized intermittently for batch processing.

With current designs of space vehicles, it would be difficult to rendezvous and make intentional contact with small planetary bodies without any type of propulsive exhaust during the touchdown phase or to land on the object's surface (which is probably tumbling) with a negligible impact velocity. Because these planetary bodies may be rich in volatiles and because some have accreted under the influence of extremely low gravitational forces—associated more with swarms of particles than with significant mass—it is probable that any impact with their surfaces will result in the ejection of materials. Furthermore, temperatures associated with a terrestrial spacecraft are relatively high (near 300 K), and as a result the spacecraft could cause sublimation of significant quantities of water ice or any number of other volatiles that may be frozen in the object's surface at the extreme vacuum conditions of the space environment. Depending on the thermal coupling between the spacecraft and the asteroid or comet surface during encounter, it would be possible to liberate large quantities of these volatiles (owing to the extremely low surface pressures) and to simultaneously scour rocks and other debris from the surface. Hence, there are a variety of processes that will result in the ejection of material from the surfaces of asteroids and comets, besides those processes associated with the deliberate extraction and collection of raw materials. Both deliberately produced and inadvertent ejecta can pose a serious hazard for space systems because they can become projectiles. However, because fracturing processes will necessarily be involved in the removal of material from asteroids, the need to capture and collect ejecta will be a primary concern.

Many problems need further study and experimental validation before a serious asteroid resource collection mission can be mounted, including how to work in the long-duration dusty environment that will probably be produced by any type of ejecta production operation. All missions to the surfaces of asteroids and comets will be greatly affected by microgravity considerations. The following specific developments, many of which will incorporate microgravity issues, are needed to support such missions:

- Accurate predictions of the local debris swarms that will exist in the vicinity of targets of opportunity;
- Material ejection and capture strategies that are proven experimentally, using hardware systems that can produce predictable and manageable surface material releases and that maximize raw material collection and minimize potential hazards to spacecraft systems;
- Experimentally verified techniques and hardware systems that will permit spacecraft devices to rendezvous with and become attached to the surface of a tumbling asteroid or comet;
- Systems that can ensure that spacecraft will not become entangled with lines, cables or nets that are used to attach hardware or communication devices to tumbling microgravity objects;
- Attach/detach coupling systems that can permit hardware units to "walk" on a tumbling microgravity surface;
- Rock climbing equivalent elements that can be used to secure and then release processing units in such a way that forces can be exerted on those surfaces that are sufficient to remove and/or manipulate raw materials;
- Development of fluid and phase separation systems that can exploit the irregular but potentially significant radial acceleration forces that are associated with the tumbling motions of typical asteroids or comets;
- Raw material batch processing systems that can extract useful resource materials and dispose of the resulting wastes in an environment with a negligible atmosphere and virtually no gravity; and
- Systems that can attach to tumbling objects and then process sufficient raw material into rocket propellant to first despin the object and then propel the object toward a desired target.

Mining Helium-3

The future demand for electrical power when carbon-based fuels become depleted will require the development of new generating technologies. The deuterium/helium-3 fusion reaction is recognized as an attractive, environmentally friendly source of power, but the terrestrial supply of helium-3 is very limited. There is a large quantity (10^9 kg) of helium-3 in the lunar regolith at a very low concentration (20-45 ppm). Very large quantities of lunar regolith would need to be processed to gather helium-3 in the quantities (hundreds of kilograms) needed for terrestrial power generation. Wittenberg et al. (1986) have analyzed and discussed this concept as a means of producing electrical energy for the entire country without causing environmental degradation.

The origin of the lunar helium-3 is the solar wind volatiles (SWVs) that are emitted from the fusion reactions in the Sun, which emits a stream of ionized elements that consist principally (96 percent) of hydrogen and a smaller amount of helium and trace amounts of other elements. The absence of a lunar atmosphere and the feebleness of the lunar magnetic field allow the SWVs to strike the lunar surface almost unimpeded and penetrate into the regolith to an estimated depth of 50 to 300 Å (Warhaut et al., 1979). Small meteor impacts constantly expose new surface of regolith to the solar wind and churn up the buried, helium 3 saturated grains so that the helium-3-concentration is nearly constant to a depth of 2.4 m (Swindle et al., 1990).

The trapped gases in the regolith are released upon heating (Pepin et al., 1970). As mentioned above, large volumes of regolith must be processed to recover useful quantities of helium-3. For example, to continuously fuel a 1000 MW fusion power plant would require 106 kg of helium-3 per full power year (Wittenberg et al., 1991; Sviatoslovsky and Jacobs, 1988). Two strategies have been proposed for the mining operation: (1) continuous area mining with linear travel of the gathering-processing equipment and transportation of the mined gases in sealed containers to a central processing station (Cameron, 1992) and (2) continuous spiral mining, where the gathering-processing equipment spirals out from a central hub to which the mined volatiles are piped for processing (Schmitt, 1992).

The collection of helium-3 from the outer planets has been proposed as a viable process (Lewis, 1997). Hydrogen and helium make up significant fractions of the masses of the outer planets; Jupiter is about 95 percent hydrogen and helium, Saturn is about 90 percent, and Uranus and Neptune are about 50 percent. In the cold atmosphere of all four planets, the heavier elements have condensed and precipitated out, so that all of their atmospheres contain helium-3 at a level of about 45 ppm. Using a probe that can be placed in the atmosphere of Uranus, which is the most accessible planet, a system of pumps and cryocoolers could process the atmosphere to stepwise-extract gases based on their liquefaction temperature and behavior with temperature to obtain pure helium-3. A payload of 10 tons could be transported to Earth with rocket engine performance only very slightly improved over that available in 1965. It is estimated that the energy that could be produced from the helium-3 would be 20,000 times that expended to gather it in this operation. The helium-3 reserve on Uranus is a staggering 16 trillion tons, so that if this gathering process could be implemented, Earth's energy needs would be adequately met for many generations to come. Lunar mining of helium-3 would involve very large scale degassing of the regolith and collection of the evolved gases while working in an ultrahigh vacuum. Chemical separation of the compressed gases is based on their liquefaction temperatures.

The reduced lunar gravity would be expected to affect the selection and design of gathering-processing equipment in ways similar to those discussed in prior sections on lunar processing. In addition, the large quantity of regolith that would have to be processed to yield useful quantities of helium-3 ensures that such an operation, and its associated design issues, would be complex.

References

Agosto, W.N. 1985. Electrostatic concentration of lunar soil minerals. Pp. 453-464 in Lunar Bases and Space Activities of the 21st Century. W.W. Mendell, ed. Houston, Tex.: Lunar and Planetary Institute.

Ash, R.L., W.L. Dowler, and G. Varsi. 1978. Feasibility of rocket propellant production on Mars. Acta Astronautica 5:705.

Ash, R.L., M.L. Stancati, J.C. Niehoff, and V. Cuda, Jr. 1981. Outer planet satellite return missions using in situ propellant production. Acta Astronautica 8:511-512.

Beck, T.R. 1992. Metals production. Pp. III.41-III.53 in Proceedings of the Lunar Materials Technical Symposium, 3rd Annual Symposium of the University of Arizona/NASA Space Engineering Research Center.

Belaventsev, Ju. E., et al. 1991. A system for oxygen generation from water electrolysis aboard the manned space station Mir. Pp. 477-479 in Proceedings of the 4th European Symposium on Space Environmental and Control Systems. European Space Agency (ESA) SP-324. Noordwijk, Netherlands: European Space Agency.

Byron, S.R., and H. Apter. 1992. Model of gas composition and plasma properties in sealed CO_2 lasers. J. Appl. Phys. 71:85-94.

Cameron, E.N. 1992. Helium resources of Mare Tranquillitatis. Technical Report WCSAR-TR-AR3-9207-1. Madison: University of Wisconsin.

Carrier, W.D., III, and J.K. Mitchell. 1990. Geotechnical engineering on the Moon. Pp. 51-58 in de Mello Volume: A Tribute to Prof. Dr. Victor F.B. de Mello. E. Blucher, ed. São Paulo, Brazil: Editora Edgard Blücher.

Chamberlain, P.G., L.A. Taylor, E.R. Podneiks, and R.J. Miller. 1993. A review of possible mining applications in space. Pp. 51-68 in Resources of Near-Earth Space. J.S. Lewis, M.S. Matthews, and M.L. Guerrieri, eds. Tucson and London: University of Arizona Press.

Cooper, D.W. 1982. Gas-particle separation. Section 9.2 in Handbook of Multiphase Systems. G. Hestroni, ed. Washington, D.C.: Hemisphere Publishing.

Cusick, R.J. 1974. Space station prototype Sabatier reactor design verification testing. ASME Paper No. 74-INAS-58. New York: American Society of Mechanical Engineers.

Dalton, C., and E. Hohmann, eds. 1972. Conceptual Design of a Lunar Colony. NASA Report No. CR-129164. Houston, Tex.: National Aeronautics and Space Administration.

DeLa'O, K.A, T.C. Eisele, D.B. Kasul, W.I. Rose, and S.K. Kawatra. 1990. Separation of lunar ilmenite: Basalt and regolith. Pp. 177-186 in Engineering, Construction and Operations in Space II, Vol. I: Proceedings of Space 90, Albuquerque, April 22-26, 1990. S.W. Johnson and J.P. Wetzel, eds. New York: American Society of Civil Engineers.

Dodge, F.T. 1990. Fluid management in low gravity. P. 369 in Low-Gravity Fluid Dynamics and Transport Phenomena: Progress in Astronautics and Aeronautics, Vol. 130. J.N. Koster and R.L. Sani, eds. New York: American Institute of Aeronautics and Astronautics.

Dodge, F.T., and D.D. Kana. 1989. Liquid dynamics in space vehicles. Technology Today, March.

Fegley, B., Jr., and T. Swindle. 1993. Lunar volatiles: Implications for lunar resource utilization. Pp. 367-426 in Resources of Near-Earth Space. J.S. Lewis, M.S. Matthews, and M.L. Guerrieri, eds. Tucson and London: University of Arizona Press.

Fraas, F. 1962. Electrostatic separation of granular materials. U.S. Bureau of Mines Bulletin 603.

Freeman, M.P. 1982. Separation mechanisms for liquid suspensions. Section 9.3 in Handbook of Multiphase Systems. Washington, D.C.: Hemisphere Publishing.

Gibson, M.A., and C.W. Knudsen. 1990. Automation and control potential for Carbotek lunar oxygen production process. P. 199 in Engineering, Construction, and Operations in Space II: Proceedings of Space 90, Albuquerque, New Mexico, April 22-26, Vol. 1. S.W. Johnson and J.P. Wetzel, eds. New York: American Society of Civil Engineers.

Gibson, M.A., C.W. Knudsen, and A. Roeger III. 1990. Development of the Carbotek process for lunar oxygen production. Pp. 357-367 in Engineering, Construction, and Operations in Space II: Proceedings of Space 90, Albuquerque, New Mexico, April 22-26, Vol. 1. S.W. Johnson and J.P. Wetzel, eds. New York: American Society of Civil Engineers.

Greeley, R., R. Leach, B. White, J. Iversen, and J. Pollack. 1980. Threshold windspeeds for sand on Mars: Wind tunnel simulations. Geophys. Res. Lett. 7:121-124.

Hepp, A.F., G.A. Landis, and C.P. Kubiak. 1994. A chemical approach to carbon dioxide utilization on Mars. Pp. 799-818 in Resources of Near-Earth Space. J.S. Lewis, M.S. Matthews, and M.L. Guerrieri, eds. Tucson and London: University of Arizona Press.

Hestroni, G., ed. 1982. Handbook of Multiphase Systems. Washington, D.C.: Hemisphere Publishing.

Hudson, J.P., and P. Squires. 1973. Evaluation of a recording continuous cloud nucleus counter. J. Appl. Meteorol. 12:175-183.

Humphries, R., K. Mitchell, J. Reuter, R. Carrasquillo, and B. Beverly. 1991. Life support and internal thermal control system design for space station Freedom. Pp. 23-27 in Proceedings of the 4th European Symposium on Space Environmental and Control Systems. European Space Agency (ESA) SP-324. Noordwijk, Netherlands: European Space Agency.

Kelly, E.G., and D.J. Spottiswood. 1982. Introduction to Mineral Processing. New York: John Wiley & Sons.

Klosky, J.L. 1997. Behavior of Composite Granular Material and Vibratory Helical Anchors. Ph.D. dissertation, Department of Civil, Environmental, and Architectural Engineering, University of Colorado.

Kolecki, J., and G.B. Hillard. 1992. Pp. 18-20 in Electrical and Chemical Interactions at Mars Worksohp, Vol. 1. NASA Conference Publication 10093. Cleveland, Ohio: NASA Lewis Research Center.

Landis, G.A., C. Baraona, C. Scheiman, and D. Brinker. 1997. Mars array technology experiment and dust accumulation and removal technology. Space Photovoltaics Research and Technology 1997, Cleveland, June 10-12.

Lewis, J.S. 1997. Mining the Sky. Reading, Mass.: Addison-Wesley.

Lewis, J.L., and M.L. Hutson. 1993. Asteroidal resource opportunities suggested by meteorite data. Pp. 523-542 in Resources of Near-Earth Space. J. Lewis, M.S. Matthews, and M.L. Guerrieri, eds. Tucson and London: University of Arizona Press.

Lewis, J.L., D.S. McKay, and B.C. Clark. 1993. Using resources from near-Earth space. Pp. 3-16 in Resources of Near-Earth Space. J. Lewis, M.S. Matthews, and M.L. Guerrieri, eds. Tucson and London: University of Arizona Press.

Lindroth, D.P., and E.R. Podneiks. 1987. Electromagnetic energy applications in lunar mining and construction. Proc. Lunar Planet. Sci. Conf. 18: 365-373.

Martin, R.B., N. Lance, R.J. Cusick, and A.T. Linton. 1983. Space station prototype Sabatier reactor design verification testing. SAE Paper No. 831110. Warrendale, Pa.: Society of Automotive Engineers.

Matsumoto, S., K. Shimiza, T. Yoshida, Y. Kai, H. Mano, K. Takagi, and H. Sato. 1990. Excavation system for lunar base construction. Pp. 325-334 in Engineering, Construction, and Operations in Space II, Vol. I: Proceedings of Space 90, Albuquerque, April 22-26. S.W. Johnson and J.P. Wetzel, eds. New York: American Society of Civil Engineers.

Mueller, P.J. 1998. Transportation of hydrogen to Mars for in situ propellant production. Jet Propulsion Laboratory Contractor Final Report No. 961007. Pasadena, Calif.: Jet Propulsion Laboratory.

Nathan, M.P., F. Barnes, H.Y. Ko, and S. Sture. 1992. Mass and energy tradeoffs of axial penetration devices on lunar soil simulant. Pp. 441-457 in Engineering, Construction, and Operations in Space III, Vol. 1: Space '92, Proceedings of the Third International Conference, May 31-June 4, Denver, Colorado. New York: American Society of Civil Engineers.

National Aeronautics and Space Administration (NASA). 1997. Human Exploration of Mars: The Reference Mission of the NASA Mars Exploration Study Team. S.J. Hoffman and D.I. Kaplan, eds. Houston: National Aeronautics and Space Administration.

National Research Council (NRC), National Materials Advisory Board. 1981. Comminution and Energy Consumption. NMAB-364. Washington, D.C.: National Academy Press.

Ness, Jr., R.O., B.D. Runge, and L.L. Sharp. 1990. Process design options for lunar oxygen production. Pp. 368-377 in Engineering, Construction, and Operations in Space II: Proceedings of Space 90, Albuquerque, April 22-26, Vol. 1. S.W. Johnson and J.P. Wetzel, eds. New York: American Society of Civil Engineers.

Okumura, M., T. Ueno, and Y. Ohashi. 1998. Regolith covering method for habitation module in an early phase of lunar base construction. Pp. 616-621 in Space 98: Proceedings of the Sixth International Conference and Exposition of Engineering, Construction, and Operations in Space, April 26-30, Albuquerque. Reston, Va.: American Society of Civil Engineers.

Pepin, R.O, L.E. Nyquist, D. Phinney, and D.C. Black. 1970. Rare gases in Apollo 11 lunar material. Pp. 1443-1454 in Proceedings of the Apollo 11 Lunar Science Conference. New York: Pergamon.

Ramohalli, K.N.R., and K.R. Sridhar. 1991. Extraterrestrial materials processing and related transport phenomena. AIAA Paper No. 91-0309. New York: American Institute of Aeronautics and Astronautics.

Rapp, D. 1998. A review of Mars ISPP Technology. Report JPL D-15223. Pasadena, Calif.: Jet Propulsion Laboratory.

Reynolds, W.C., and H.M. Satterlee. 1966. Liquid propellant behavior at low and zero G. The Dynamic Behavior of Liquids in Moving Containers. H.N. Abramson, ed. NASA SP-106. Washington, D.C.: National Aeronautics and Space Administration.

Richter, R. 1981. Basic investigation into the production of oxygen in a solid electrolyte. AIAA Paper No. 81-1175. New York: American Institute of Aeronautics and Astronautics.

Sabatier, P., and J.B. Senderens. 1902. Comptes Rendus Academe des Sciences (Paris) 134:514-689.

Schmitt, H.H. 1992. Interlune concept for helium-3 fusion development. Pp. 805-814 in Engineering, Construction, and Operations in Space III: Space '92, Proceedings of the Third International Conference, Denver, Colorado, May 31-June 4. W.Z. Sadeh, S. Sture, and R.J. Miller, eds. New York: American Society of Civil Engineers.

Schrunk, D., M. Thangavelu, B. Cooper, and B. Sharpe. 1998. Physical transportation on the moon: The lunar railroad. Pp. 347-353 in Space 98: Proceedings of the Sixth International Conference and Exposition of Engineering, Construction, and Operations in Space, April 26-30, Albuquerque. Reston, Va.: American Society of Civil Engineers.

Seglin, L. 1975. Methanation of Synthesis Gas. Advances in Chemistry Series, Vol. 146. Washington, D.C.: American Chemical Society.

Senior, C.L. 1993. Lunar oxygen production by pyrolysis. Pp. 179-197 in Resources of Near-Earth Space. J. Lewis, M.S. Matthews, and M.L. Guerrieri, eds. Tucson and London: University of Arizona Press.

Sharp, W.R., J.P.H. Steele, B.C. Clark, and E.R. Kennedy. 1990. Mining and excavating systems for a lunar environment. Pp. 294-304 in Engineering, Construction, and Operations in Space II: Proceedings of Space 90, Albuquerque, April 22-26, Vol. 1. S.W. Johnson and J.P. Wetzel, eds. New York: American Society of Civil Engineers.

Sridhar, K.R., and S.A. Miller. 1994. Solid oxide electrolysis technology for ISRU and life support. Space Technol. 14(5):339.

Sridhar, K.R., and B.T. Vaniman. 1995. Oxygen production on Mars using solid oxide electrolysis, 25th International Conference on Environmental Systems. SAE Paper No. 951737. Warrendale, Pa.: Society of Automotive Engineers.

Sridhar, K.R., and R. Foerstner. 1998. Regenerative CO/O_2 solid oxide fuel cells for Mars exploration. AIAA Paper No. 98-0650. Reston, Va.: American Institute of Aeronautics and Astronautics.

Stancati, M.L., J.C. Niehoff, W.C. Wells, and R.L. Ash. 1979. Remote automated propellant production: A new potential for round trip spacecraft. AIAA Paper No. 79-0906. New York: American Institute of Aeronautics and Astronautics.

Sullivan, T.A., D. Linne, L. Bryant, and K. Kennedy. 1995. In-situ-produced methane and methane/carbon monoxide mixtures for return propulsion from Mars. J. Propulsion Power 11:1056-1062.

Sviatoslovsky, I.N., and M. Jacobs. 1988. Mobile helium-3 mining and extraction system and its benefits toward lunar base self-sufficiency. Pp. 310-321 in Engineering, Construction, and Operations in Space: Proceedings of Space 88. S.W. Johnson and J.P. Wetzel, eds. New York: American Society of Civil Engineers.

Swindle, T.D., C.E. Glass, and M.M. Poulton. 1990. Mining lunar soils for ^3He. University of Arizona/NASA Space Engineering Research Center TM90/1. Tucson: NASA Space Engineering Research Center for Utilization of Local Planetary Resources.

Szabo, B., F. Barnes, S. Sture, and H.Y. Ko. 1994. Effectiveness of vibrating bulldozer and plow blades on draft force reduction. Proceedings of the Winter Meeting of the American Society of Agricultural Engineers. No. 941535. St. Joseph, Mich.: American Society of Agricultural Engineers.

Taylor, L.A., and W.D. Carrier III. 1993. Oxygen production on Mars using solid oxide electrolysis. Pp. 69-108 in Resources of Near-Earth Space. J. Lewis, M.S. Matthews, and M.L. Guerrieri, eds. Tucson and London: University of Arizona Press.

Vaniman, D.T., and G.H. Heiken. 1990. Getting lunar ilmenite from soils or rocks? Pp. 107-116 in Engineering, Construction, and Operations in Space II: Proceedings of Space 90, Albuquerque, April 22-26, Vol. 1. S.W. Johnson and J.P. Wetzel, eds. New York: American Society of Civil Engineers.

Vuskovic, L., R.L. Ash, Z. Shi, S. Popovic, and T. Dinh. 1997. Radio-frequency-discharge reaction cell for oxygen extraction from Mars atmosphere. SAE Paper No. 972499. Warrendale, Pa.: Society of Automotive Engineers.

Vuskovic, L., R.L. Ash, S. Popovic, T. Dinh, and A. Van Orden. 1998. Radio-frequency plasma assisted production of carbon monoxide/oxygen propellant directly from Mars atmosphere. AIAA/ASME/SAE/ASEE 34th Joint Propulsion Conference and Exhibit, Cleveland, July 12. AIAA Paper No. 98-3304. Reston, Va.: American Institute of Aeronautics and Astronautics.

Warhaut, M., J. Kiko, and T. Kirsten. 1979. Microdistribution patterns of implanted rare gases in a large number of individual lunar soil particles. P. 1531 in Proceedings of the 10th Lunar Planetary Science Conference. New York: Pergamon.

Wempfen, S.P. 1973. Mine costs and control. Pp. 13-17 in SME Mining Handbook, Vol. 2. I. Given, ed. New York: Society for Mining, Metallurgy, and Exploration.

Wittenberg, L.J., J.F. Santarius, and G.L. Kulcinski. 1986. Lunar source of He-3 for commercial fusion power. Fusion Technology 10:167.

Wittenberg, L.J., E.N. Cameron, G.L. Kulcinski, S.H. Ott, J.F. Santarius, G.I. Sviatoslavsky, I.N. Sviatoslavsky, and H.E. Thompson. 1991. Technical Report WCSAR-TR-AR3-9107-1. Madison: University of Wisconsin.

Wood, M. 1992. Oxygen generation by static feedwater electrolysis for Space Station Freedom. Pp. 127-137 in Proceedings of the International Conference on Life Support and Biospherics. Huntsville, Ala.: National Aeronautics and Space Administration.

Wu, D., R.A. Outlaw, and R.L. Ash. 1996. Extraction of oxygen from CO_2 using glow-discharge and permeation techniques. J. Vac. Sci. Technol. A 14:408-414.

Wydeven, T. 1988. A survey of some regenerative physico-chemical life support technology. NASA Technical Memorandum 101004. Moffett Field, Calif.: National Aeronautics and Space Administration.

Zhao, Y., and F. Shadman. 1993. Production of oxygen from lunar ilmenite. P. 149 in Resources of Near-Earth Space. J. Lewis, M.S. Matthews, and M.L. Guerrieri, eds. Tucson and London: University of Arizona Press.

Zubrin, R.M., D.A. Baker, and O. Gwynne. 1991. Mars direct: A simple, robust, and cost effective architecture for the space exploration initiative. AIAA Paper 91-0326. New York: American Institute of Aeronautics and Astronautics.

Zubrin, R.M., B. Frankie, and T. Kito. 1997. Mars in-situ resource utilization based on the reverse water gas shift. AIAA Paper No. 97-2767. Reston, Va.: American Institute of Aeronautics and Astronautics.

III.F CONSTRUCTION AND MAINTENANCE

Introduction

It is an underlying assumption in this discussion that the need for power, even if it is large, will be met locally. Potential power sources and generation systems are discussed in previous sections and are not covered again here. Much thought has been given to the incorporation of novel manufacturing processes that would uniquely benefit from the extraterrestrial microgravity environment and the hard vacuum (10^{-9} to 10^{-12} torr) in the lunar environment and in space. However, to date, no examples have been found of products that could be advantageously manufactured in space for commercial use on Earth (NRC, 1992).

It is clear that the HEDS program must plan to use local resources for life support since it will not always be possible to transport the needed facilities or replacement parts from a terrestrial base. While it is acceptable to provide all the needed supplies for the nearby International Space Station, this approach would become expensive on a lunar base and for missions to Mars or beyond, and in addition, the one-way transit time of hundreds of days would make such resupply impossible. It is clear that unexpected component failures can provide challenges to the success of a mission. It is also obvious that a spare parts kit, no matter how extensive, cannot anticipate every possible emergency. Direct manufacturing, an important new technology that builds metal or ceramic piece parts by computer-controlled, step-by-step deposition rather than by the machining of bulk feedstock, may have some dependence on gravity. An alternative would be to use a universal, compact machine shop that could process parts ranging from a wristwatch gear to an antenna mount and that would have no direct dependence on gravity.

The problems that might be encountered at fractional gravity in lunar and Martian inhabitation are discussed here in relation to site preparation and habitat construction. In considering materials handling and transport technologies, it becomes evident that the direct use of unmodified terrestrial equipment would involve many shortfalls in performance at fractional gravity. For example, as discussed above with respect to mining operations, the reduced traction in transport vehicles and the reduced friction available to secure tethers into the regolith make it desirable to develop new structural designs and innovative processes. Granular materials such as the lunar and Martian regolith exhibit cohesion and are arrangements of rigid particles in frictional contact. Because gravity contributes to the normal stress of interaction between particle surfaces and frictional forces are typically proportional to the normal (hence gravitational) forces, the behavior of granular material is strongly dependent on gravity. Dust management as gravity nears zero is of concern since, in the absence of gravitational settling, one needs a positive filtration technology capable of dealing with a large range of particulate sizes and densities, both during space travel and to cope with tenacious lunar fines and Martian dust.

Construction will be facilitated by the ability to use local materials to fabricate concrete (Lin and Bhattacharja, 1998), which will be useful for many applications such as securing tethers at reduced gravity and general construction. There will also be a need to refine metals from local ores and perform manufacturing operations to provide needed articles and to generate replacement parts. There will also be a need to provide comfortable facilities, such as abodes and habitats protected from radiation, where the occupants can work and dwell in reasonable comfort, safely isolated from the harsh environment.

Site Preparation

The ability to establish outposts on extraterrestrial surfaces represents a serious challenge for future HEDS missions. Initial site preparation for all but touch-and-go surface exploration missions will require that the landing site be prepared to support research stations, spacecraft landing sites, habitats shielded from hazardous radiation, surface transportation systems, power generating stations, and an energy distribution infrastructure. Nearly all contemplated site development scenarios assume that significant portions of these preparation activities will be accomplished robotically, before humans arrive. Based on present knowledge, fairly detailed mission designs can be developed for lunar and, to a lesser extent, Mars missions, but it is also logical to consider the technologies that might permit site development on other extraterrestrial bodies such as asteroids and the Galilean satellites of Jupiter. Furthermore, even though it seems natural to imagine conventional earth-moving and excavation equipment operating robotically on these surfaces, gravitational and other environmental differences ensure that that will not be the case. Even the simple task of covering a structure with regolith involves a number of unresolved engineering issues. Because of our more detailed knowledge of the lunar surface, it is instructive to frame the reduced gravity research issues in that context.

Lunar soil characteristics at depths greater than 0.7 m are not well known (Klosky et al., 1998, Carrier et al., 1991). While a few core samples were taken down to depths of 3 m using rotary drilling devices, the samples were undoubtedly altered during the collection process. However, it is likely that site preparation and resource extraction activities will require excavation operations to depths on the order of 5 m, which is, incidentally, the estimated thickness of Martian regolith needed for passive shielding from radiation (Hepp et al., 1994). This will require the ability to perform bedrock mining as well as to handle and process regolith. Both soil composition and soil density variation with depth will change from one site to another, but Carrier et al. (1991) found that the lunar soil appears to be compacted more than terrestrial soil just beneath the surface. In addition, if site preparation studies are shifted to locations near the permanently shadowed polar craters, where the Lunar Prospector has indicated significant quantities of water ice are present, consideration must be given to the manipulation of soil containing water ice in an environment that is considered to be a hard vacuum on Earth. An appropriate area of study would be the development of instruments and gravity-dependent soil models that could address soil strength, stiffness, and density on the Moon, Mars, and other extraterrestrial bodies. Creation and control of dust during excavation and soil placement operations will be a major concern (Colwell et al., 1998). Since much of the site preparation activity (and subsequent mining activity) will involve operating small, low-power, robotic machines, their ability to perform automated tasks reliably in a dusty environment, unattended for long periods of time, is critical.

Because of the high value of hydrogen and oxygen molecules for propellant and life support outside Earth's gravitational field, attention should be directed toward developing water extraction sites on a range of low-gravity surfaces whose compositions range from dirty ice to permafrost. However, the presence of water ice can profoundly influence the types of soil manipulation operations that can be accomplished. Under some conditions, the water ice can fill pore spaces in the soil, decreasing the friction angle of the soil and thus decreasing the energy required for excavation. Conversely, permafrost (which is probable at some locations at Mars) is extremely difficult to excavate or even penetrate, and subsequent to any site preparation operations on permafrost surfaces, design considerations associated with maintaining the permafrost in a solid state must be addressed. Because of the low pressures that exist—even on Mars—the excavation process will probably generate sufficient heat to liberate water. Aside from the problems of processing water ice in a vacuum, Perkins and Madson (1996a) described a number of basic geotechnical issues in some detail.

There is a need for smaller-scale soil excavation tools capable of moving cubic meters of material per day (as opposed to their terrestrial counterparts, which move hundreds of cubic meters of material per day). These tools must be very low in mass and consume only small amounts of power during initial site preparation operations. These requirements are very different from those for their terrestrial counterparts, where mass and power consumption are variables that can be optimized to improve reliability and efficiency. Unfortunately, the reduced gravitational acceleration at all contemplated HEDS sites translates to reduced frictional forces, which ultimately control the magnitudes of forces that can be exerted in excavation operations. Some work has been done on possible systems (Szabo et al., 1994; Boles et al., 1997), but to date research in this area has been sporadic and has failed to properly address the scaling factors that relate to low-gravity operations. Owing to the nonlinearities in soil behavior and consistency of stress fields (Ko, 1988), it is probably more desirable to construct 1/6 scale models of lunar equipment and operate them in Earth's gravity than to attempt to conduct tests in reduced gravity on a KC 135 flight. Future excavation research should focus on properly scaling the direct and indirect gravitational parameters, which control both the characteristics of the in situ material being excavated and the forces required to effect excavation.

The following important issues relating to preparation and native material handling remain to be addressed:

- There is a clear need for better soil mechanics information from greater depths on both the Moon and Mars before a long-term human presence can be established. As part of this effort, information could also be gathered on the availability of various resources for mining at the chosen site. Augered rather than push-in technologies show promise for introducing probes to significant depths with small machines (Klosky et al., 1998).
- The energy and force needed to excavate, haul, and place extraterrestrial soils are a central design parameter needed to define the equipment required for site preparation. Research into more efficient methods of excavation using lightweight, low-power equipment is called for. This research would need to properly account for scaling factors related to operations and forces in low-gravity environments.
- A method of locomotion and tractive efficiency for equipment remain important questions. Terrestrial soil-handling equipment operators have long found that continuous-loop (tracked) vehicles, such as bulldozers and backhoes, have significant advantages over wheeled vehicles in long-term operation. Further, the isolation of the drive mechanism from the surface has proven to greatly decrease maintenance. Some of these issues were discussed recently by Costes and Sture (1998).
- The behavior of ice-laden lunar and Martian soils will need to be addressed. Before this can occur, we must obtain practical estimates of the concentration of this ice and determine whether it is continuous in the pore spaces of the soils. Then, the effect of this ice on excavation forces, foundation elements, and slope stability will need to be evaluated.
- Procedures for the installation of foundation elements and reliable estimation of their load-bearing capacity will be required very early in the deployment of extraterrestrial bases (Perkins and Madsen, 1996b). Helical plate anchor/foundations have been suggested as a reusable and efficient alternative for this application (Klosky, 1997; Klosky et al., 1995, 1996, 1998). Further evaluation of the load-bearing capacity and installation energy requirements for this and other foundation types is also appropriate.
- A method for detecting and removing or otherwise dealing with large rocks in excavation areas needs to be discovered. Terrestrial methods, typically involving blasting or otherwise applying massive energy impulses, will not be practical. This is a very significant problem and has not been adequately addressed as yet.

Construction

For cost reasons, equipment that is transported to space will have to be as light as is feasible to perform a given task and, therefore, the design must provide stability and stiffness. On Earth, gravity plays an important role directly by providing a restoring force following mechanical vibration or movement and indirectly in that since structures are made massive enough to withstand compressive gravity loads, they are automatically stable and stiff. For HEDS structures, structural stability and stiffness are issues to be addressed in the earliest stages of design. Preliminary architectural studies will uncover the specific research issues of structural dynamics that NASA should investigate.

Structures must be suitably anchored to provide the desired relationship with the surface. For anchoring on Earth, gravitational body force and friction are important factors in most designs; in very low gravity, as, for example, on an asteroid, these factors are reduced or absent, but still the anchoring must be able to reliably attach structures to fixed positions. Wind forces must be withstood in an environment such as that of Mars, where ubiquitous dust devils result in rather strong short-duration wind loads, even though the local atmospheric pressures are low. Even when wind forces are small, they can be important when acting on structures that are large for operational reasons but that, for cost reasons, are of low strength and stiffness, as discussed above.

Large area and volume of habitat, while difficult to provide, will be desirable because equipment and crew will need to be isolated from the environment to a degree unusual on Earth, and the crew will need room for their operational work and to maintain mental and physical health. Dust intrusion must be prevented, particularly since in reduced gravity, dust will not settle as rapidly as it does on Earth.

Inside atmosphere must be maintained across a wall or membrane separating it from extreme conditions on the outside, where very low pressure and very high or very low (perhaps variable) temperatures would exist. Diffusional loss of inside gases must be prevented, appropriate thermal insulation must be provided, and pressure loading of the wall must be withstood. All these issues will be troublesome for the large, low-mass structures that might be contemplated for HEDS colonies. Ideally, the habitat should provide for comfortable living in a shirtsleeve atmosphere.

The construction methods for such structures will utilize equipment and processes that are described in other sections of this Chapter for materials handling, transport, bedrock mining, welding, and concrete use for anchoring and joining.

In-depth research is needed on the design requirements and possibilities for all potential station locations, taking into account the relevant environmental factors, including low gravity, and with the health, safety, and happiness of crew being given paramount importance. This research should pay particular attention to structural analysis, dynamic as well as static, and to specific construction methods.

Concreting

Weight considerations are a strong argument against the transportation of structural building materials from Earth to bodies such as the Moon or Mars. The use of in situ building materials is therefore highly desirable, and the production of concrete is often cited (Lin and Battacharja, 1998) as a particularly attractive approach for a number of reasons, including the enormous experience base that engineers could bring to it. The various aspects of concrete production and use are described in detail in a textbook by Mindess and Young (1981) and are briefly summarized here. Concrete is made up of an aggregate embedded in a cement, or bonding agent. Aggregate can contain a variety of materials but is usually composed of rock of some type. There are also various types of cement, but a type known as portland cement is used commonly in construction. For this reason, the following discussion refers to portland cement for the purposes of illustration. (Note, however, that there is a family of portland cements differing in composition.)

Production of Portland Cement

The central process in the production of portland cement is the heating of a mixture of sources of calcium carbonate and silica in a kiln at 1400 to 1600 °C. In this temperature range, calcium silicates are formed. The raw materials are processed before burning to obtain a feed that is thoroughly pulverized and homogeneous, which ensures a product that is uniform in composition. Failure to do this would result in cement with irregular properties and performance. Similarly, the control of the burning process is critical. This heat treatment, taking place in a rotary kiln, is called clinkering. It consists of a number of steps, as shown in Figure III.F.1. The inclination of the kiln, together with its rotation of 60 to 200 rph, causes the feed to move slowly along the length of the kiln. Material may remain in the kiln between 20 min and 2.5 h depending on the type of kiln and the specific production method. The material emerging from the kiln is called clinker. It is conveyed to ball mills, where it is finely ground with a small amount of gypsum and then stored until needed.

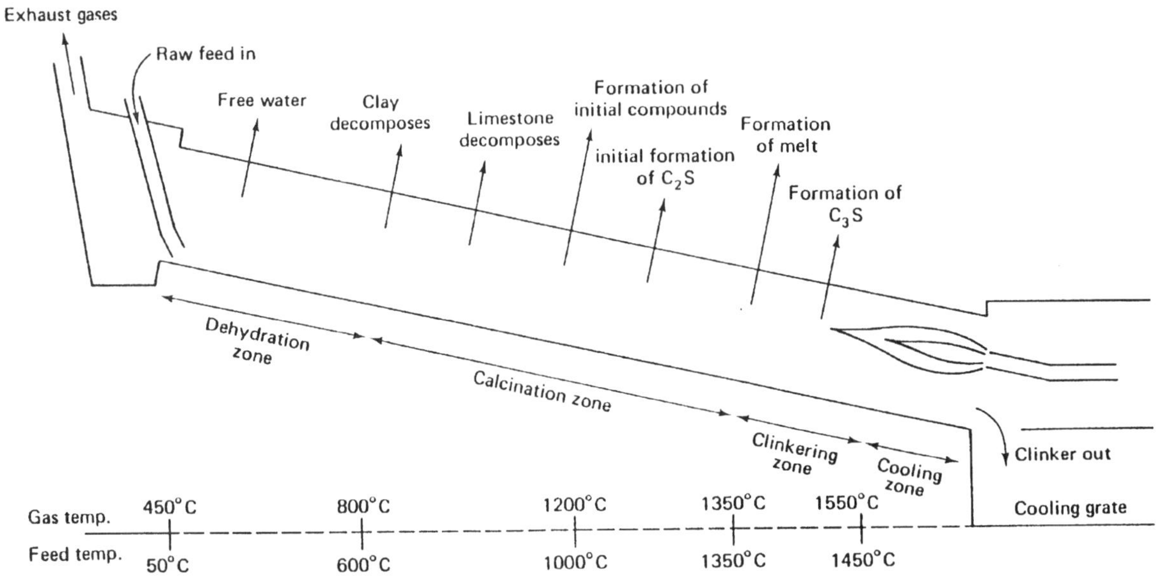

FIGURE III.F.1 Schematic outline of conditions and reactions in a typical cement rotary kiln (dry process). SOURCE: Mindess and Young (1981).

Aggregates

Aggregates make up 70 to 80 percent of the volume of concrete and therefore have a profound influence on its properties. They are granular materials mostly coming from rock, crushed stone, gravel, and sand, although other materials such as slag can be used. For extraterrestrial use, aggregate will no doubt be an in situ resource. Since aggregate greatly affects the properties of concrete, the potential in situ sources of aggregate should be characterized and their effect on concrete properties understood.

Batching, Mixing, and Placement

Homogeneity is absolutely necessary to ensure uniform and adequate performance of the concrete. There are well-defined tolerances on the amounts of cement, water, aggregates, and admixtures, and thorough and uniform mixing is a requirement.

Hydration of Cement and Curing

The various compounds that make up cement undergo chemical reactions when the cement is mixed with water. These reactions, referred to as hydration, are responsible for the hardening of the concrete. The rate of hydration is therefore directly related to the rate of hardening of the concrete and the rate at which its properties, such as compressive strength, develop. Furthermore, the hydration reactions are exothermic, so that the concrete increases in temperature as it hardens. The temperature increase will be a function of the hydration reaction rate of each of the compounds, the amount of each compound in the concrete, and the rate of heat loss to the surroundings. Concrete must be properly cured to develop its optimum properties. The most critical parameter is the amount of water present. An adequate supply of moisture must be present to achieve as much hydration as possible. In principle, there is enough water in concrete to ensure complete hydration if the water to cement weight ratio is at least 0.42. However, water is lost by evaporation or absorption by aggregates, framework, and subgrade. Once enough moisture is lost to reduce the internal relative humidity to about 80 percent, hydration will stop. Strength development will also stop, and the concrete will not achieve its potential.

Another critical parameter is the temperature of curing. Curing at low temperatures gives an initially low rate of strength development but can ultimately result in a higher compressive strength. Hydration will occur at temperatures down to 263 K, and the heat of hydration, along with adequate insulation, should protect the concrete from freezing in the early stages. At ambient temperatures where insulation is inadequate, external heating is required. In extraterrestrial environments such as the Moon, where surface temperatures vary from 80 to 390 K, and Mars, where surface temperatures vary from 130 to 300 K, curing processes are obviously a concern.

Summary of Gravity Impacts

The steps described above for the manufacture of concrete are varied and are fairly challenging for environments such as the Moon and Mars. They include the production of cement, followed by the production of concrete and its handling and placement. Cement production involves quarrying, grinding and blending, clinkering, and ball milling. Concrete production involves batching of aggregate and cement, mixing, transportation, placement, and curing.

The production process for cement is greatly affected by gravity level throughout. First, it is most important to have a feed that is thoroughly pulverized and homogeneous. Since grinding and blending operations on Earth are basically driven by gravity, in reduced gravity they will need considerable modification. Likewise, the terrestrial clinkering process is gravity-driven. It depends on a number of fundamental reaction steps, which in turn depend on the rate of fall through the various zones of the rotary kiln. Furthermore, the reaction rates are controlled by the rates of heat and mass transport, which can themselves be gravity-dependent.

All terrestrial batching and mixing processes are also dependent on gravity in that they all make use of free fall. Hence, the dynamics (or kinetics) of these processes in extraterrestrial environments are critical if concreting in those environments is to be viable.

Likewise, all placement methods on Earth are reliant on gravity by virtue of its control of sedimentation and buoyancy. After placement, the concrete is worked to eliminate voids and entrapped air as well as to consolidate it into corners and around any reinforcing steel. Again, the dynamics of these processes and the resulting concrete integrity and properties are likely to differ at extraterrestrial gravity levels.

Hydration and curing are not directly affected by the gravity level, but they are certainly affected by the local environment. Specifically, the temperature increase during hydration has a direct effect on the final properties of the concrete. This increase is dependent on the rate of heat loss to the surroundings, which is, in turn, dependent on ambient conditions. Likewise, as already mentioned, the temperature of curing is a critical parameter; again, ambient conditions are important and countermeasures for temperature extremes are necessary.

Direct Manufacturing

An important new technology, direct manufacturing, is being actively developed at a number of government, university, and industry laboratories. There are many variants of direct manufacturing, and they go by a number of names. All systems have in common that they involve a three-dimensional rendering and the production of a complex physical form by the continuous, layer-by-layer buildup of metals, ceramics, or polymers. The subsystems are, basically, a powder delivery subsystem, a mechanical subsystem to drive the building of the part, a laser, and a control subsystem. The powder delivery subsystem consists of a compressed gas supply, a powder feeder, and a cyclone mixer. The powder is fed through a nozzle into a gettered glove box, where the objects are made. A multiaxis mechanical subsystem is used to manipulate the object under the laser beam. A microcomputer control subsystem drives the mechanical stage and the laser.

These computer controlled layer deposition techniques allow the direct production of high value replacement parts without use of conventional casting, forging, and machining. The ability to produce new shapes at will lends itself to rapid, flexible, customized production and offers considerable potential for the extraterrestrial production of spare parts.

While many variations of the technology are being studied, some typical processes are briefly described. In selective laser sintering (SLS) (Bergan, 1998), a layer of metal powder is deposited on a surface and a laser beam directed by the computer numerical control program fuses and consolidates individual powder particles in selected regions. Only the particle surfaces are fused, so that complex geometry control is maintained. The interior of each shell is fused in a second heating to achieve full density, with the sintered shell acting as a mold or forming die. Promising results have been achieved with titanium, Inconel 625, and mild steel/nickel alloys.

In another example (W.H. Hofmeister, private communication, 1998; Keicher, 1999), powders between 50 and 100 μm are entrained in a gas flow. The powders are delivered coaxially with a laser beam to a molten pool on the workpiece. The laser and powder feed traverse the workpiece to build parts in layers. Beam diameters from 0.25 to 0.5 mm have been used with layer depths from 0.2 to 1.00 mm. Linear traverse speeds from 15 to 45 mm/s have been demonstrated. The metal volumes deposited have been on the order of cubic centimeters per minute. The cooling rates in this process are as high as 10^5 °C/s, so that highly refined microstructures are produced, comparable to those made by other rapid solidification processes. A number of materials, including stainless steels, tool steels, nickel-based superalloys, and titanium alloys, have been successfully processed. Multiaxis laser control has been used to form complex geometries with this process.

Complex ceramic parts have been successfully produced using deposits of ceramic-loaded polymers. For example (Danforth et al., 1998), the molten polymer is extruded out of a 250 to 635 μm diameter nozzle, directed by a computer program, to a platform where the polymer freezes. Another process (Brady and Halloran, 1998) to achieve ceramic structures uses successive layers of ceramic-loaded polymers that are ultraviolet-curable so that layer patterns can be defined by stereolithography. In both cases, the polymer is removed and the ceramic is densified in subsequent heatings.

Though these processes are in their infancy, the potential advantages to the NASA program are obvious. In the area of fabricating a prototype part or parts in limited numbers, successful implementation could drastically reduce the cost and lead time for procurement. In remote locations such as the Moon or Mars, direct fabrication from computer numerical control programs could be used to produce items on location, reducing reliance on spare parts inventories. One of the original applications of this technology was in aircraft turbine repair (W.H. Hofmeister, private communication, 1998), demonstrating that the equipment is also capable of laser welding and repair of critical structural items.

Further technology development and actual implementation of the technology require that considerable research be done in the area of microstructural control. The research is necessary to learn how to control the process to allow tailoring the microstructure of each part manufactured, thus ensuring that the resulting properties are appropriate for the desired application. The weld pools involved in the buildup of layers contain very large thermal gradients. As a result, both surface-tension-driven flows and gravity-driven flows can be large, leading to significant effects on the microstructure, which must be understood. Also, the powder feed is delivered by forced convection, and the powder particles not captured by incorporation into the process must be recovered for reuse. Current recovery methods depend on gravitational settling, so alternative methods will be necessary in reduced gravity.

Fabrication of Components and Structural Elements from Raw or Processed Materials

The success of a mission to unexplored destinations can depend on the ingenuity with which local resources are utilized to meet unexpected challenges. Such a challenge could come from the unanticipated failure of a mechanical component, which would require the fabrication of a replacement part to repair the problem. As discussed above, it would not be practical to carry a complete spare parts inventory, nor would an extensive collection of spares necessarily fulfill every emergency need. A different approach has been described whereby a universal, compact machine shop with a very broad capability that might even extend to repairing itself would be included in the spacecraft or at the base (Stryker, 1987). This machine "shop" would be able to generate replacement parts as small as a wristwatch gear or as heavy as an antenna mount. Such a machine was developed by a mechanical engineer in the 1950s for personal use and was first described in 1974 by Urwick. Its operation is based on the common geometries of three machines: the lathe, the horizontal milling machine, and the horizontal boring machine. The machine is commercially available[10] and has been used by the Royal Navy and various research institutes as a general-purpose machine tool. It is completely modular and can be disassembled and reassembled. The design allows for working on a wide range of part sizes. Tiny parts can be manufactured by bringing the movable machine components close together. The largest parts may require partial disassembly of the machine, and their handling can be assisted by use of robotic positioning.

In kit form, the machine weighs 300 kg and occupies a volume of 1 m^3. Operation requires minimal on-site skill since the machine's activities can be controlled by an onboard computer or remotely by an operator elsewhere on the space station or on Earth. The latter approach offers the advantage that the machining procedure can be measured, evaluated, and planned by an expert stationed at an identical machine.

The feedstock would initially be aluminum or a suitable alloy from unneeded external vehicle tanks. In situ mining and refining activities could also generate iron-nickel-titanium alloys and other suitable feedstock metals and ceramics.

Such a universal machine could provide the wide range of capabilities needed to sustain a space station. It would be able to mill, shape, saw, and grind metal parts and even to generate special-purpose vises, collets, chucks, faceplates, and clamps that were not available in the kit of spare replacement parts.

The mechanical and thermal stability of the machining operations present no unusual problems, but operation in a hard vacuum would require appropriate lubrication of sliding and mating surfaces to prevent cold welding. The machine's operation is not sensitive to gravity, but the generated filings, cuttings, etc., must be collected at reduced gravity so that a clean environment can be maintained. At zero gravity, the manufacturing should occur in an isolated atmosphere to prevent contamination of the adjacent spaces.

In addition to machining, it is useful to note here some other common metal-working processes that may eventually be needed on HEDS missions, including extrusion, rolling, drawing, forging, bending, and pressing. None of these processes are expected to be gravity-dependent and so are not discussed further.

Casting in Reduced Gravity

Casting is a process in which a molten material is allowed to freeze or solidify, usually in a mold, to produce a solid object of the desired shape. The liquid is introduced into the mold by pouring or by pressure, as in die casting (Scully, 1988). Containerless solidification is also possible if a uniform crystal structure is desired, free of contamination from a container, but the resulting object will probably require machining as this process allows only a limited range of shapes (Shong et al., 1987; Hofmeister et al., 1987; Naumann and Elleman, 1986). Molds or dyes of complicated shapes would usually have to be made in situ, as it would be impossible to anticipate or transport all the ones that might be needed. In some cases, this could be done using sand molds and lost wax or similar techniques. The performance of the product is directly determined by the microstructure resulting from the casting process.

The large body of knowledge on casting metals in terrestrial gravity is treated extensively in a handbook by ASM (1988). There is also a substantial body of work demonstrating that the microstructure of castings in microgravity differs from that in terrestrial gravity (Curreri and Stefanescu, 1988). Finally, there is considerable research in progress (e.g., Glicksman et al., 1987, 1995a-c; Abbaschian, 1996; Bassler et al., 1995) exploring the fundamentals of solidification in a microgravity environment. At present, however, there is insufficient understanding of these fundamentals to allow predictions of the detailed effect of gravity level on the microstructure of a casting. Microgravity experiments continue to yield surprises. For example, solidification of eutectic alloys in microgravity (Larson and Pirich, 1982) shows a closer spacing of finer rods than in Earth gravity.

In general, the solidification process and the resulting microstructure are affected by gravity levels. The effect is ultimately due to differences in the strength of density-induced convection in the liquid phase. These differences affect the distribution of temperature, solute, and suspended particles or bubbles, which in turn affect the solidified microstructure. Moreover, casting operations may perform differently in reduced gravity. For example, many

[10]Anthony Croucher Ltd., Alton, Hampshire, England.

casting operations depend on the gravity feed of liquid by way of risers as part of the design of the mold. At reduced gravity, such feeds would be less effective. Conduction/convection furnaces may also have different operating characteristics in microgravity, as treated in a study by Lenski and Piller (1987).

Sintering

Sintering is an important manufacturing process for making near-net-shape parts from powder in the solid state. It is discussed, along with the reduction of a material to powder, in *Metals Handbook* (ASM, 1984). The powder is poured or injected into a mold or dye, with or without the aid of a binder. Filling the dye is extremely important, particularly for parts with complicated geometry, and it relies on gravity feed unless injection molding is used. If injection molding is used, the powder may be introduced as a slurry or paste, in which case the part is usually subject to light machining after sintering. The powder compact is then heated to a temperature below the melting point of the solid, with or without pressure, to produce the consolidated part. The process may take place entirely in the solid state of the material or may be facilitated by the presence of a liquid phase in the solid particle interstices; in the latter case, it is called liquid-phase sintering (LPS).

Sintering offers advantages over casting that include its capability to (1) use high-melting-point materials, (2) produce porous materials as used in self-lubricating bearings, and (3) use mixed powders, whose separate liquids are immiscible, to produce materials that cannot be formed by casting. In all three cases, however, molds or dyes are required. As with casting, molds or dyes of complicated shapes would usually have to be made in situ, as it would be impossible to anticipate and carry all the ones that might be needed.

During the sintering process, densification of the aggregate of solid particles takes place by the formation of connecting necks between particles and the concomitant reduction of pore volume during heating below the melting point of the solid. The driving force is surface energy reduction and—if the aggregate has been compacted or is under pressure—plastic and elastic energy reduction. In the case of solid-phase sintering, material transport occurs by diffusion on the surface of the solid particles, volume diffusion occurs in the interior of the solid particles (including high diffusivity paths), and vapor transport occurs in the pores; in LPS, there are additional processes of flow of the interparticle liquid phase, diffusive transport in the liquid phase, and local melting/freezing and dissolution/precipitation at the liquid/solid interface. Particle reorientation and plastic deformation or viscous flow of the solid phase may also play a role. Each of these processes is dependent on the temperature and the particle size; for example, surface diffusion dominates at smaller scales and lower temperatures and volume diffusion dominates under the opposite conditions.

Sintering is not only an important technique for making precision parts but it is also considered to be potentially important for fabricating building material brick from lunar regolith (Allen et al., 1992, 1994; Pletka, 1993). Solid-state sintering is slow and leads to very uneven heating owing to the low thermal conductivity of the regolith. Pletka (1993) has described an LPS process in which the liquid phase may derive from the glassy silicates in the regolith itself or from reactions that occur in the material when heated and which thus requires no additive. The advantage of sintering over casting is that lower temperatures are sufficient. The disadvantage is that the material must be comminuted and/or sieved to small particle sizes (typically $\sim 100\,\mu m$) for the sintering rates to be reasonable.

The diffusion transport processes that occur during solid-state sintering are not affected to any significant degree by the level of gravity. The spatial distribution of particles, however, is affected by it. In Earth's gravity, particles settle, forming a skeleton characterized by an average coordination number (Yang and German, 1991). In microgravity, an aggregate of independent particles would not form a compact unless pressure is applied, with the effective coordination number depending on the pressure.

Similarly, the distribution of particles in LPS is affected by gravity level. If the volume fraction of particles is low for LPS under microgravity, the particles tend to agglomerate toward the center, surrounded by liquid (Kohara, 1994; German, 1995), rather than settling toward the bottom as they do in Earth's gravity. This agglomeration has been interpreted as being driven by the reduction of surface and interface energy that can occur when particles coalesce to form grain boundaries at their junctions (German, 1995); the process is analogous to the coalescence of liquid drops brought into contact.

On a microstructural level, LPS involves solidification and is therefore affected by gravity level by virtue of density-induced convection and sedimentation in the freezing liquid, as was discussed in the section on casting. For example, it is found that materials formed by LPS in microgravity may be more porous (German et al., 1995) than those formed in Earth's gravity, presumably because the bubbles formed by outgassing are not eliminated by buoyancy migration; the migration of liquid-filled pores is also affected by gravity level (Heaney et al., 1995). The coarsening of particle sizes that occurs during sintering has been studied in microgravity, mostly to test theoretical models that do not include gravitational effects. Again, there are some surprises related to the behavior of pores in LPS (German et al., 1995).

The sintering operation requires a mold or dye with equipment (e.g., injection equipment) to fill it, a furnace capable of operating at sintering temperatures (typically above 1000 °C), and an atmosphere regulating system. As pointed out, molds would usually have to be made in situ. This poses no problem for simple shapes like bricks but is a serious limitation for complicated shapes unless they can be made by a lost wax or similar technique.

Composite Materials

On Earth, increasing use is being made of composite materials. These materials are combinations of a matrix and a dispersion. Broadly, they are classified as one of three basic types, depending on their matrix: polymer matrix composites (PMCs), metal matrix composites (MMCs), and ceramic matrix composites (CMCs). The utility of these materials stems from the synergism achieved by combining different materials into a single entity (Eckold, 1994; Schaffer et al., 1995; Callister, 1997).

It is not anticipated that exploration plans over the next few decades would include the extraterrestrial manufacture of matrix materials and particulate/fiber materials because of its complexity and the drain on resources that such activities would entail, including the demand for manufacturing capability. Instead, if composites were needed, judiciously chosen materials could be part of the cargo. These raw materials could then be used for repair, maintenance, or replacement.

The main manufacturing methodologies for polymer matrix composites are hand lay-up, filament winding, and pultrusion. These do not appear to involve gravity effects in a major way, although in hand lay-up, spray processes typically rely on free fall to distribute fiber in the mold for making the final product. Therefore, development of techniques for use in reduced gravity would be needed. For example, confinement of sprayed material such as resin, catalyst, and particulate would be necessary.

Metal matrix composites can be cast to shape using an intermediate feedstock but, alternatively, these composites can be forged to shape. Other options are production by hot processing or casting to shape using pressurized feeding of liquid metal into a mold cavity containing fiber preform. Of these alternatives, casting to shape could be affected considerably by reduced gravity through its effects on flow, convection, buoyancy, and sedimentation. The process would be a likely candidate for experimental work in reduced gravity and microgravity.

Products from ceramic matrix composites can be fabricated by pressing, hot or cold, and by sintering of prepregs as composite feedstock. In these cases, gravity level is not a factor.

Joining Methods in Space

Joining structural members in space is important for both construction and repair. Methods usually considered are mechanical joining, adhesive bonding, and welding or soldering (including brazing). However, mechanical joining requires special design (e.g., provisions for O rings) to assure pressured seals, and the high polymers used for adhesive bonding are subject to degradation in space owing to outgassing and radiation damage. We therefore focus here on welding, which may be used for repair (e.g., to patch holes caused by micrometeors) as well as for construction.

Welding entails the fusion of the base metals at the junction. It may be done either with or without a welding rod; in the latter case, it is called autogenous welding. In brazing and soldering, only the solder and not the base metal is melted; above 450 °C the process is called brazing, below 450 °C, soldering.

The history of welding in space is described in AWS (1991). It starts with Russian experiments in 1969 on Soyuz 6, followed by a number of subsequent Russian experiments and tests. The first American trials occurred in 1973 on Skylab with the welding of three metals (stainless steel, an aluminum alloy, and high purity tantalum); the results were examined at Battelle and NASA. In 1984, Russian cosmonauts spent 3 hours welding outside Salyut 7 using a handheld electron beam gun designed by the E.O. Paton Electric Welding Institute in Kiev. In 1986, two cosmonauts constructed a large truss in EVA off Salyut 7. This showed that quality welds could be done with little prior training and that the VHT (versatile handgun tool) performed well in space; however, there were dangers to the welder from emitted X rays. Numerous underwater welding experiments have been carried out at Marshall Space Flight Center in a neutral buoyancy tank. Welding experiments scheduled for the October 1997 shuttle flight STS-87 were postponed and have still not taken place.

The EVA welding environment is characterized by microgravity ($10^{-6}\ g_0$) with jitter, hard vacuum (modified by outgassing from the vehicle), meteoroids and debris, sunlight and ionizing radiation, atomic oxygen, and large thermal gradients (near the Sun/shade boundary). Technical conditions or limitations for welding are limited power sources (a few kilovolts for a few minutes) and the paucity of nondestructive testing methods for space. For some welding methods the associated health hazards make robotic welding very desirable.

Gravity is not a dominant factor in the welding process itself, even in Earth's gravity, as indicated by the fact that welding is routinely done upside down. Under microgravity conditions, the weld pool dynamics are completely dominated by capillary and electromagnetic forces. Thus, even though gravity-induced convection and sedimentation are absent, Marangoni-induced convection (due to the dependence of surface tension on temperature and composition) may be strong; added to this are electromagnetic stirring forces due to the welding current. The shape of the weld pool that moves in concert with the welding rod is determined by the interplay among these forces in a way that is not entirely understood. The shape in turn affects the weld quality. A cusp shape at the trailing edge produces a seam that is generally detrimental to the material properties since impurities tend to segregate there.

Since welding involves continuous solidification of the trailing edge of the moving molten zone, gravity level will have some effect on the resulting microstructure, as it does in all solidification processes. Typical microstructural variables in welded material that are affected are the grain size, distribution of phases, distribution of inclusions, and porosity and cracks. Some characteristics of the microstructure of welds conducted in microgravity as opposed to terrestrial gravity (Nance and Jones, 1993) are smaller grain size despite the slower cooling rates (possibly due to nucleation on suspended particles), increased porosity in some cases (possibly due to lack of buoyancy forces on bubbles), and a more uniform distribution of inclusions throughout the weld (again possibly due to lack of sedimentation forces).

References

Abbaschian, R. 1996. In-situ monitoring of crystal growth using MEPHISTO. Pp. 45-87 in Second United States Microgravity Payload: One Year Report. P.A. Curreri and D.E. McCauley, eds. Huntsville, Ala.: NASA Marshall Space Flight Center.

Allen, C.C., J.A. Hines, D.S. McKay, and R.V. Morris. 1992. Sintering of lunar glass and basalt. Pp. 1209-1218 in Engineering, Construction, and Operations in Space III: Space '92, Proceedings of the Third International Conference, Vol. II. W.Z. Sadeh, S. Sture, and R.J. Miller, eds. New York: American Society of Civil Engineers.

Allen, C.C., J.C. Graf, and D.S. McKay. 1994. Sintering bricks on the moon. Pp. 1220-1229 in Engineering, Construction, and Operations in Space IV: Proceedings of Space '94. R.G. Galloway and S. Lokaj, eds. New York: American Society of Civil Engineers.

American Society for Metals (ASM). 1984. Metals Handbook, 9th Ed. Metals Park, Ohio: American Society for Metals.

American Society for Metals (ASM). 1988. Metals Handbook, 9th Ed., Vol. 15. Metals Park, Ohio: ASM International.

American Welding Society (AWS). 1991. Proceedings of Welding in Space and the Construction of Space Vehicles by Welding, cosponsored by the American Welding Society-USA and the E.O. Paton Electric Welding Institute-USSR.

Bassler, B.T., W.H. Hofmeister, and R.J. Bayuzick. 1995. Examination of solidification velocity determination in bulk undercooled nickel. Proceedings of the 1994 Materials Research Society (MRS) Fall Meeting. Warrendale, Pa.: Materials Research Society.

Bergan, P. 1998. Potential Navy applications for selective laser sintering. P. 5 in Naval Research Reviews, Vol. L. Washington, D.C.: Government Printing Office.

Boles, W., W. Scott, and J. Connolly. 1997. Excavation forces in reduced gravity environment. J. Aerospace Eng. 10(2):99-103.

Brady, G.A., and J.W. Halloran. 1998. Solid freeform fabrication of ceramics via stereolithography. P. 39 in Naval Research Reviews, Vol. L. Washington, D.C.: Government Printing Office.

Callister, W.D., Jr. 1997. Materials Science and Engineering—An Introduction, 4th Ed. New York: John Wiley & Sons.

Carrier, W.D., G.R. Olhoeft, and W. Mendell. 1991. Physical properties of the lunar surface. Pp. 476-567 of Lunar Sourcebook. G. Heiken, D. Vaniman, and B.M. French, eds. New York: Cambridge University Press.

Colwell, J.E., M. Horanyi, A. Sickafoose, S. Roberston, and R. Walch. 1998. Dynamic dust in photoelectron layers near surfaces in space. Proceedings of the Fourth Microgravity Fluid Physics and Transport Phenomena Conference, NASA Lewis Research Center. Cleveland, Ohio: National Aeronautics and Space Administration.

Costes, N.C., and S. Sture. 1998. A mobility concept for martian exploration. Pp. 301-318 in Space 98: The Sixth International Conference and Exposition on Engineering, Construction, and Operations in Space. R.G. Galloway and S. Lokaj, eds. Reston, Va.: American Society of Civil Engineers.

Curreri, P.A., and D.M. Stefanescu. 1988. Low-gravity effects during solidification. P. 147 in Metals Handbook, 9th Ed., Vol. 15. Metals Park, Ohio: ASM International.

Danforth, S.C., A. Safari, M. Jafari, and N. Landrana. 1998. Solid freeform fabrication (SFF) of functional advanced ceramic components. P. 27 in Naval Research Reviews, Vol. L. Washington, D.C.: Government Printing Office.

Eckold, G. 1994. Design and Manufacture of Composite Structures. New York: McGraw-Hill.

German, R.M. 1995. Grain agglomeration in solid-liquid mixtures under microgravity conditions. Met. Mater. Trans. B 26:649.

German, R.M., R.G. Iacocca, J.L. Johnson, Y. Liu, and A. Upadhyaya. 1995. Liquid-phase sintering under microgravity conditions. J. Met. 47(8):46-48.

Glicksman, M.E., E. Winsa, R.C. Hahn, T.A. Lograsso, R. Rubinstein, and M.E. Sellick. 1987. Isothermal dendritic growth. P. 37 in Materials Processing in the Reduced Gravity Environment of Space: Proceedings of the 1986 Fall Materials Research Society (MRS) Meeting, Vol. 87. R.H. Doremus and P.C. Nordine, eds. Warrendale, Pa.: Materials Research Society.

Glicksman, M.E., M.B. Koss, L.T. Bushnell, and J.C. LaCombe. 1995a. The isothermal dendritic growth experiment: Implications for theory. P. 633 in Modeling of Casting, Welding, and Advanced Solidification Processes VII. M. Cross and J. Campbell, eds. Warrendale, Pa.: The Minerals, Metals & Materials Society.

Glicksman, M.E., M.B. Koss, L.T. Bushnell, J.C. LaCombe, and E.A. Winsa. 1995b. Dendritic growth of succinontrile in terrestrial and microgravity conditions as a test of theory. Iron and Steel Institute of Japan (ISIJ) International 35(6):1216

Glicksman, M.E., M.B. Koss, L.T. Bushnell, J.C. LaCombe, and E.A. Winsa. 1995c. Dendritic growth in terrestrial and microgravity conditions. P. 13 in Fractal Aspects of Materials: Materials Research Society (MRS) Symposia Proceedings, Vol. 367. F. Family, P. Meakin, B. Sapoval, and R. Wool, eds. Warrendale, Pa.: Materials Research Society.

Heaney, D.F., R.M. German, and A. Griffo. 1995. Gravitational effects on low solid-volume fraction liquid-phase sintering. J. Mater. Sci. 30:5808-5812.

Hepp, A.F., G.A. Landis, and C.P. Kubiak. 1994. A chemical approach to carbon dioxide utilization on Mars. Pp. 799-818 in Resources of Near-Earth Space. J. Lewis, M.S. Matthews, and M.L. Guerrieri, eds. Tucson and London: University of Arizona Press.

Hofmeister, W., M.B. Robinson, and R.J. Bayuzick. 1987. Undercooling of bulk high temperature metals in the 100 meter drop tube. P. 149 in Materials Processing in the Reduced Gravity Environment of Space: Proceedings of the 1986 Fall Materials Research Society (MRS) Meeting, Vol. 87. R.H. Doremus and P.C. Nordine, eds. Warrendale, Pa.: Materials Research Society.

Keicher, D.M. 1999. Direct fabrication via the Laser Engineered Net Shaping (LENS^{TM}) process. Forum on Rapid Manufacturing, January 25-26, San Francisco.

Klosky, J.L. 1997. Behavior of Composite Granular Materials and Vibratory Helical Anchors, Ph.D. dissertation. University of Colorado at Boulder.

Klosky, J.L., S. Sture, H.-Y. Ko, and F. Barnes. 1995. Augered foundations for lunar operations. Proceedings of the Third International Symposium on Mine Mechanization and Automation. Golden, Colo.: Colorado School of Mines.

Klosky, J.L., S. Sture, H.-Y. Ko, and F. Barnes. 1996. Vibratory excavation and anchoring tools for the lunar surface. Pp. 903-909 in Engineering, Construction, and Operations in Space V: Proceedings of the Fifth International Conference on Space 96, Vol. 1. S.W. Johnson, ed. New York: American Society of Civil Engineers

Klosky, J.L., S. Sture, H.-Y. Ko, and F. Barnes. 1998. Helical anchors for combined anchoring and soil testing in lunar operations. Pp. 489-494 in Space 98: The Sixth International Conference and Exposition on Engineering, Construction, and Operations in Space. R.G. Galloway and S. Lokaj, eds. Reston, Va.: American Society of Civil Engineers.

Ko, H.-Y. 1988. Summary of the state-of-the-art in centrifuge testing. Pp. 11-19 in Centrifuges in Soil Mechanics. W.H. Craig et al., eds. Rotterdam, Netherlands: AA Balkema Publishers.

Kohara, S. 1994. Study on Liquid Phase Sintering. Tokyo: Science University of Tokyo.

Larson, D.J., and R.G. Pirich. 1982. Influence of gravity driven convection on the directional solidification of Bi/MnBi eutectic composites. P. 523 in Materials Processing in the Reduced Gravity Environment of Space. G.E. Rindone, ed. Warrendale, Pa.: Materials Research Society.

Lenski, H., and J. Piller. 1987. The influence of natural convection on heating characteristics of space furnaces. P. 313 in Materials Processing in the Reduced Gravity Environment of Space: Proceedings of the 1986 Fall Materials Research Society (MRS) Meeting, Vol. 87. R.H. Doremus and P.C. Nordine, eds. Warrendale, Pa.: Materials Research Society.

Lin, T.D., and S. Bhattacharja. 1998. Lunar and Martian resources utilization: Cement. Pp. 592-600 in Space 98: The Sixth International Conference and Exposition on Engineering, Construction, and Operations in Space. R.G. Galloway and S. Lokaj, eds. Reston, Va.: American Society of Civil Engineers.

Mindess, S., and J.F. Young. 1981. Concrete. Englewood Cliffs, N.J.: Prentice-Hall.

Nance, M., and J.E. Jones. 1993. Welding in space and low-gravity environments. P. 1020 in Metals Handbook, Vol. 6, Welding, Brazing and Soldering. Metals Park, Ohio: ASM International.

National Research Council (NRC), Space Studies Board. 1992. Toward a Microgravity Research Strategy. Washington, D.C.: National Academy Press.

Naumann, R.J., and D.D. Elleman. 1986. Containerless processing technology. P. 294 in Material Science in Space. B. Feuerbacher, H. Hamacher, and R.J. Naumann, eds. New York: Springer-Verlag.

Perkins, S.W., and C.R. Madson. 1996a. Mechanical and load-settlement characteristics of two lunar soil simulants. J. Aerospace Eng. 9(1):1-11.

Perkins, S.W., and C.R. Madsen. 1996b. Scale effects of shallow foundations on lunar regolith. Pp. 963-972 in Engineering, Construction, and Operations in Space V: Proceedings of the Fifth International Conference on Space 96, Vol. 2. S.W. Johnson, ed. New York: American Society of Civil Engineers.

Pletka, B.J. 1993. Processing of lunar basalt materials. P. 325 in Resources of Near-Earth Space. J.S. Lewis, M.S. Matthews, and M.L. Guerrieri, eds. Tucson and London: University of Arizona Press.

Schaffer, J.P., A. Saxena, S.D. Antolovich, T.H. Sanders, Jr., and S.B. Warner. 1995. The Science and Design of Engineering Materials. Chicago: Richard D. Irwin.

Scully, L.J.D. 1988. Die casting. P. 286 in Metals Handbook, 9th Ed., Vol. 15. Metals Park, Ohio: ASM International.

Shong, D.S., J.A. Graves, Y. Ujiie, and J.H. Perepezko. 1987. Containerless processing of undercooled melts. P. 17 in Materials Processing in the Reduced Gravity Environment of Space: Proceedings of the 1986 Fall Materials Research Society (MRS) Meeting. R.H. Doremus and P.C. Nordine, eds. Warrendale, Pa.: Materials Research Society.

Stryker, J.M. 1987. A job shop for space manufacturing. Pp. 158-163 in Proceedings of the Eighth Princeton/AIAA/SSI Conference. B. Faughnan and G. Maryniak, eds. Washington, D.C.: American Institute of Aeronautics and Astronautics.

Szabo, B., F. Barnes, H.-Y. Ko. 1994. Effectiveness of vibrating bulldozer and plow blades on draft force reduction. Proceedings of the Winter Meeting of the American Society of Agricultural Engineers. No. 941535. St. Joseph, Mich.: American Society of Agricultural Engineers.

Yang, S.-C., and R.M. German. 1991. Gravitational limit of particle volume fraction in liquid-phase sintering. Met. Mater. Trans. A 22:786.

III.G MATRICES OF SUBSYSTEMS, PROCESSES, AND PHENOMENA

Tables III.G.1 through III.G.3 summarize information from the preceding sections regarding which common subsystems and processes are likely to be affected by changes in gravity level. The tables identify the specific phenomena that are most likely to play an important role in the operation of a given subsystem when the gravity level is altered. These phenomena are discussed in greater detail in Chapter IV.

TABLE III.G.1 Phenomena Associated with the Common Subsystems Likely to Be Affected by Gravity Level

Subsystem/ Variant	**Phenomenon**																		
	Capillarity	**Wetting**	**Marangoni Flows**	**Two-Phase Flows**	**Phase Separation and Distribution**	**Mixing**	**Flow in Porous Media**	**Convection**	**Boiling Heat Transfer**	**Condensation Heat Transfer**	**Multiphase Flow and Heat Transfer**	**Evaporation Heat Transfer**	**Phase-Change Heat Transfer**	**Solidification**	**Multiphase System Dynamics**	**Combustion**	**Pyrolysis**	**Granular Flow**	**Structural Dynamics**
Storage tanks																			
Gas								•											•
Liquid	•	•	•	•	•	•		•	•			•		•					•
Cryogenic	•	•	•	•	•			•	•	•	•	•			•				•
Pumps																			
Condensate											•								
Liquid line															•				
Microdevices																			
Compressors																			
Rotary											•								
Adsorption											•								
Piping																			
Gas-phase									•	•	•								•
Liquid-phase					•			•	•		•								•
Two-phase	•	•	•		•	•		•	•	•	•				•				•
Radiators																			
Solid-state																			•
Gas-phase								•											•
Two-phase					•	•		•		•	•			•	•				•
Heat pipes																			
Capillary pumped loop	•	•	•		•		•			•	•	•		•	•				
Simple			•		•		•			•	•	•		•	•				
Fans and blowers								•											
Evaporators																			
Boilers			•		•	•			•				•		•				
Vaporizers		•	•		•			•				•	•						
Liquifiers																			
Condensers					•					•					•				
Distillations units								•											

TABLE III.G.1 Continued

	Phenomenon																		
Subsystem/ Variant	**Capillarity**	**Wetting**	**Marangoni Flows**	**Two-Phase Flows**	**Phase Separation and Distribution**	**Mixing**	**Flow in Porous Media**	**Convection**	**Boiling Heat Transfer**	**Condensation Heat Transfer**	**Multiphase Flow and Heat Transfer**	**Evaporation Heat Transfer**	**Phase-Change Heat Transfer**	**Solidification**	**Multiphase System Dynamics**	**Combustion**	**Pyrolysis**	**Granular Flow**	**Structural Dynamics**
Filters/separators																			
Gas/solid					•						•						•		
Gas/liquid	•		•		•	•					•								
Liquid/liquid	•				•	•					•								
Liquid/solid	•				•	•					•								
Vortex separators					•						•								
Rotating drum separators					•														
Spargers					•	•					•								
Valves and actuators											•								•
Heaters								•											
Catalyst beds							•												
Seals											•								
Heat exchangers																			
Gas/gas								•											
Gas/liquid								•		•			•						
Gas/solid								•					•						
Fluidized-bed							•												
Fire extinguishers						•		•								•	•		
Smoke detectors																•	•		

TABLE III.G.2 Phenomena Associated with the Materials Handling Equipment Likely to Be Affected by Gravity Level

Equipment	Phenomenon																	
	Capillarity	Wetting	Marangoni Flow	Free-Surface Flow	Phase Separation and Distribution	Mixing	Flow in Porous Media	Convection	Boiling Heat Transfer	Condensation Heat Transfer	Multiphase Heat Transfer	Evaporation Heat Transfer	Phase-Change Heat Transfer	Solidification	Multiphase System Dynamics	Combustion	Pyrolysis	Granular Flow
Screens	•	•		•														
Hoppers																		•
Excavators																		•
Conveyers																		•
Drillers																		•
Bulldozer																		•
Anchor																		•
Trucks																		•
Cranes																		•
Bucket scoop																		•
Winch																		•
Rotating drum or slide charging unit																		•
Electrostatic generator																		
Gravity collection bins																		•

TABLE III.G.3 Phenomena Associated with Various Material Processes Likely to Be Affected by Gravity Level

	Phenomenon																	
Process	**Capillarity**	**Wetting**	**Marangoni Flow**	**Free-Surface Flow**	**Phase Separation and Distribution**	**Mixing**	**Flow in Porous Media**	**Convection**	**Boiling Heat Transfer**	**Condensation Heat Transfer**	**Multiphase Flow and Heat Transfer**	**Evaporation Heat Transfer**	**Phase-Change Heat Transfer**	**Solidification**	**Multiphase System Dynamics**	**Combustion**	**Pyrolysis**	**Granular Flow**
Crushing/grinding																		•
Settling				•	•													•
Sieving							•											•
Transport[a]					•						•							•
Sintering (LPS)	•	•							•									
Casting									•					•				
Welding	•	•	•						•					•				

[a]Includes such bulk material transport processes as ore transport and slurry flow in pipes.

IV

Phenomena of Importance in Reduced Gravity

In Chapter III various HEDS technologies are explored and the phenomena likely to affect their operation in reduced gravity are identified. In this chapter, those phenomena and their dependence on, or importance in, reduced gravity are discussed, along with research that would be needed to develop an adequate database as well as predictive models for better characterizing the phenomena. The material in this chapter and in Chapter V provides the basis for the research recommendations outlined in Chapter VI.

IV.A GENERAL CONSIDERATIONS

Introduction

In microgravity research, one is concerned with how, in various circumstances, the relationships or balances among various fundamental phenomena depend on the presence or absence of gravity and, consequently, how particular processes of technical importance depend on gravity level.

Some phenomena can therefore be isolated for study if gravity is eliminated. Moreover, knowledge of the underlying physics can often be obtained when it is not masked by phenomena that are induced by terrestrial gravity. This basic scientific goal is emphasized in the NASA microgravity research program (Woodward, 1998).

Understanding and mitigating the technological consequences of low gravity must be one of NASA's primary research goals. If the HEDS program is to be successful, however, its research goals must differ from NASA's purely scientific goals. In particular, the emphasis must be on applied research, which is the focus of this report. Indeed, the report identifies the kinds of enabling technologies required to meet the goals of the HEDS program.

The following paragraphs provide an overview of the basic effects of a reduction in gravity; those effects are then treated in more detail in subsequent sections of the chapter. The emphasis is on how the interplay of gravity and other, competing basic physical processes finds expression in various dimensionless combinations of physical parameters. Either explicitly or implicitly, these dimensionless groups will govern the interaction of microgravity with the specific phenomena discussed in this chapter. The overview concludes with some general thoughts about the need for microgravity research to characterize the relationships among these groups.

Gravity and Density Difference

First, the general importance of the combination of gravity with density difference should be noted. Gravity is a body force per unit mass (a virtual acceleration, according to Newtonian mechanics). Therefore, the gravity-force-density is the product of gravity and density (ρ), and fluid motion is affected by gravity if the density is nonuniform. Accordingly, one expects any change of gravity level coupled with a density difference (whether preexisting or due to the flow process) to affect the phenomena of interest. In other words, a product $g\Delta\rho$ will be of basic significance, where $\Delta\rho$ is a characteristic parameter describing the density difference.

Clearly, the larger the density difference the larger will be the effect of a change in gravity level. The largest density differences normally encountered are those due to the presence of different phases, especially gas and liquid, or gas and solids. Thus, one must be especially concerned with multiphase flows. Smaller, but often important, density differences also occur as the result of thermal expansion of a single phase, as will be discussed in a later section of this chapter.

Frequently, when gravity is changed, physical phenomena that depend on gravity will change in ways that can be characterized as a change in stability. Stability is an especially useful concept for engineering purposes, since one normally intends a system to stay unchanged over time or to maintain a steady motion. There are many classical books treating stability; for fluids, the work of Chandrasekhar (1961) and of Yih (1965) abounds in stability problems involving gravity.

A common issue for the fluid systems discussed in Chapter III is a change of local stability when gravity is reduced or vanishes. In Earth gravity, a liquid/vapor system will tend to stratify stably, with vapor above and liquid below; then, if the interface is displaced it will tend to return. If gravity is absent, the restoring force is absent and the position of the interface is indeterminate. This is the situation in a cryogenic storage tank. If the situation in Earth gravity is initially the reverse, with vapor below the liquid, then the situation is unstable, and the vapor will tend to rise as a consequence of buoyancy. This instability enables a simple boiler to function. As gravity is reduced, this helpful instability is reduced, and boiler performance is impaired. Clearly, either stability or instability associated with gravity on Earth may be desirable, depending on the function considered. In some cases, such as the dispersal of particulates or droplets, neutral stability would be desired, and in such cases, reduced gravity could be helpful.

Gravity-Density Coupling in Various Basic Processes

Various basic flow processes may now be related to buoyancy as expressed by the gravity-density difference product, and typical dimensionless groups may thus be identified. It will be seen that if such groups are numerically very large or very small, dominance of one process over another is implied. More importantly, such groups are parameters of any theoretical or computational model for system behavior, with effects depending on numerical coefficients, or correlations, derived from the appropriate theories or experiments.

- *Flow forced by pressure difference.* If the velocity of buoyant rise (or fall) through a distance L owing to a given density difference ($\Delta\rho$) is compared with the velocity produced by a given pressure difference, Δp, acting as the only other force on the fluid, the comparison can be expressed in terms of the dimensionless group $g\Delta\rho L/\Delta p$. If this parameter[1] is much less than unity, then buoyancy is insignificant. This is certainly true in rocket exhausts, for example, where the pressure drop is very large.
- *Capillarity, wetting, and Marangoni flows.* The role of buoyancy in comparison with that of capillarity, or surface tension (σ), can be expressed by the dimensionless group $g\Delta\rho L^2/\sigma$ when surface tension acts on a curved interface of scale L. This group is the "static" Bond number (Ostrach, 1982) based on density difference. One sees that for a given surface tension coefficient (σ), capillarity becomes more important relative to buoyancy if either

[1] With pressure drop expressed in terms of the flow velocity it produces, this parameter is the reciprocal, squared, of the Froude number as given by Yih (1965) or the "densimetric" Froude number in the hydraulics literature.

g or L becomes small and less important if L is large, unless g is especially small or zero. In Marangoni flows, it is the gradient of σ that is important (Ostrach, 1982), so in the above group, σ would be replaced by a surface-tension difference $\Delta\sigma$, which in turn varies with temperature or composition. This topic is discussed in more detail in Section IV.B.

- *Diffusion.* Transport by buoyancy may compete with molecular diffusion in affecting composition, temperature, or some other local property. If the diffusion coefficient, D, is known for the property difference of interest, the competition is governed by the dimensionless group $g\Delta\rho L^3/\rho D^2$. If the diffusing quantity is thermal energy, the foregoing group becomes $g\Delta\rho L^3\rho c_p{}^2/k^2$, where k is thermal conductivity and c_p is specific heat. This subject is discussed more fully in Section IV.B.
- *Viscosity.* Similarly, buoyant transport of momentum may compete with viscosity, according to the group $g\Delta\rho L^3/\rho\nu^2$, in which ν is the kinematic viscosity coefficient of the fluid.[2]
- *Chemical transformation and combustion.* Obviously, chemical change results in density change, and therefore an interaction with gravity can be expected, as discussed above. Consequences of chemical change in low gravity are also discussed in Section IV.F.
- *Electromagnetic forces.* Electric or magnetic fields may exert body forces whose effects can be compared with the gravitational body force, $g\Delta\rho$.
- *Vibration.* Time dependence of motion, or unsteady motion in general, introduces yet another degree of freedom of fluid or solid motion, and the accelerations involved clearly compete with the virtual acceleration represented by gravity. Thus, the importance of gravity during vertical vibration of a container with a liquid/vapor interface (Yih, 1965) will be governed by the group $g\Delta\rho/\rho\varepsilon\omega^2$, where ε is the amplitude of the imposed vibration. The higher the imposed frequency ω, the less important buoyancy becomes relative to forces due to acceleration.

The dynamic behavior of machines or structures also depends on gravitational loading, as a simple pendulum illustrates. Unsteadiness in competition with gravity is an important topic called g-jitter, where gravity itself appears to fluctuate owing to imposed accelerations (Woodward, 1998) arising from crew motions, rocket firings, and so on. Acoustics in the presence of gravity furnishes still another important example of time dependence. In general, if a process is time-dependent, new parameters, and hence additional dimensionless groups, will need attention.

- *Phase change.* In a subsequent section, the physical process of solidification and melting is discussed, and it is made clear that the role of buoyancy, and hence of gravity level, is an interactive one: the phase change itself depends on buoyant transport, which in turn depends on the degree of phase change already achieved.
- *Granular behavior.* In Section IV.G, problems of particulate or granular flows are described. Density and gravity are of course coupled in such flows, and particle interactions with each other and with liquid or gaseous media will furnish examples of the static and dynamic phenomena with which buoyancy must be compared.

Gravity Regime Boundaries

It is obvious from the foregoing discussion that particular combinations of processes must be identified for specific problems. For example, inertia, viscosity, heat conduction, and capillarity may be simultaneously important, along with buoyancy, in some technically important phenomena. Therefore, many specific dimensionless groups will be needed to describe complex flow regimes of interest (Ostrach, 1982).

Nevertheless, the crude outline given above shows that, generally, gravity level, g, finds itself in a group that includes density difference and some positive power (n) of length scale, that is, $g\Delta\rho L^n$. Thus, one may infer that when gravity is reduced, any effect of that reduction will be amplified by large density difference (especially that due to phase difference) and also by large scale.

---

[2]This group may be recognized as the classical Grashof number (Eckert and Drake, 1972), to which the previous two groups are related through the Schmidt number (ν/D) and the Prandtl number ($\rho c_p\nu/k$).

Gravity reduction will be significant only in relation to phenomena competing with buoyancy or settling processes. Therefore, in any given reduced-gravity situation, one would like to know whether the dimensionless-group coefficients or correlations used in Earth gravity still apply, or whether a different set of models or correlations needs to be developed. That is, one asks what numerical levels of the various dimensionless groups applicable to a specific technical situation represent critical boundary zones between regimes of essentially terrestrial and essentially microgravity behavior. This is the problem posed by fractional gravity environments, which is where a large proportion of HEDS operations are expected to be carried out.

Research Issues

It would be useful for NASA to develop a catalog of the regime-change zones for the dimensionless parameters of all relevant fundamental phenomena. This would presumably entail reviewing and extending the experiments performed in microgravity. This effort would provide the basis for assessing the computational design capabilities in the future, capabilities essential for comprehensive and credible designs of efficient, reliable systems and components for HEDS.

Clearly the number and the ranges of relevant parameters are very great. Therefore, the development of the suggested catalog would require effort carried out over a long period of time, focused on a great variety of difficult issues and problems.

References

Chandrasekhar, S. 1961. Hydrodynamic and Hydromagnetic Stability. International Series Monographs on Physics. Oxford: Clarendon Press.
Ostrach, S. 1982. Low-gravity fluid flows. Annu. Rev. Fluid Mech. 14:313-346.
Woodward, D. 1998. NASA's Microgravity Research Program. NASA Report TM 1998-208418. Huntsville, Ala.: NASA Marshall Space Flight Center.
Yih, C.S. 1965. Dynamics of Nonhomogeneous Fluids. London: Macmillan.

IV.B INTERFACIAL PHENOMENA

Interfacial or capillary phenomena generally refer to the broad field of surface-tension-related phenomena (Maxwell, 1878; Gibbs, 1878; de Gennes, 1985; Haynes and Langbein, 1987). The terminology derives from the surface-tension-induced rise (fall) of liquid in a capillary tube for contact angles less than (greater than) 90°. These phenomena are not directly influenced by gravity level, but they become increasingly important in determining the configuration and movement of liquid as gravity level is reduced, and they may become dominant in microgravity (Ostrach, 1982). In some cases, the effects may be utilized to compensate for the loss of gravity in the management of liquids in microgravity; examples are wicked structures as in heat pipes (Faghri, 1995; also see Figure III.A.5 of this report), capillary pumped loops, and vane structures as in cryogenic storage tanks (Dodge, 1990). Because of their dominance and practical importance in reduced gravity, the following interfacial phenomena are discussed in this section: static and dynamic capillary configurations, wetting, and the Marangoni effect (arising from gradients of the surface free energy).

Capillary Equilibrium and Dynamic Forms

Most of the problems and work described in this section assume a uniform surface tension and therefore omit Marangoni effects, but in many practical situations these must be included; they are discussed later on. The dimensionless Bond number measures the ratio of (gravity forces)/(capillary forces) and is given by $B = \rho g L^2/\sigma$, where ρ and σ are, respectively, the density and surface tension of the liquid, g is the gravitational strength, and L is a characteristic length; for example, the ratio of height to radius of a liquid in a capillary tube is inversely proportional to B. Similarly, the Bond number enters into a description of equilibrium shapes determined by the competition of gravitational and capillary forces, as exemplified by sessile and pendant drop shapes (Antar and Nuotio-Antar, 1993; Haynes and Langbein, 1987; Concus and Flinn, 1990).

There is a large literature devoted to static equilibrium capillary shapes in the absence of gravity; these are shapes that minimize the area of the surface of a mass of liquid subject to boundary conditions of a fixed volume and given contact angles at the perimeter of the surface (Antar and Nuotio-Antar, 1993; Haynes and Langbein, 1987; Concus and Flinn, 1990; Carter, 1998). In some cases (as for a box with a shallow layer of partially wetting liquid), there are multiple solutions: the liquid may cover a face of the box or collect along the edges and/or corners (Martinez et al., 1987). In addition, there is the question of the stability of equilibrium shapes, as illustrated by the Rayleigh instability of a cylinder or the stability of liquid bridges (Koster, 1990), that arise in containerless processing of materials.

In addition to static problems, there are many dynamic problems in capillary theory exemplified by Bénard convection, the oscillations of liquid drops or bubbles, resonances between the applied forces or accelerations (e.g., *g*-jitter), and capillary modes of motion of a mass of liquid in a container (Zhang and Vinals, 1997), including so-called sloshing problems (Antar and Nuotio-Antar, 1993). Bénard convection, arising from the heating from below of a layer of liquid with a free surface and leading to a pattern of hexagonal convection cells, is often dominated by Marangoni convection (Antar and Nuotio-Antar, 1993) rather than the buoyancy-driven convection analyzed by Rayleigh. In particular, Bénard cells have been studied under microgravity conditions.

The stability and dynamics of capillary-dominated configurations of liquids are expected to play an important role in the management of liquids and in the boiling/condensation process in heat exchangers under reduced gravity conditions (Westbye et al., 1995). As mentioned above, they also underlie the operation of heat pipes, capillary pumped loops, microgrooved heat pipes, and veined structures.

Surface-tension-driven free surface flows arise in nature, science, and technology. An important technological application of free surface flows is the laser printer, and HEDS applications include the liquid-droplet surface radiator and the electrostatic liquid film radiator. It is recognized that drops generally result from the motion of free surfaces. The dynamics and breakup of drops and of other free surface flows, including theoretical developments and experimental work, have recently been reviewed and unresolved problems have been outlined (Eggers, 1997; Stone, 1994).

Research Issues

A body of classical knowledge and current results is available, as indicated above, but there are many unsolved problems of capillarity-determined liquid configurations involving practical boundary conditions (e.g., vessel and conduit shapes) that require additional modeling and experiments. Dynamic problems involving resonances between imposed vibrations and accelerations (e.g., *g*-jitter) and capillary modes of liquid masses are important because of the possibility of uncontrolled excursions of the liquid mass; such problems need to be extended. In addition, the inclusion of Marangoni effects will require computer modeling of complex fluid flows as well as a greatly improved knowledge of the Marangoni parameters (e.g., temperature and composition dependence of the surface tension).

Wetting

Wetting is a phenomenon in which one condensed phase spreads over the surface of a second condensed phase (Adamson, 1982; Findenegg and Telo de Gama, 1987). If the spreading stops at some equilibrium configuration with the surfaces of both phases exposed, it is called partial wetting; this is illustrated in Figure IV.B.1 for the case of a drop of liquid L on the surface of a rigid solid S, both in contact with the vapor V. The contact angle θ is determined in the classical description by the balance of surface tensions (indicated in the figure) described by the Young equation:

$$\sigma_{SV} = \sigma_{LS} + \sigma_{LV} \cos\theta.$$

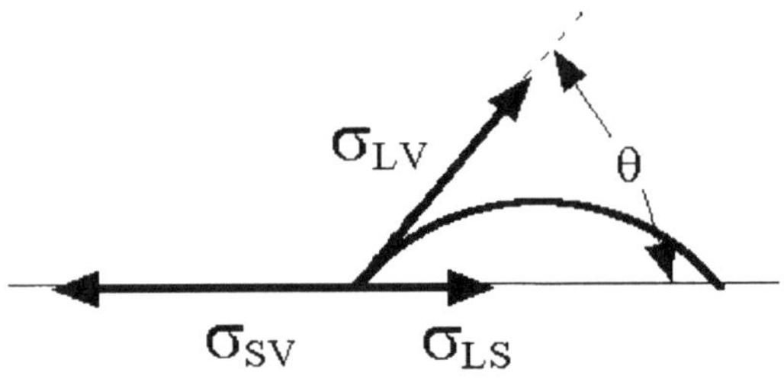

FIGURE IV.B.1 A liquid drop in equilibrium with a rigid solid surface and the vapor; interface free energies are labeled by the adjacent phases.

If the spreading does not stop, which occurs when $\sigma_{SV} \geq \sigma_{LS} + \sigma_{LV}$, then complete wetting is said to occur; the liquid phase spreads indefinitely on (i.e., it wets) the solid surface as long as it is thick enough to be described as a macroscopic liquid. In general, the two phases may be liquid or solid. A related phenomenon is the partial or complete penetration of a second phase into a grain boundary between two crystals. Although wetting is a capillary or surface phenomenon not significantly affected by gravity level, it becomes increasingly important when the gravity level is reduced, which makes it one of several phenomena that dominate events under microgravity conditions, and for this reason it is included in this report.

Wetting, partial or complete, underlies such important technologies as soldering and welding (Nance and Jones, 1993); liquid-phase sintering (German et al., 1995); the operation of wicks in capillary pumped loops (Anatar and Nuotio-Antar, 1993) and vanes in the cryogenic storage of liquids (Dodge, 1990); heat pipes (Faghri, 1999; Peterson et al., 1998); boiling/condensation heat transfer, including the rewetting of a hot surface (Westbye et al., 1995); and lubrication. Usually, small contact angles or complete wetting are desired in these techniques.

To understand the meaning of complete wetting (de Gennes, 1985; Dietrich, 1988) in the context of the previous illustration, let $W = \sigma_{SV} + \sigma_{LV} - \sigma_{LS}$ be the work per unit area required to separate the liquid from the solid. Then, the condition for complete wetting becomes $W \geq 2\sigma_{LV}$, which means that the adhesion of the liquid to the solid is greater than the cohesion of the liquid. Complete wetting is favored for liquids of low surface energy in contact with solids to which they are strongly attracted. A film of oil or grease that interferes with adhesion will favor partial wetting with a large contact angle.

Extensions and modification of the Young description are necessary to account for several important features of wetting that include (1) hysteresis of the contact angle, (2) dynamics of wetting, and (3) breakdown of the macroscopic description on sufficiently small scales. Hysteresis of the contact angle (Decker and Garoff, 1997; Ramé and Garoff, 1996) refers to the greater value of the contact angle obtained from measurement if the contact line is advancing (extending the liquid) than if it is receding; the difference between the advancing and receding contact angles makes it possible for a drop on a tilted solid surface in 1 g_0 to be stationary (e.g., raindrops on a window pane). The extent of hysteresis is affected by the microscopic topography and condition of the surface.

The dynamics of wetting is exemplified by dependence of the geometry of the liquid configuration in the immediate vicinity of a moving contact line on the velocity of the contact line; in particular, the effective contact angle depends on that velocity. This is particularly true of the ebullition cycle during boiling, where the dynamic contact angle is required to adequately model the nucleation of bubbles on the heated wall. There is also evidence that a spreading liquid drop on a solid surface is preceded by a thin film (Marsh et al., 1993). The dynamics of wetting has become an important area of investigation (Decker and Garoff, 1997; Ramé and Garoff, 1996).

The thermodynamic description of wetting breaks down when the thickness of the liquid becomes comparable to or less than the correlation length in the liquid, i.e., when the liquid in a thin film or filament can no longer be described in macroscopic thermodynamic terms. This can occur if the liquid is imbibed into a porous structure (as in a wick) whose pore dimensions are comparable to or smaller than the correlation length in the liquid (Wiltzius et al., 1989).

Two additional developments in wetting science in recent years are the following: (1) A phenomenon known as premelting (Frenken and van der Veen, 1985) can occur in which the surface layer of a solid loses translational long-range order below the melting point of the solid. As the melting point is approached, the thickness of the

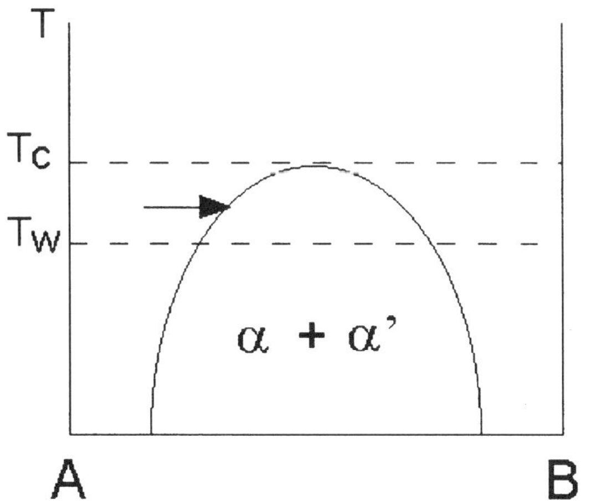

FIGURE IV.B.2 Miscibility gap with critical and wetting temperatures T_c and T_W.

layer diverges to become a bulk liquid layer that completely wets the solid at the melting point. Premelting has been confirmed in several systems. (2) A theory of Cahn (1977), now confirmed in several systems, shows that in a binary system with a critical temperature T_c, as shown schematically in Figure IV.B.2, there is a temperature $T_W < T_c$ in the two-phase region above which one of the two phases (say α, rich in A) will be completely wetted by the other (α′, rich in phase B). Further, as the two-phase field is approached from the single-phase A side (shown by the arrow), an adsorbed layer of B on the A phase occurs and increases in thickness until at the two-phase boundary, wetting of α by α′ occurs.

Research Issues

Technologies that depend on wetting generally require good or complete wetting of a fluid on solid surfaces (i.e., low or zero contact angle). Based on this requirement and the preceding discussion, areas that would profit from research are the following: (1) the hysteresis effect, which can inhibit the spread of the wetting liquid, (2) the dynamics of wetting, which determines the rapidity with which wetting or rewetting will take place, (3) the description of wetting in porous materials when the scale of the pores is comparable to or smaller than the correlation length in the liquid, so that a bulk description of the liquid is no longer appropriate, and (4) development of the molecular theory of wetting, which would enlarge the knowledge base on material combinations and conditions for good wetting and on wetting or tensioactive agents (Schrader and Loeb, 1992; Eustathopoulos et al., 1998).

Marangoni Effect

The Marangoni effect (Hondros, 1998; Antar and Nuotio-Antar, 1993; Legros et al., 1987, 1990; Ostrach, 1982) commonly refers to liquid convection caused by surface tension gradients at the free surface of a liquid or at the interface between two liquids. Because this dependence of surface tension on position arises in the presence of temperature or composition gradients along the surface when the surface tension depends on the temperature and/or on composition, the effect may also be called the thermosolutal, thermocapillary, or solutocapillary effect. When the surface tension is a function of position along a surface or interface, there is a resultant force on an element of the surface or interface. Since the net force on the element must vanish to avoid essentially infinite acceleration of the atomically thin element, a velocity gradient perpendicular to the surface is generated that supplies the counterbalancing viscous force. The result is a thermosolutal-induced convection in the fluid. The Marangoni effect occurs in the absence of gravity and is a dominant cause of convection in microgravity. A special publication of the Philosophical Transactions of the Royal Society, London (Hondros et al., 1998) is devoted entirely to the Marangoni effect and provides a comprehensive look at what is known about this phenomenon, especially as it affects materials processing. It has been suggested (Ostrach, 1977a,b, 1982) that, for historical reasons, the term Marangoni instability should be used to describe those effects resulting from gradients initially normal to an interface. Terms such as thermocapillary and solutocapillary can be applied to effects resulting from gradients of the corresponding variables initially parallel to the interface. However, the term Marangoni effect (or flow) is commonly applied to the generality of flows induced by any gradient that produces a variation of surface tension with position on the interface, whether directly or through the effect of perturbations, and this is the usage that has been adopted in this report.

As an example of the Marangoni effect, consider a single component liquid with a liquid/vapor surface lying in the xy plane with a surface tension σ. Suppose there is a temperature gradient $\partial T/\partial x$ at the surface. Then, the force on a surface element in the x direction per unit length along y is $\partial\sigma/\partial x = (\partial\sigma/\partial T)(\partial T/\partial x)$. To balance this, there must be a velocity gradient in the z direction, normal to the surface, of magnitude given by $\mu\partial v/\partial z = (\partial\sigma/\partial T)(\partial T/\partial x)$, where μ is the coefficient of viscosity. This velocity gradient serves as a boundary condition that generates a convection pattern in the fluid. It is clear from this simple expression that the induced velocity gradient is proportional to the temperature gradient ($\partial T/\partial x$) and inversely proportional to the viscosity. The actual flow in any given system may be very complicated because the convection affects the surface gradients of temperature and composition which, in turn, generate the convection. This coupling of Marangoni-induced convection with the surface tension via the surface temperature and composition can lead to oscillatory flows (Verlarde, 1998). In the case of a liquid/solid interface, there is no Marangoni effect since the solid exerts forces that balance interface tension gradients (and cause a zero velocity at the interface).

In general, the dependence of σ on position arises from both temperature gradients and composition gradients (the latter is exemplified by the tearing of wine in a glass), since σ generally depends on both variables; often the effects are mixed, as in the combustion or gasification of liquid drops (Zhang et al., 1996; Aharon and Shaw, 1996). In a multicomponent system, temperature gradients may have both a direct effect on σ and an indirect effect through the temperature dependence of the surface composition. Thus, whereas $\partial\sigma/\partial T$ is negative for a pure component, it may be positive in a multicomponent system, because as T increases, σ may increase if the adsorption of a surface-active component decreases sufficiently; a notable example is sulfur in stainless steel above 40 ppm (Mills et al., 1998). Thus, in multicomponent systems, the Marangoni effect can have either sign (defined as the sign of $\partial\sigma/\partial T$).

The Marangoni effect can lead to the migration of drops and bubbles (Verlarde, 1998). For example, consider a liquid drop with $\partial\sigma/\partial T < 0$. If it is placed on a plate with a temperature gradient, it will move toward the cold end. If it is suspended in a fluid with a temperature gradient, however, it will move toward the hot end (assuming the same sign for the Marangoni coefficient); in the latter case, the return flow along the drop's center line pushes the nose of the drop forward toward the hot end. Similar phenomena happen with bubbles and can have a strong effect on pool and forced convective boiling heat transfer.

Marangoni convection usually dominates gravity-induced convection in weld pools in Earth gravity; it is undiminished in microgravity. When the sign of the effect is negative, the liquid surface is pulled toward the cooler outer edges of the pool, and when it is positive, the reverse is true. The shape of the weld pool is affected (Mills et al., 1998). The Marangoni effect dominates gravity-induced convection in the Bénard instability arising from heating a liquid layer from below, provided the layer is not too deep. The threshold ΔT required to initiate the flow that was calculated by Rayleigh based on gravity-induced convection is 10^4 to 10^5 times larger than that found experimentally by Bénard (Legros et al., 1987); the discrepancy was attributed to the Marangoni effect, which causes warm liquid rising to the top center of a cell to be pulled outward by surface tension to the cooler cell edges, where it then sinks. As the gravity level is reduced, Marangoni convection increasingly dominates gravity-induced convection.

The relative strengths of Marangoni convection and gravity-induced convection are quantified by the dimensionless Marangoni (Ma) and Rayleigh (Ra) numbers:

$$Ma = \sigma_T \beta L^2/\rho \nu k$$

$$Ra = g\alpha\Delta T L^3/\nu k$$

(Legros et al., 1990; Verlarde, 1998), where, in the first formula, $\sigma_T = \partial\sigma/\partial T$, $\beta = \nabla T$, ρ is the density, ν is the kinematic viscosity, k is the thermal diffusivity, and L is a characteristic distance, and in the second formula, g is the gravitational constant, α is the thermal expansion coefficient, ΔT is a temperature difference over a characteristic distance L, and ν and k are defined as above. To initiate surface-tension-induced Bénard flow, that is, flow due to Marangoni instability, a critical value of $Ma \approx 80$ is required, and to initiate buoyancy-induced Bénard flow, a critical vale of $Ra \approx 1{,}100$ is required. Other characteristics of the flow (e.g., oscillatory and turbulent) are determined by these and other dimensionless numbers of fluid dynamics.

As indicated by the above examples, Marangoni effects are ubiquitous wherever liquid/fluid interfaces are subject to temperature and composition gradients. The effects become dominant in reduced gravity, as in the stirring of a weld pool, the migration of liquid in spills, fire control, and two-phase fluid transport, and in capillary-operated devices such as heat pipes or capillary pumped loops (referred to in the subsection on wetting). Moreover, multicomponent mixtures may have significantly higher critical heat fluxes than occur for single-component boiling, and this observed increase in critical heat flux is apparently due to the Marangoni-induced flows.

Research Issues

Based on the previous discussion, the research areas of particular relevance are (1) the modeling and experimental study of the complex convection flows induced by the Marangoni effect, (2) the experimental determination of the parameters that enter into the Marangoni and other relevant dimensionless numbers (Egry et al., 1998), (3) investigation of tensioactive agents (Eustathopoulos et al., 1998; Verlarde, 1998) to control the magnitude and sign of the effect, and (4) inclusion of thermal and concentration gradients to assess the merits of designs where the Marangoni-induced flow of fluids can be useful or detrimental.

References

Adamson, A.W. 1982. Physical Chemistry of Surfaces, 4th Ed. New York: Interscience.
Aharon, I., and B.D. Shaw. 1996. Phys. Fluids 8:1820.
Antar, B.N., and V.S. Nuotio-Antar. 1993. Liquid gas capillary surfaces. Fundamentals of Low Gravity Fluid Dynamics and Heat Transfer. Boca Raton, Fla.: CRC Press.
Cahn, J.W. 1977. J. Chem. Phys. 66:3667.
Carter, W.C. 1988, The forces and behavior of fluids constrained by solids. Acta Metall. 36(8):2283.
Concus, P., and R. Flinn. 1990. Capillary surfaces in microgravity. P. 183 in Low-Gravity Fluid Dynamics and Transport Phenomena: Progress in Astronautics and Aeronautics, Vol. 130. J.N. Koster and R.L. Sani, eds. New York: American Institute of Aeronautics and Astronautics.
de Gennes, P.G. 1985. Wetting: statics and dynamics. Rev. Mod. Phys. 57:827.
Decker, E., and S. Garoff. 1997. The need for new experimental and theoretical models. J. Adhes. 63:159.
Dietrich, S. 1988. P. 1 in Phase Transitions and Critical Phenomena, Vol. XII. C. Domb and J.L. Lebowitz, eds. New York: Academic Press.
Dodge, F.T. 1990. Fluid management in low gravity. P. 369 in Low Gravity Fluid Dynamics and Transport Phenomena: Progress in Astronautics and Aeronautics, Vol. 130. J.N. Koster and R.L. Sani, eds. New York: American Institute of Aeronautics and Astronautics.
Eggers, J. 1997. Nonlinear dynamics and break-up of free-surface flows. Rev. Mod. Phys. 69:865-929.
Egry, I., M. Langen, and G. Lohöfer. 1998. Measurements of thermophysical properties of liquid metals relevant to Marangoni effects. Philos. Trans. R. Soc. London, Ser. A 356(1739):845.
Eustathopoulos, N., J.P. Garandet, and B. Drevet. 1998. Influence of reactive solute transport on spreading kinetics of alloy droplets on ceramic surfaces. Philos. Trans. R. Soc. London, Ser. A 356(1739):871.
Faghri, A. 1995. Heat Pipe Science and Technology. Washington, D.C.: Taylor and Francis.
Faghri, A. 1999. Recent advances in heat pipe analysis and simulation. Annual Review of Heat Transfer, Vol. 8. C.-L. Tien, ed. New York: Begell House.
Findenegg, G.H., and M.M. Telo de Gama. 1987. Wetting and adsorption phenomena. P. 191 in Fluid Sciences and Materials Science in Space. H.U. Walter, ed. New York: Springer-Verlag.
Frenken, J.W.M., and J.F. van der Veen. 1985. Phys. Rev. Lett. 54:134.

German, R.M., R.G. Iacocca, J.L. Johnson, Y. Liu, and A. Upadhyaya. 1995. Liquid-phase sintering under microgravity conditions. J. Met. 47(8):46-48.

Gibbs, J.W. 1878. On the equilibrium of heterogeneous substances. Republished (1961) in The Scientific Papers of J. Willard Gibbs, Vol. 1. Mineola, N.Y.: Dover, p. 55.

Haynes, J.M., and D. Langbein. 1987. Fluid statics and capillarity. P. 53 in Fluid Sciences and Materials Science in Space. H.U. Walter, ed. New York: Springer-Verlag.

Hondros, E.D. 1998. Introduction: significance of capillary driven flows in materials processing. Philos. Trans. R. Soc. London, Ser. A 356(1739):815.

Hondros, E.D., M. McLean, and K.C. Mills, eds. 1998. Marangoni and interfacial phenomena in materials processing. Philos. Trans. R. Soc. London, Ser. A 356(1739):811-1061.

Koster, J.N. 1990. P. 369 in Low-Gravity Fluid Dynamics and Transport Phenomena: Progress in Astronautics and Aeronautics, Vol. 130. J.N. Koster and R.L. Sani, eds. New York: American Institute of Aeronautics and Astronautics.

Legros, J.C., A. Sanfeld, and M. Verlarde. 1987. Fluid dynamics. P. 109 in Fluid Sciences and Materials Science in Space. H.U. Walter, ed. New York: Springer-Verlag.

Legros, J.C., O. Dupont, P. Queeckers, and S. Van Vaerenbergh. 1990. Thermohydrodynamic instabilities and capillary flows. P. 207 in Low-Gravity Fluid Dynamics and Transport Phenomena: Progress in Astronautics and Aeronautics, Vol. 130. J.N. Koster and R.L. Sani, eds. New York: American Institute of Aeronautics and Astronautics.

Marsh, J.A., S. Garoff, and E.B. Dussan. 1993. Dynamic contact angles and hydrodynamics near a moving contact line. Phys. Rev. Lett. 70:2778.

Martinez, I., J.M. Haynes, and D. Langbein. 1987. P. 53 in Fluid Sciences and Materials Science in Space. H.U. Walter, ed. New York: Springer-Verlag.

Maxwell, G.J.C. 1878. Capillary action. Encyclopedia Britannica, 9th Ed. New York: Encyclopedia Britannica.

Mills, K.C., B.J. Keene, R.F. Brooks, and A. Shirali. 1998. Marangoni effects in welding. Philos. Trans. R. Soc. London, Ser. A 356(1739):911.

Nance, M., and J.E. Jones. 1993. Welding in space and low-gravity environments. P. 1020 in ASM Handbook, Vol. 6: Welding, Brazing and Soldering. Metals Park, Ohio: ASM International.

Ostrach, S. 1977a. Motion induced by capillarity. Pp. 571-589 in Physicochemical Hydrodynamics: V.G. Levich Festschrift, Vol. 2. London: Advanced Publications.

Ostrach, S. 1977b. Convection due to surface-tension gradients. Pp. 563-570 in Committee on Space Research (COSPAR) Advances in Space Research, Vol. 19. M.J. Rycroft, ed. Oxford and New York: Pergamon.

Ostrach, S.A. 1982. Low-gravity flows. Annu. Rev. Fluid Mech. 14:313.

Peterson, G.G., L.W. Swanson, and F.M. Gerner. 1998. Micro heat pipes. Pp. 295-337 in Microscale Energy Transport. C.-L. Tien, A. Majumdar, and F.M. Gerner, eds. New York and Philadelphia: Taylor and Francis.

Ramé, E., and S. Garoff. 1996. Microscopic and macroscopic dynamic interface shapes and the interpretation of dynamic contact angles. J. Colloid Interface Sci. 177:234.

Schrader, M.E., and G.L. Loeb, eds. 1992. Modern Approaches to Wettability. New York: Plenum Press.

Stone, H.A. 1994. Dynamics of drop deformation and breakup of viscous fluids. Annu. Rev. Fluid Mech. 26:26-65.

Verlarde, M.G. 1998. Drops, liquid layers and the Marangoni effect. Philos. Trans. R. Soc. London, Ser. A 356(1739):829.

Westbye, C.S., M. Kawaji, and B.N. Antar. 1995. Boiling heat transfer in the quenching of a hot tube under microgravity. J. Thermophys. Heat Transfer 9:302.

Wiltzius, P., S.B. Dierker, and B.S. Dennis. 1989. Wetting and random-field transition of binary liquids in a porous medium. Phys. Rev. Lett. 62(7):804.

Zhang, B.L., J.M. Card, and F.A. Williams. 1996. Combust. Flame 105:267.

Zhang, W., and J. Vinals. 1997. Pattern formation in weakly damped Faraday waves. J. Fluid Mech. 336:301.

IV.C MULTIPHASE FLOW

Both single and multiphase[3] fluid flows may be used in microgravity environments. While single-phase flows can behave somewhat differently in space (owing, for example, to the absence of natural-convection-induced flows), they can be reliably calculated in most cases.

The microgravity issues associated with single-phase flows are primarily those related to heat-transfer-induced density changes. These phenomena are discussed in Section IV.D, which is concerned with heat transfer phenomena. Multiphase flows are inherently more complicated than single-phase flows. Because of differences in phasic density and inertia, multiphase flows may exhibit pronounced phase separation and distribution. Indeed,

[3]The simultaneous flow of several phases or components, for example, vapor/liquid flows and solid/fluid flows.

multiphase flows normally configure themselves into distinctive flow regimes in which the various phases are nonuniformly distributed across the duct through which they flow. Significantly, the modeling of turbulence, pressure drop, heat transfer, and stability must take flow regimes into consideration. Moreover, surface-tension-induced (e.g., Marangoni) forces and surface phenomena are likely to be much more important in space than they are on Earth. It should be noted that virtually all flow-regime-specific phenomena will be influenced by gravity level during both normal forced-flow operating conditions and various postulated accidents.

Although NASA has so far been able to design around the need for the extensive use of multiphase systems and processes in its missions, the agency is well aware of the fact that there are numerous issues and concerns facing space system designers attempting to utilize multiphase flow and heat transfer processes and systems in space (McQuillen, 1999). However, it is inevitable that such systems and processes will be needed for HEDS proposed interplanetary missions and extraterrestrial colonies. Examples, which are discussed in Chapter III, include Rankine cycle power plants, boilers, evaporators, condensers, multiphase thermal buses, electrolysis units, and various other life support and materials processing systems for which performance and weight are important design considerations. In addition, there is a need to better understand multiphase particulate/fluid systems (e.g., dust transport and deposition) in reduced gravity if extraterrestrial colonies are to be established.

Unfortunately the current state of knowledge concerning how multiphase systems behave in reduced and microgravity environments is inadequate to support NASA's proposed missions and goals. Indeed, if one is to have confidence in the performance of multiphase systems and processes in space and at extraterrestrial sites, then a well-coordinated research and development effort will be required to provide the needed mission-enabling technology.

To understand why this effort is so crucial, it is important to note that multiphase flow and heat transfer technology is a mature one that has been widely used on Earth during the last century. Nevertheless, it is a field that has been empirically based. Unfortunately, many of the design rules and correlations that are valid on Earth are invalid for microgravity applications. One reason for this is that the flow regimes (i.e., how the phases configure) are quite different on Earth and in space. Moreover, natural circulation and buoyancy are suppressed in space while they play an important role in the performance of many multiphase systems on Earth.

It will not be economically viable to develop multiphase systems for space in the same way as these systems have been developed on Earth. That is, it will not be practical to test different configurations in space until an acceptable design is achieved. Rather, physically based analytical models (which can be used in computer codes for design purposes) should be developed to take into account all relevant aspects of reduced-gravity phenomena. These analytical models could then be used to optimize designs and scale up small scale microgravity data to full scale.

Phase Separation and Distribution

It is well known (Heppner et al., 1975) that buoyancy plays an important role in the phase distributions that lead to the development of the various flow regimes and that, as shown in Figure IV.C.1, the regimes can be very different in microgravity environments.

This subsection focuses on some multiphase flow phenomena that are considered to be of vital importance to the HEDS technologies. Many important and challenging problems in multiphase flow and heat transfer have to do with multidimensional (i.e., three-dimensional) phenomena, in particular, phase separation and distribution phenomena.

When a flowing multiphase mixture (vapor/liquid or solid/fluid) changes direction or is subject to other types of accelerations, the phases may separate nonuniformly. This is because of the relatively large differences in the inertia of each phase for sufficiently large dispersed particles or bubbles. A good example of this phenomenon can be seen in Figure IV.C.2, which illustrates phase separation for a bubbly gas/liquid mixture in a piping Tee (Hwang et al., 1988).

Because the gas phase has a lower density, and thus a lower momentum flux, than the liquid phase, it has an easier time changing direction from the inlet section into the side branch of the Tee. Thus the position of the dividing streamline for the gas (δ_G) is farther into the incoming stream than that for the liquid phase (δ_L). Hence

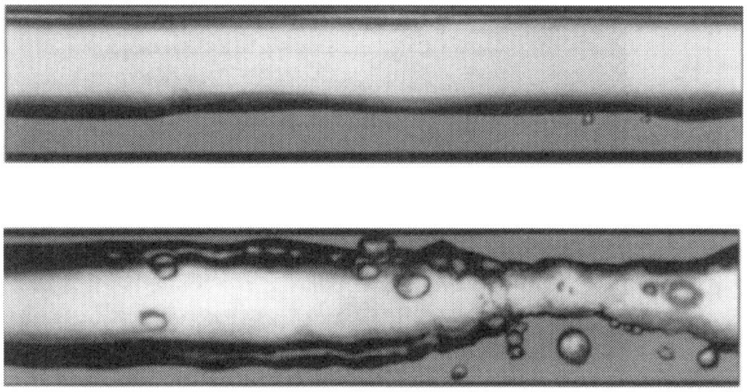

FIGURE IV.C.1 The effect of gravity on flow regimes. Image courtesy of NASA.

the zone of influence of the gas is larger than that of the liquid phase. As a consequence, a preferential phase separation of gas into the side branch occurs such that the volume fraction of gas is higher in the side branch and lower in the run (downstream of the branch) than it was at the inlet of the Tee.

Similar phase separation phenomena may occur when there is a change in direction or acceleration of a multiphase mixture. In particular, phase separation should be expected in any phase change equipment that is geometrically complex (e.g., plena).

One important source of the acceleration force on a multiphase mixture is gravity. Because of gravity, pronounced phase separation regularly occurs in horizontal ducts on Earth, in which the heavier phase will concentrate and flow in the lower part of the duct. In contrast, gravitational phase separation will not occur in microgravity environments. This, of course, will have a pronounced effect on the flow regime (i.e., the phasic configuration) and thus on the overall behavior of the flow.

Reliable multidimensional models of multiphase flow will be required to predict phase separation in microgravity and reduced-gravity environments. Fortunately, models of this type have recently been developed and used on Earth. In particular, physically based, multidimensional two-fluid models have been developed (Drew and Passman, 1998) by ensemble-averaging the conservation equations of mass, momentum, and energy, and these models were numerically evaluated using computational fluid dynamics (CFD) techniques (Lahey and Drew, 2000). It appears that multidimensional CFD models of this type can also be developed for microgravity conditions. However, this will require carrying out parametric and separate-effects experiments in reduced-gravity environments to develop the constitutive relations required for closure of the two-fluid model.

It should be stressed that NASA must be able to reliably calculate phase separation if multiphase systems and processes are to be used in space, because, as noted previously, significant phase separation may occur in multiphase systems (e.g., in pipes, boilers, and condensers) every time the flow accelerates or changes direction.

It is also important to note in Figure IV.C.2 that the gas, which is extracted through the side branch of the Tee, depends on the inlet phase distribution. That is, it depends on the phasic distribution within the zones of influence. It has been known for some time that pronounced lateral phase distribution may occur on Earth (Serizawa, 1974) and in microgravity multiphase conduit flows (Heppner et al., 1975; Colin et al., 1991; Bousman and Dukler, 1993; McQuillen et al., 1998).

Figure IV.C.3 shows a void fraction (i.e., the gas volume fraction, (α) distribution measured on Earth by Serizawa (1974) for dispensed air/water upflow in a pipe. It can be seen that the gas concentrates near the wall ($r/R = \pm 1$) of the pipe for low inlet gas flows (i.e., qualities, <x>) but near the pipe's center line for higher inlet gas flows. Phase distribution phenomena of this type have been reported for many conditions, including laminar flows (Antal et al., 1991) and flows in complex geometry conduits (Lopez de Bertodano et al., 1990).

Lateral phase distribution occurs because of the various lateral forces on the dispersed (i.e., noncontinuous)

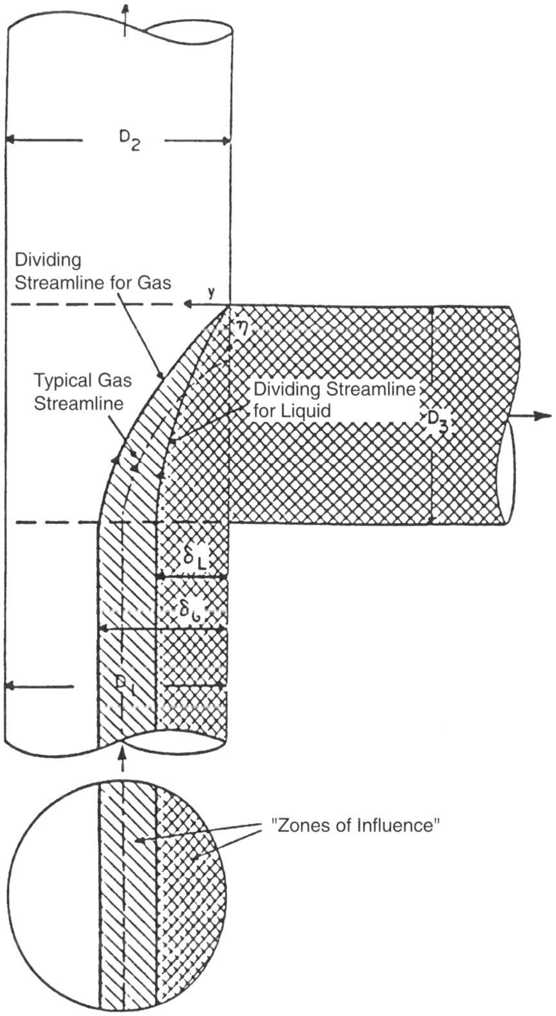

FIGURE IV.C.2 Phase separation in a piping Tee. SOURCE: Lahey (1992). Reprinted with permission from Elsevier Science Publishers.

and continuous phases. This phenomenon is significant because it can lead to flow regime transition, and the thermal-hydraulic characteristics (e.g., pressure drop and heat transfer) of a flowing multiphase mixture are strongly dependent on flow regime and, thus, gravity (Dukler et al., 1988; Jayawardena et al., 1997; Chen et al., 1991; Zaho and Rezkallah, 1993; Bousman et al., 1996; McQuillen et al., 1998).

In principle it is possible to analyze multiphase flows using Lagrangian (interface tracking) volume of fluid or level set numerical formulations, or using direct numerical simulations. However, predictions of this type are prohibitively time-consuming and expensive. The current state of the art is CFD analysis of multiphase systems based on Eulerian or Eulerian/Lagrangian formulations of multidimensional two-fluid models (Lahey, 1992).

The interfacial transfer laws required for closure or two-fluid models must be flow-regime-specific and can be derived using a combination of analytical and numerical models and suitable experiments. In particular, both separate-effects experiments, which isolate the various important physical phenomena (e.g., lift and drag), and integral experiments, in which various phenomena interact, will be required to properly assess and verify the multidimensional two-fluid model. Reduced-scale experiments of this type will need to be performed in micro-gravity to assure their relevance. The resulting two-fluid model can be evaluated using CFD techniques. Indeed, this approach is now widely used for the analysis of single and multiphase flows on Earth.

Significant progress has been made on the ability to predict phase distribution phenomena using multidimen-

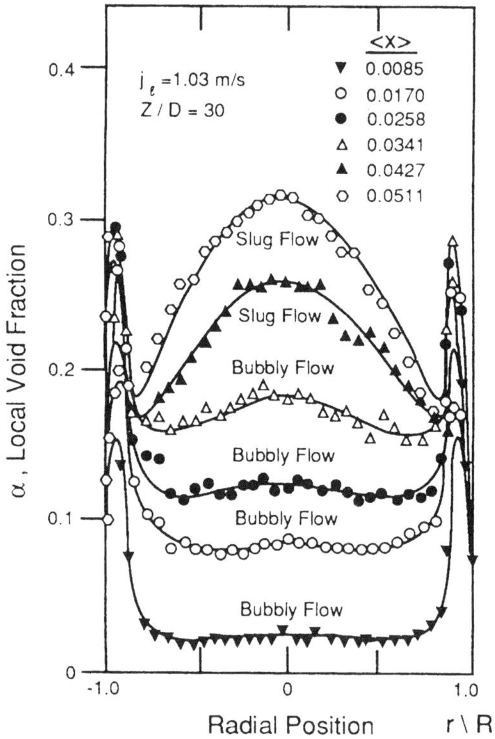

FIGURE IV.C.3 Radial void distribution for 1 g_0 and bubbly upflow (Serizawa, 1974), where α is the local void fraction and r/R is the relative radial position. SOURCE: Lahey (1992). Reprinted with permission from Elsevier Science.

sional, two-fluid CFD models (Lahey and Drew, 2000). This has required the analytical modeling of flow-regime-specific interfacial transfer laws (e.g., lift and virtual mass) and multiphase turbulence. Moreover, detailed experiments were required on Earth for the development and assessment of these flow-regime-dependent models. These efforts resulted in the development by the U.S. Department of Energy-Naval Reactors program of a multidimensional, four-field, two-fluid CFD model, which has been extensively used by the U.S. Navy. While these computational models are not directly applicable to reduced gravity, they can be extended for this application once interfacial closure laws are developed for microgravity that include all relevant physical phenomena.

Developing models will require a better understanding of multiphase fluid mechanics and of interfacial mass, momentum, and energy transfer mechanisms in reduced gravity environments. Proper quantification of these models, through experiments and analysis, should give NASA the ability to reliably calculate phase distribution and separation, as well as phasic mixing, in simple and complex geometries. In particular, it should be possible to predict the various flow regimes, as well as multiphase pressure drop, flow regime transitions, and boiling/condensing heat transfer (Lahey, 1996).

To better appreciate the current state of the art, let us consider some comparisons of a multidimensional, two-fluid model with some multiphase data taken on Earth. Because of its importance in many commercial processes, the bubbly flow regime has been chosen to demonstrate the model's predictive capabilities. (However, similar predictions are possible for other flow regimes.) As can be seen in Figures IV.C.4 through IV.C.8, a properly formulated three-dimensional, two-fluid CFD model is capable of predicting a wide variety of adiabatic multiphase flows on Earth. In particular, Figures IV.C.4 through IV.C.6 show that for bubbly upflow in a pipe, the local void fraction, turbulent intensities, and Reynolds stress (i.e., turbulent shear stress) are well predicted. Similarly, Figure IV.C.7 shows that the same model can also predict bubbly downflows in a pipe in which the void fraction distribution is completely different from that for upflow (see Figure IV.C.4). Similarly, Figure IV.C.8 shows that the same model also gives excellent predictions for bubbly upflow in complex geometry conduits (in this case, an isosceles triangle).

Moreover, as can be seen in Figures IV.C.9 and IV.C.10, when the condition of the boiling surface (i.e., the nucleation site density) is properly characterized, Freon (R-113) and SUVA (R-134a) subcooled boiling data are

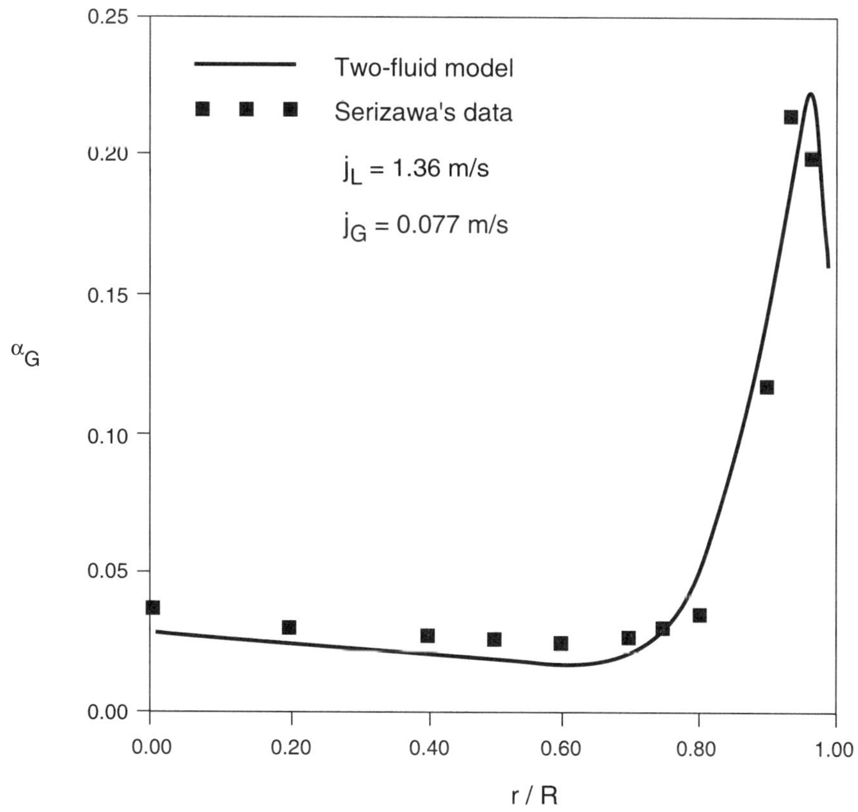

FIGURE IV.C.4 Comparison of model predictions with Serizawa's bubbly upflow data (1974): void fraction distribution at 1 g_0 (α_G is the local void fraction and r/R is the relative radial position). SOURCE: Lahey and Drew (2000, in press). Courtesy of R.T. Lahey, Jr.

also well predicted using the same multidimensional, two-fluid CFD model. The results shown in Figure IV.C.10 are particularly significant since they show that a properly formulated two-fluid model can accurately predict two-phase pressure drop and the developing phase distributions for several different flow regimes (Lahey, 1996). That is, the heated channel was, in effect, a "once-through evaporator" in which subcooled liquid entered and saturated annular flow exited.

It should be noted that all of the above examples are for conditions of forced (i.e., pumped) flows in which buoyancy-driven natural circulation plays no role. It is significant that pronounced phase distribution is seen for forced flow conditions, since only multiphase flows of this type are suitable for microgravity conditions (that is, as discussed previously, natural circulation will not be effective in microgravity environments).

The applications for these predictive capabilities are extensive, ranging from the design and analysis of boilers, evaporators, condensers, reprocessing systems, electrolysis units, Rankine cycle power plants, and propulsion systems. Unfortunately, very little research of the type needed to support the development of multiphase CFD predictive capabilities for reduced and/or microgravity applications has been conducted by NASA in the past.

Other Multiphase Phenomena

It should be noted that some of the phenomena needed in the closure laws (e.g., surface phenomena such as wetting and surface-tension-induced forces) will also have important applications in the analysis of other pro-

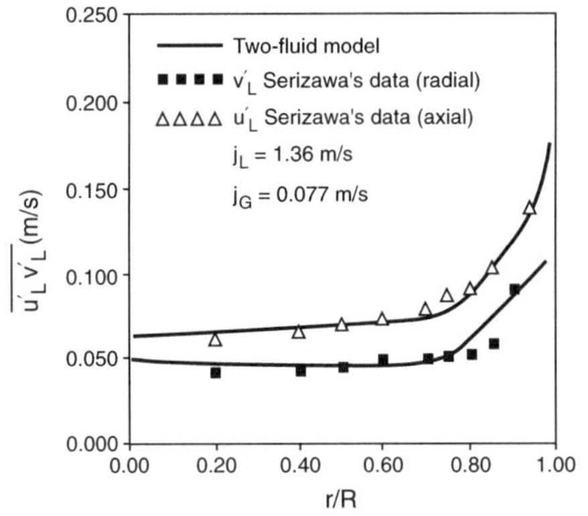

FIGURE IV.C.5 Comparison of model predictions with Serizawa's bubbly upflow data (1974): velocity fluctuation distributions at 1 g_0 ($\overline{u'_L}$ and $\overline{v'_L}$ are the local liquid turbulent fluctuations in the axial and radial directions, respectively, and r/R is the relative radial position). SOURCE: Lahey and Drew (2000, in press). Courtesy of R.T. Lahey, Jr.

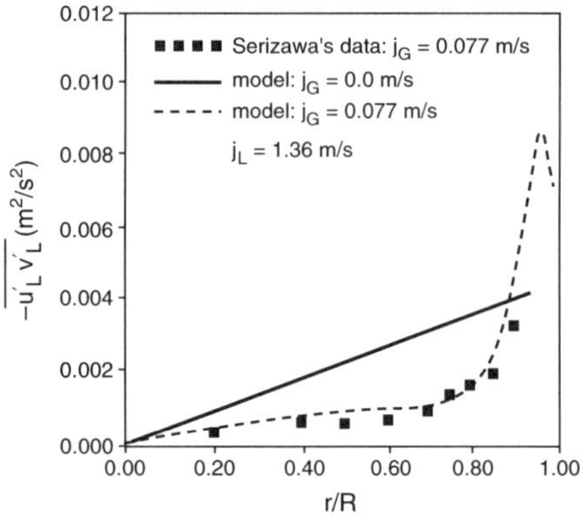

FIGURE IV.C.6 Comparison of model predictions with Serizawa's bubbly upflow data (1974): Reynolds stress distribution at 1 g_0 ($\overline{u'_L v'_L}$ is the local liquid Reynolds stress and r/R is the relative radial position). SOURCE: Lahey and Drew (2000, in press). Courtesy of R.T. Lahey, Jr.

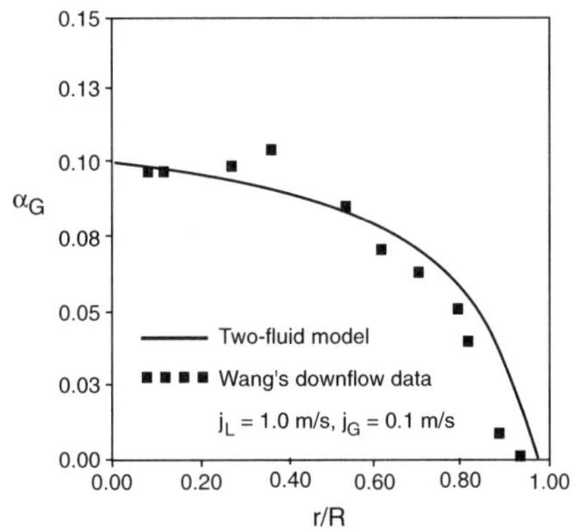

FIGURE IV.C.7 Comparison of model predictions with Wang's bubbly downflow data (Wang et al., 1987): void fraction distribution at 1 g_0 (α_G is the local void fraction and r/R is the relative radial position). SOURCE: Lahey and Drew (2000, in press). Courtesy of R.T. Lahey, Jr.

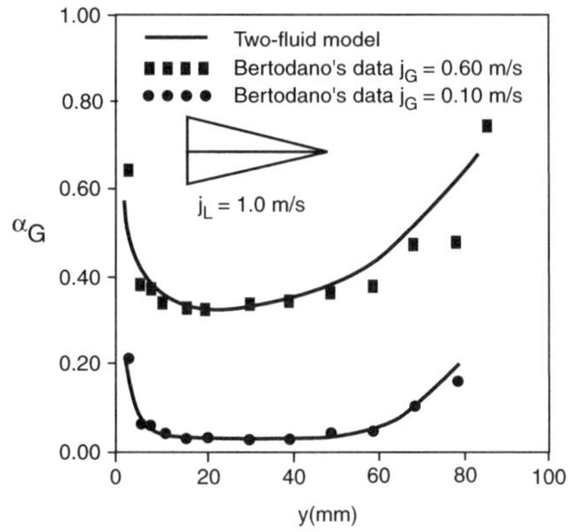

FIGURE IV.C.8 Comparison of model predictions with Lopez de Bertodano's bubbly upflow data (1993): void fraction distribution at 1 g_0 (α_G is the local void fraction and y is the lateral position). SOURCE: Lahey and Drew (2000, in press). Courtesy of R.T. Lahey, Jr.

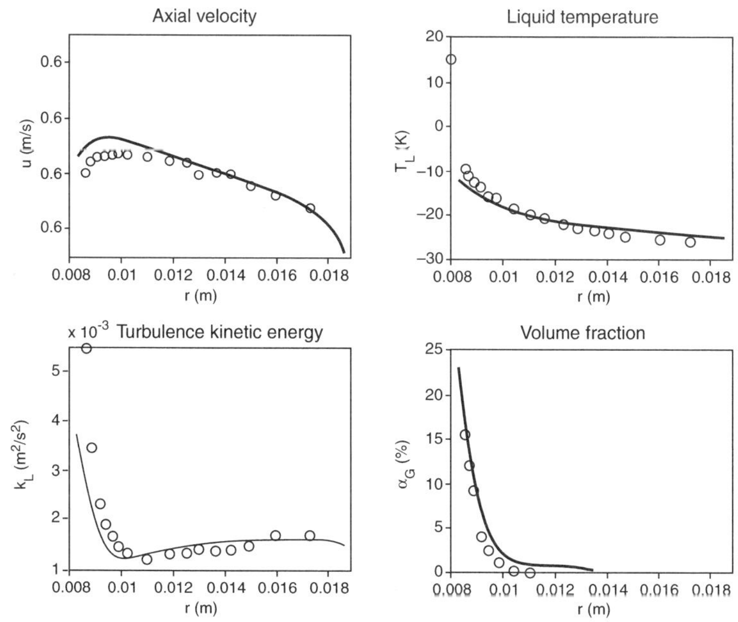

FIGURE IV.C.9 Predictions for subcooled boiling R-113 upflow in an annulus at 1 g_0, with data of Velindandla et al. (1995) (u is the local liquid velocity, T_L is the local liquid temperature, α_G is the local void fraction, k_L is the local turbulent kinetic energy of the liquid phase, and r is the radial position). SOURCE: Lahey (1996). Reprinted with permission from Begell House.

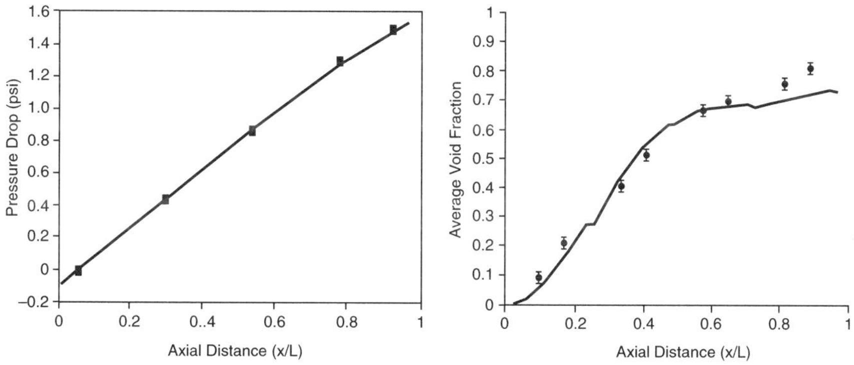

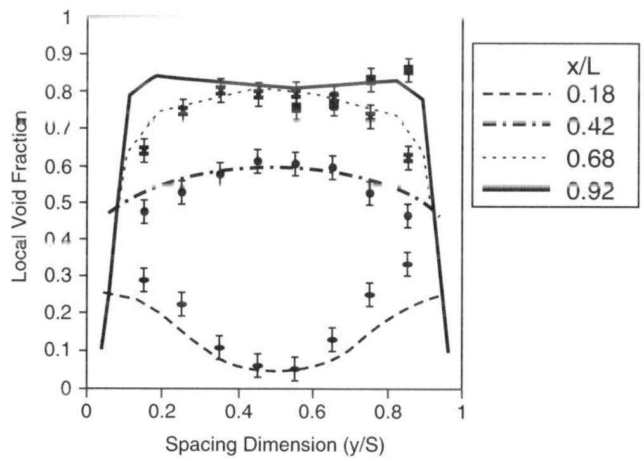

FIGURE IV.C.10 Predictions of subcooled boiling SUVA upflow between heated parallel plates at 1 g_0. SOURCE: Lahey (1996). Reprinted with permission from Begell House.

cesses, such as welding, that will be needed in space. These phenomena are discussed at greater length in Section IV.B.

While this discussion has focused on vapor/liquid systems, it should be mentioned that similar phenomena have been seen in solid/fluid systems (Alajbegovic et al., 1994). Indeed, multidimensional two-fluid CFD models have been developed and successfully applied to solid/fluid flows (Alajbegovic et al., 1999). Moreover, a similar CFD approach that includes aerosol physics (e.g., coagulation and thermophoresis) could also be used. Thus, if desired, multidimensional aerosol (e.g., smoke) and dust transport and deposition models could be developed for space and extraterrestrial application. The theory of aerosol transport and deposition is a fairly mature one (Fuchs, 1964), and numerous aerosol transport codes have been developed, benchmarked against experiments, and applied to the transport and deposition of aerosols (e.g., Croff, 1980). Missions to distant moons, asteroids, or planets such as Mars will require that we develop more soil-specific knowledge of the particulate mechanics associated with dust transport/deposition, filtration, and accumulation, as well as techniques for the cleaning of vital life support systems (e.g., solar cells).

In other cases, multidimensional, multiphase CFD models could also be developed and used to analyze the flow of solid/fluid systems, including models for particle transport and deposition, regolith transport and processing, particulate flow in hoppers, and flow through porous media. However, the appropriate closure laws will be different from those used for vapor/liquid systems (i.e., particle-to-particle interactions will be very significant), and appropriate particle transport experiments will be needed to establish flow behavior in reduced gravity and for atmospheric conditions relevant to the extraterrestrial site.

In other applications—for example those associated with powder metallurgy processes, direct metal deposition techniques for rapid, one-of-a-kind manufacture in space, and for some pyrolysis processes—ad hoc phenomenological model development and supporting experiments may be sufficient. In any event, it appears that much of the research that will be needed can be done on Earth, and NASA may be able to take advantage of the ongoing particulate mechanics research programs being sponsored by the National Science Foundation.

Research Issues

In summary, significant phase separation and distribution should be expected when multiphase systems are used in space. Fortunately, it appears that the multidimensional, multiphase CFD models that have been developed for vapor/liquid flows on Earth can be extended to accommodate reduced-gravity environments. To accomplish this, however, a well-focused research program would be required to obtain a better understanding of multiphase phenomena in reduced gravities, to build a database for model assessment, and to establish the roles of various dimensionless parameters involving gravity.

Mixing

Chaotic mixing may occur in both single- and two-phase fluids as a result of turbulence (Ottino, 1990; Lahey and Drew, 2000). In addition, fluid/fluid mixing may be used to produce materials that have special desirable properties.

In particular, plastic blends with novel fibrous or multilayer film microstructures have been produced recently by inducing chaotic motion within multiphase melts (Zumbrunnen, 1998), and it is unclear if they will be affected by gravity. The microstructures have been associated with enhancements in physical properties such as toughness, strength, and electrical conductivity. Owing to the repeated stretching and folding inherent in chaotic motion, initially large minor-phase bodies are eventually transformed into films, which may then subdivide into fibers owing to interfacial instabilities. Studies of chaotic mixing have been done by Ottino (1990) and Ottino et al. (1992). Similar microstructures might also be formed in metallic systems having immiscibility gaps in phase diagrams at elevated temperatures. However, on Earth, phase separation due to gravity has prevented attempts to create these microstructures in metallic systems. Results with plastics demonstrate that it may be possible to create metallic alloys with fibrous or multilayer film microstructures if processing is performed in a microgravity environment.

It is well known that during the flow of mixtures of immiscible liquids (Nadler and Mewes, 1997), droplet entrainment and mixing may cause emulsification of the continuous phase, which leads to significantly higher viscosities and pressure drops. There has been virtually no research on these important phenomena in microgravity environments.

Single-phase turbulence is a vast field that has occupied the attention of many outstanding scientists and engineers for over 100 years. As a consequence, modern single-phase CFD codes regularly involve the computation of turbulence phenomena using so-called k-ε and τ–ε models, where k is the turbulent kinetic energy, ε is the dissipation rate, and τ are the Reynolds stresses (Wilcox, 1998). Moreover, models are available that allow large eddy simulation and direct numerical simulation of single-phase turbulence, although analysis of this type is currently time-consuming and very expensive (Wilcox, 1998).

In contrast, understanding of turbulence phenomena in multiphase flows is much less advanced. Nevertheless, it is known (Theofanous and Sullivan, 1982; Lance and Bataille, 1991; Lopez de Bertodano et al., 1994a) that for dilute dispersed particulate and bubbly flows, the shear-induced and particle-induced Reynolds stresses can be superimposed. The shear-induced Reynolds stress can be evaluated using a two-phase k-ε or τ–ε model (Lahey and Drew, 2000), and a particle-induced Reynolds stress was derived by Nigmatulin (1979).

It has been found (Lopez de Bertodano et al., 1994b) that the ability to predict lateral phase distribution strongly depends on turbulence modeling. As a consequence, research on flow-regime-specific multiphase turbulence is important. Unfortunately, virtually no research on the effect of gravity on multiphase turbulence has been conducted to date. This must be rectified if reliable multidimensional, multiphase CFD models are to be developed for applications in space.

Research Issues

In summary, experimental and analytical research focused on measuring and modeling flow-regime-specific multiphase turbulent phenomena in reduced-gravity environments is required if reliable predictions are to be made of the multiphase flow and heat transfer phenomena expected to occur in many of the power, propulsion, and life support systems and subsystems to be used for HEDS missions.

Multiphase Systems Dynamics

Multiphase flow also exhibits important global phenomena, which can affect system/subsystem operation and performance. For example, it is well known that liquid handling and storage is an important issue for NASA, and attempts have been made in the past to address it with wicking structures installed in tanks, for instance. Nevertheless, more research is needed, particularly for flashing cryogenic liquids and for the cavitation and/or the release of dissolved gases in liquids, in support of proposed future deep-space missions that may involve the harvesting and storage of water from asteroids, or on colonized extraterrestrial sites, and the handling and treatment of liquids associated with waste management and food production.

Other important global thermal-hydraulic phenomena are associated with system dynamics, in particular, the stability of multiphase systems. Such systems may exhibit static instabilities (Lahey and Podowski, 1989) such as the following:

- Excursive (i.e., Ledinegg) instabilities,
- Flow regime relaxation, and
- Geysering/chugging.

In addition, there may be dynamic instabilities such as these (Lahey and Podowski, 1989):

- Pressure-drop oscillations,
- Density-wave oscillations (DWOs), and
- Flow-regime-induced oscillations.

Of these instabilities, the most important from the point of view of NASA's HEDS goals are excursive (Ledinegg) instabilities, pressure-drop oscillations, and DWOs.

Excursive Instabilities

It is known (Lahey and Moody, 1995) that a boiling system may undergo Ledinegg-type flow excursions if

$$\frac{\partial(\Delta p_{system})}{\partial w} < \frac{\partial(\Delta p_{ext})}{\partial w}$$

where w is the flow rate and Δp_{ext} and Δp_{system} are, respectively, the external pressure change (e.g., due to a pump) and the irreversible pressure loss in the system.

Four cases are shown in Figure IV.C.11, where the boiling channels head-flow curve is given by the S-shaped curve and the externally impressed channel pressure is given in cases 1 through 4. Cases 1 and 4, for a positive displacement pump and a high-head centrifugal pump, respectively, are stable. In contrast, cases 2 and 3, for a parallel channel and a low-head centrifugal pump, respectively, are unstable. In particular, the unstable cases have multiple operating states, and CHF may occur for states 2 and 2′. It should be noted that this stability problem may be mitigated by inlet orificing and/or increasing system flow rate (w). However, these fixes can cause a large penalty in terms of increased system pressure drop and reduced boiling heat transfer effectiveness.

Pressure-Drop Instabilities

The inertia associated with an accumulator's dynamics may cause over/under shoots, which drive system oscillations. As can be seen in Figure IV.C.12, the accumulator can convert a Ledinegg-type flow excursion in a boiling loop into an equally undesirable periodic oscillation between points 2′ and 3′ (Lahey and Podowski, 1989).

Density-Wave Oscillations

Density-wave oscillations (DWOs) are caused by the lag introduced into a flowing multiphase system by the finite speed of propagation of density waves (Lahey and Podowski, 1989). To understand DWO phenomena in a boiling or condensing channel we can consider the channel to be a negative feedback control system (Lahey and Podowski, 1989), in which the inlet liquid velocity is induced by the dynamics of the diabatic system (i.e., the boiling or condensing channel). Similar instability phenomena may occur in electronic and electromechanical systems.

Figure IV.C.13 is a schematic of a typical U-tube condenser, and Figure IV.C.14 is the block diagram of the analytical model representing this phase-change system. Similar models and block diagrams can be derived for boiling channels (Lahey and Podowski, 1989).

It is significant to note in Figure IV.C.15 that gravity (g) and inlet velocity (j) appear to have a pronounced effect on DWO in boiling channels (Lahey and Podowski, 1989). That is, variations of the inverse Froude number (Fr^{-1}) can cause a significant change in the linear stability boundary. In particular, it appears that stability is increased as the inlet velocity is increased and/or gravity is reduced. The results shown in Figure IV.C.15 are theoretical results in which homogeneous flow was assumed. It is unlikely that this flow regime will occur in reduced-gravity environments, and thus more detailed analysis and/or experimental confirmation are needed. Nevertheless, gravity can be expected to have a strong influence on channel stability.

Significantly, nonlinear supercritical and subcritical Hopf bifurcations (e.g., limit cycles) and chaotic oscillations can also occur in boiling/condensing channels (Achard et al., 1985). Chaotic (i.e., nonperiodic) instabilities may also occur in phase change systems (Clausse and Lahey, 1991; Chang and Lahey, 1997). Finally, it should be noted that DWOs may occur not only in channels in which phase change takes place but also in phase-change loops (Lahey and Podowski, 1989).

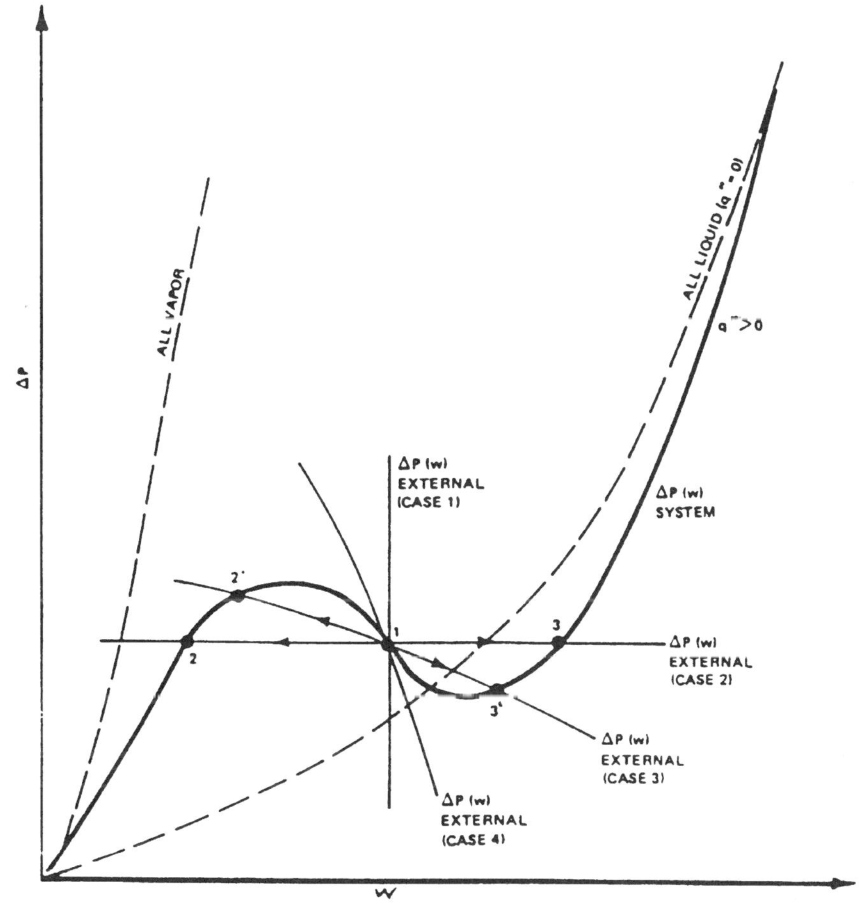

FIGURE IV.C.11 Excursive instabilities in a boiling channel (where Δp is the pressure drop and w is the flow rate). SOURCE: Reprinted with permission of the American Nuclear Society from R.T. Lahey, Jr., and F.J. Moody, *The Thermal-Hydraulics of a Boiling Water Nuclear Reactor*, Second Edition, p. 329. Copyright © 1993 by the American Nuclear Society, La Grange Park, Illinois.

Multiphase instabilities in diabatic channels and loops must be avoided since the amplitude of the flow excursions and oscillations can be quite large and dangerous to the integrity of the phase change equipment. Indeed, Ledinegg instabilities have destroyed commercial boilers on Earth (Profos, 1962). Unfortunately, NASA has performed little research to date on multiphase instabilities in reduced-gravity environments.

There are also other potentially important multiphase phenomena that should be studied, including counter current flow limitations, mixed convection and buoyancy (particularly for the emergency and/or long-term cooling of space-based nuclear power plants), the performance of lubrication films and pumps (e.g., cavitation will be more likely to reduce net positive suction heads), and inertial or condensation-induced loads (e.g., liquid-hammer-type loads). It is also important to note that inertial loads and flow unsteadiness may lead to the flow-induced vibration of piping systems and spacecraft structures. This is particularly true for multiphase flows, where inherent unsteadiness in some flow regimes (e.g., slug flow) may create a forcing function for structural vibrations. These topics are considered in Section V.A.

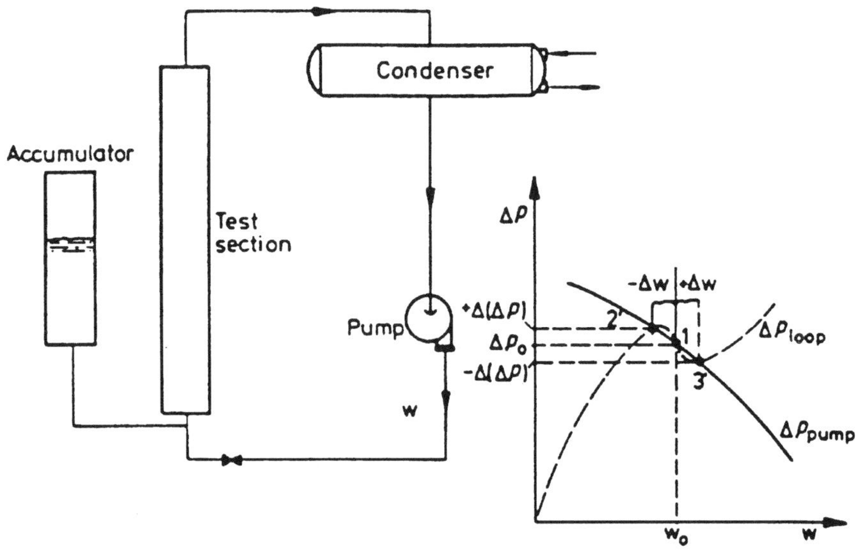

FIGURE IV.C.12 Pressure-drop instabilities in a boiling loop. SOURCE: Lahey and Podowski (1989). Reprinted by permission of Hemisphere Publishing Corporation.

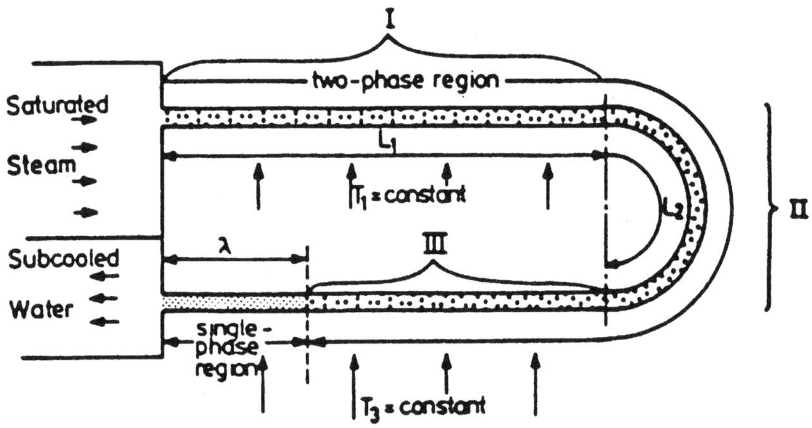

FIGURE IV.C.13 Schematic of a U-tube condenser. SOURCE: Lahey and Podowski (1989). Reprinted by permission of Hemisphere Publishing Corporation.

Research Issues

In summary, multiphase systems can exhibit global (systemwide) phenomena that are sensitive to gravity level. In particular, both excursive and oscillatory system instabilities are expected to be gravity-dependent. Unfortunately, very little research of this type has been done by NASA. It appears that an experimental and analytical research program focused on, at a minimum, Ledinegg (i.e., excursive) pressure drop and density-wave instabilities is needed if reliable multiphase systems are to be used on HEDS missions.

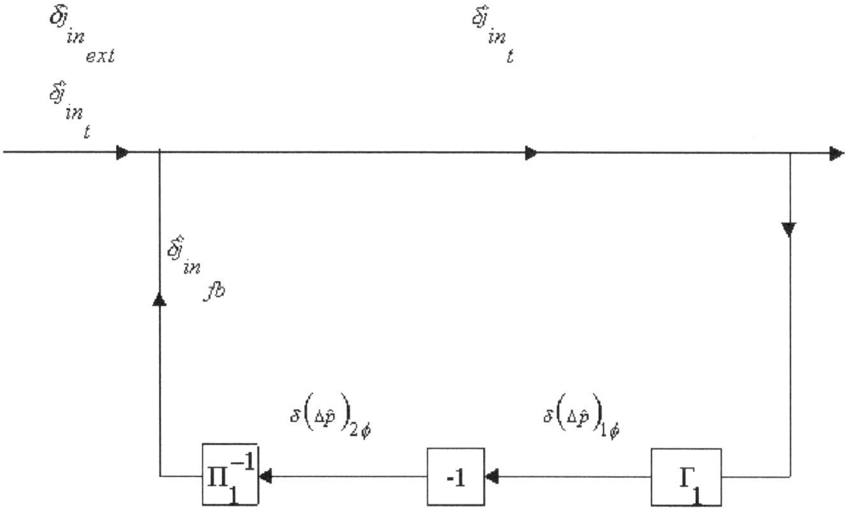

FIGURE IV.C.14 Block diagram for a parallel channel in a U-tube condenser (Γ_1 is the single-phase pressure-drop-to-inlet-velocity ratio and Π_1 the two-phase pressure-drop-to-inlet velocity ratio). Courtesy of R.T. Lahey, Jr.

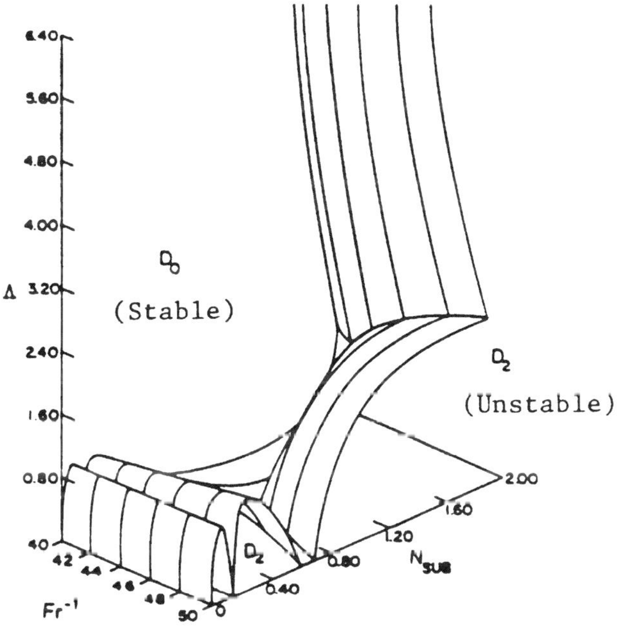

FIGURE IV.C.15 Three-dimensional marginal linear stability surface ($\tilde{j} = 0.3$; $42.0 \leq Fr^{-1} = gL_H/j^2_\lambda \leq 50.0$) where j and L_H are the inlet velocity and length of the heated channel, respectively, *g* is the acceleration of gravity, and Fr is the Froude number. SOURCE: Lahey and Podowski (1989). Reprinted by permission of Hemisphere Publishing Corporation.

Flow in Porous Media

Problems involving single- and multiphase flow, heat transfer, and multicomponent mass transport in porous media arise in a number of scientific, geophysical, and engineering disciplines. Important technological applications include geothermal energy exploration, enhanced oil recovery, chemical and drying processes, capillary-assisted thermal management technologies for space exploration, and numerous others (Woodling and Morel-Seytox, 1976; Wang and Cheng, 1997; Adler and Brenner, 1998; Masuoka, 1999). For example, porous media in a capillary pumped loop, which is used for thermal management in microgravity, determines the available pumping head for transport of heat in the loop. Changes in the wick (porous medium) properties to increase the transport capacity of the loop may affect (positively or adversely) the overall loop performance and system response. Following a depriming or dryout of the structure due to excess heat addition or system transients, rewetting of the capillary must occur for normal operation to resume. Delivery of nutrients to the roots of plants in growth media (i.e., porous root media) is an example of fluid flow in porous media associated with plant cultivation for life support in fractional gravity or microgravity environments (Eckart, 1996).

Two-phase devices such as heat pipes and capillary pumped loops have become key elements in the thermal control systems of space platforms. Capillary and porous structures are necessary and widely used in these devices, especially in high-heat-flux and zero-gravity applications (Faghri, 1995, 1999). As is discussed in Chapter III, porous media are also used in fuel cells, life support systems, bioregenerative food production, and chemical processing. Because of its diverse and important geophysical and technological applications, flow in porous media has received considerable research attention (Bear and Bachmant, 1990; Dullien, 1992; Nield and Bejan, 1992; de Boer, 1996; Wang and Cheng, 1997). Single- and multiphase fluid flow in saturated, partially saturated, and unsaturated porous media is of interest, and extensive discussion of the issues can be found in the references cited. Darcy's law, which describes fluid flow in porous media, has been generalized for both steady and unsteady multiphase flow in porous media (Dullien, 1992). The macroscopic-scale Darcy's law for two-phase flows in porous media has, however, been criticized from several perspectives (Wang and Cheng, 1997; Adler and Brenner, 1998), and controversy continues about the extension of the law. Data needed for system designs involving porous media are usually obtained empirically, because pore structure, effective thermophysical characteristics, transport coefficients, and other information needed as inputs to flow and transient model equations are not readily available.

A number of complex, interacting transport phenomena may take place in a nonisothermal, multiphase system. In general, flow in porous media is driven by gravitational, capillary, and viscous forces. Gravity causes phase separation and migration in the direction of the gravitational field. In microgravity, capillary and viscous forces play fundamental roles in controlling the phase distribution and, hence, multiphase flow and transport in porous media. According to Dullien (1998), surface tension, wettability, pore morphology, and displacement are historically the fundamental parameters that determine the topologies of immiscible fluids in porous media at capillary equilibrium. Steady two-phase flow at a given level of saturation depends on viscosities and fluid velocities, in addition to the parameters listed above. No fundamental experimental studies of fluid flow in porous media under isothermal or nonisothermal flow conditions have been performed in reduced or microgravity environments. The lack of experimental data constitutes a major impediment for the conceptual development of flow models in porous media.

Basic understanding of two-phase flows through porous media is of interest in many chemical processing and geophysical applications (Wang and Cheng, 1997). The early work was primarily motivated by potential applications in absorption towers employed in the chemical industry. More recently, two-phase flow in porous media with heat transfer has become a concern in light-water nuclear reactor accident scenarios in which the reactor core is severely degraded (Tung and Dhir, 1988; Chung and Catton, 1991). With respect to HEDS technologies, adiabatic two-phase flows or two-phase flows with heat transfer are relevant to life support systems, fuel cells, AMTEC, in situ resources utilization, heat pipes, and materials processing. For example, the delivery of nutrients to roots of plants in growth media is an example of fluid flow in porous media for life support in fractional or microgravity environments (see Section III.C). However, no fundamental studies of single- or two-phase flows in porous media, with or without heat addition, under reduced or microgravity conditions could be identified in the literature. Flow regimes (flow patterns), pressure drop, and dryout are also expected to be greatly influenced by gravity and surface tension. Size and shape of particles, porosity, permeability, wetting, type of fluids, etc. are expected to be important factors affecting two-phase flows in porous media.

The effect of combined thermal and gravity modulation on the onset of convective flow in porous media has not been studied and is not understood. Recently, it was shown theoretically that low-frequency *g*-jitter can have a significant effect on flow stability (Malashetty and Padmavathi, 1998), but experimental data do not appear to exist in the literature that could be used to validate the theoretical predictions. Knowledge of turbulence is important when predicting flow, heat, and mass transfer characteristics in porous media at high Reynolds numbers. Macroscopic turbulence models have been proposed (Travkin et al., 1993; Masuoka and Takatsu, 1996). Masuoka and Takatsu in their model introduced the concepts of interstitial vortex and pseudovortex to reflect the microscopic vortex behaviors intrinsic to porous media. Their macroscopic turbulence model has been subjected to experimental verification (Takatsu and Masuoka, 1998). It has been shown that the obstruction of a solid matrix not only imposes spatial restrictions on the magnitude of interstitial vortices but also induces flow distortion. The fluid parcel in the mixing zone is transported by the flow distortion with mixing length being of the order of the particle diameter. This motion produces additional mixing in the interstitial vortex. Practical porous-media turbulence models for predicting transport coefficients such as thermal dispersion and eddy diffusivities have not yet been proposed (Masuoka, 1999).

Research Issues

In summary, there are many fundamental research issues for single, multiphase, and multicomponent flows and transport in porous media that are not understood and have not been studied under fractional or microgravity conditions. These issues include the following: (1) There is a need to identify single-phase flow regimes in porous media, in both the presence and absence of forced flow, and express them as a function of operating variables. Such information is required for constructing porous media flow and transport models. (2) The simultaneous flow of gas and liquid through a fixed (packed) bed of particles in microgravity needs to be characterized. To this end, different flow regimes that exist (say, air-water systems) should be determined for a range of conditions of technological interest, and the boundaries between flow regimes should be established. (3) The effects of combined capillary and gravitational forces in partially saturated porous media and effects of mechanical surface characteristics such as roughness have not been studied and are not understood. Research is needed to relate pore-scale hydrodynamics to macroscopic-scale flow parameters and to identify flow regimes. (4) The effects of local thermal nonequilibrium, on a macroscopic scale, between the solid and fluid phases, on steady transport processes in saturated porous media need to be assessed and conditions identified for which they are important. (5) Experimental and theoretical work is needed to develop models on a micro- and macroscopic scale to rigorously address the coupling between vapor dynamics on the pore level and two-phase transport phenomena on the system level. (6) It would be highly desirable to account for pore-level fluid mechanics and to predict the development of complex flow structures and the two-phase zone at the onset of boiling and dryout when a liquid is heated in a porous medium. This is because various transient effects and flow instabilities may exist in the highly nonlinear system in which two-phase adiabatic or nonadiabatic flow occurs in a porous medium.

References

Achard, J.-L., D.A. Drew, and R.T. Lahey, Jr. 1985. The analysis of nonlinear density-wave oscillations in boiling channels. J. Fluid Mech. 155:213-232.

Adler, P.M., and H. Brenner. 1998. Multiphase flow in porous media. Annu. Rev. Fluid Mech. 20:35-59.

Alajbegovic, A., A. Assad, R.T. Lahey, Jr., and F. Bonetto. 1994. Phase distribution and turbulence structure for solid/fluid upflow in a pipe. Int. J. Multiph. Flow 20(3):453-479.

Alajbegovic, A., D.A. Drew, and R.T. Lahey, Jr. 1999. An analysis of phase distribution and turbulence in dispersed particle/liquid flows. Chem. Eng. Commun. 174:85-133.

Antal, S.P., R.T. Lahey, Jr., and J.E. Flaherty. 1991. Analysis of phase distributions in fully developed laminar bubbly two-phase flow. Int. J. Multiph. Flow 17(5):635-652.

Bear, J., and Y. Bachmant. 1990. Introduction to Modeling of Transport Phenomena in Porous Media. Dordrecht, Netherlands: Kluwer Academic Publishers.

Bousman, W.S., and A.E. Dukler. 1993. Studies of gas-liquid flow in microgravity: Void fracture pressure wrap all flow patterns. ASME Symposium Vol. 174/Fed. Vol. 175. New York: American Society of Mechanical Engineers.

Bousman, W.S., J.B. McQuillen, and L.C. Witte. 1996. Gas-liquid flow patterns in microgravity: Effects of tube diameter, liquid viscosity and surface tension. Int. J. Multiph. Flow 22(6):1035-1053.

Chang, C.J., and R.T. Lahey, Jr. 1997. The analysis of chaotic instabilities in boiling systems. Nucl. Eng. Des. 167:307-334.

Chen, I.Y., R.S. Downing, E. Keshock, and M. Al-Skarij. 1991. Measurements and correlation of two-phase pressure wrap under microgravity conditions. J. Thermophys. Heat Transfer 5:514-523.

Chung, M., and I. Catton. 1991. Post-dryout heat transfer in a multi-dimensional porous bed. Nucl. Eng. Des. 128:289-304.

Clausse, A., and R.T. Lahey, Jr. 1991. The analysis of periodic and strange attractors during density-wave oscillations in boiling flows. J. Chaos, Solitons and Fractals 1(2):167-178.

Colin, C., J. Fabre, and A.E. Dukler. 1991. Gas-liquid flow at microgravity conditions—I, dispersed bubble and slug flow. Int. J. Multiph. Flow 17(4):533-544.

Croff, A.G. 1980. ORIXEN2—A Revised and Updated Version of the Oak Ridge Isotope Generation and Depletion Code. ORNL-5621. Oak Ridge, Tenn.: Oak Ridge National Laboratory.

de Boer, R. 1996. Highlights in the historical development of the porous media theory: Toward a consistent macroscopic theory. Appl. Mech. Rev. 49:201-262.

Drew, D.A., and S.L. Passman. 1998. Theory of Multicomponent Fluids: Applied Mathematical Sciences Series, Vol. 135. New York: Springer-Verlag.

Dukler, A.E., J.A. Fabre, S.B. McQuillen, and R. Vernon. 1988. Gas-liquid flow at microgravity conditions: Flow patterns and their transitions. Int. J. Multiph. Flow 14(9):389-400.

Dullien, F.A.L. 1992. Porous Media-Fluid Transport and Pore Structure, 2nd Ed. New York: Academic Press.

Dullien, F.A.L. 1998. Capillary effects and multiphase flow in porous media. J. Porous Media 1:1-29.

Eckart, P. 1996. Spaceflight Life Support and Biospherics. Torrance, Calif.: Microcosm Press, and Dordrecht, Netherlands: Kluwer Academic Publishers.

Eckert, E.R.G., and R.M. Drake, Jr. 1972. Analysis of Heat and Mass Transfer. New York: McGraw-Hill.

Faghri, A. 1995. Heat Pipe Science and Technology. Washington, D.C.: Taylor and Francis.

Faghri, A. 1999. Recent advances in heat pipe analysis and simulation. Annual Review of Heat Transfer 8. C.-L. Tien, ed. New York: Begell House.

Fuchs, N.A. 1964. The Mechanics of Aerosols. New York: Pergamon Press.

Heppner, D.B., C.D. King, and J.W. Libble. 1975. Zero-gravity experiments in two-phase fluid flow patterns. ASME Preprint IS-ENAS-24. New York: American Society of Mechanical Engineers.

Hwang, S.T., H. Soliman, and R.T. Lahey, Jr. 1988. Phase separation in dividing two-phase flows. Int. J. Multiph. Flow 14(4):439-458.

Jayawardena, S., V. Balakataiah, and L.C. Witte. 1997. Flow pattern maps for microgravity two-phase flow. American Institute of Chemical Engineering J. 43:1637-1640.

Lahey, R.T., Jr. 1992. The prediction of phase distribution and separation phenomena using two-fluid models. Pp. 85-122 in Boiling Heat Transfer—Modern Developments and Advances. R.T. Lahey, Jr., ed. New York: Elsevier.

Lahey, R.T., Jr. 1996. A CFD analysis of multidimensional two-phase flow and heat transfer phenomena. Process, Enhanced, and Multiphase Heat Transfer: A Festschrift for A.E. Bergles. A.E. Bergles, R.M. Manglik, and A.D. Kraus, eds. New York: Begell House.

Lahey, R.T., Jr., and D.A. Drew. 2000. The analysis of two-phase flow and heat transfer using a multidimensional, four-field two-fluid model. Proceedings of the Ninth International Topical Meeting on Nuclear Reactor Thermal Hydraulics (NURETH-9), October 3-8, 1999. Nuclear Engineering and Design, in press.

Lahey, R.T., Jr., and F.J. Moody. 1995. The Thermal-Hydraulics of a Boiling Water Nuclear Reactor. ANS Monograph. LaGrange, Ill.: American Nuclear Society.

Lahey, R.T., Jr., and M.Z. Podowski. 1989. On the analysis of instabilities in two-phase flows. Multiphase Science and Technology 4:183-370.

Lance, M., and J. Bataille. 1991. Turbulence in the liquid phase of a uniform bubbly air-water flow. J. Fluid Mech. 222:95-118.

Lopez de Bertodano, M., S.-J. Lee, R.T. Lahey, Jr., and D.A. Drew. 1990. The prediction of two-phase turbulence and phase distribution phenomena using a Reynolds stress model. J. Fluids Eng. 112(1):107-113.

Lopez de Bertodano, M., R.T. Lahey, Jr., and O.C. Jones, Jr. 1993. "Phase Distribution in Complex Geometry Conduits." Nuclear Engineering and Design 141 (1 and 2).

Lopez de Bertodano, M., O.C. Jones, and R.T. Lahey, Jr. 1994a. Development of a k-ε model for bubbly two-phase flow. J. Fluids Eng. 116(1):128-134.

Lopez de Bertodano, M., R.T. Lahey, Jr., and O.C. Jones. 1994b. Phase distribution in bubbly two-phase flow in vertical ducts. Int. J. Multiph. Flow 20(3):453-479.

Malashetty, M.S., and V. Padmavathi. 1998. Effect of gravity modulation on the onset of convection in a porous layer. J. Porous Media 1:219-226.

Masuoka, T. 1999. Some aspects of fluid flow and heat transfer in porous media. Proceedings of the 5th ASME/JSME Joint Thermal Engineering Conference, March 15-19, 1999, San Diego, Calif. Paper No. AJTE 99-6304. New York: American Society of Mechanical Engineers.

Masuoka, T., and Y. Takatsu. 1996. Turbulence model for flow through porous media. Int. J. Heat Mass Transfer 39:2803-2809.

McQuillen, J. 1999. Reduced gravity gas/liquid flows. Proceedings of the Second Annual Institute for Multifluid Science and Technology (IMuST)—1999, March 18-20, Santa Barbara, California. Santa Barbara: Institute for Multifluid Science and Technology.

McQuillen, J., C. Colin, and J. Fabre. 1998. Ground-based gas-liquid flow, research in microgravity conditions: State of knowledge. Space Forum 3:165-203

Nadler, M., and D. Mewes. 1997. Flow induced emulsification in the flow of two immiscible liquids in horizontal pipes. Int. J. Multiph. Flow 23(1):55-63.

Nield, D.A., and A. Bejan. 1992. Convection in Porous Media. New York: Springer-Verlag.

Nigmatulin, R. 1979. Spatial averaging in the mechanics of heterogeneous and dispersed systems. Int. J. Multiph. Flow 5:333-385.

Ottino, J.M. 1990. The Kinematics of Mixing: Stretching, Chaos, and Transport. Cambridge: Cambridge University Press.

Ottino, J.M., F.J. Muzzio, M. Tjahjadi, J.G. Franjione, S.C. Jana, and H.A. Kusch. 1992. Chaos, symmetry, and self-similarity: Exploiting order and disorder in mixing processes. Science 257:754-760.

Profos, P. 1962. Die Regelung von Dampfanlagen. Heidelberg: Springer-Verlag.

Serizawa, A. 1974. Fluid Dynamic Characteristics of Two-Phase Flow. Ph.D. dissertation. Kyoto University, Japan.

Takatsu, Y., and T. Masuoka. 1998. Turbulence phenomena in flow through porous media. J. Porous Media 1:243-251.

Theofanous, T.G., and J.P. Sullivan. 1982. Turbulence in two-phase dispersed flows. J. Fluid Mech. 116:343-362.

Travkin, V.S., I. Catton, and L. Gratton. 1993. Single phase transport in prescribed isotropic and stochastic porous media. Pp. 43-48 in Heat Transfer in Porous Media, HTD Vol. 240. New York: American Society of Mechanical Engineers.

Tung, V.X., and V.K. Dhir. 1988. A hydrodynamic model for two-phase flow through porous media. Int. J. Multiph. Flow 14:47-65.

Velindandla, V., S. Putta, R.P. Roy, and S.P. Kaira. 1995. Velocity field in turbulent subcooled boiling flow. Proceedings of the National Heat Transfer Conference, Vol. 12, HTD Vol. 314. New York: American Society of Mechanical Engineers.

Wang, C.Y., and P. Cheng. 1997. Multiphase flow and heat transfer in porous media. Pp. 93-196 in Advances in Heat Transfer, Vol. 30. J.P. Hartnett, T.F. Irvine, Jr., Y.I. Cho, and G.A. Green, eds. San Diego, Calif.: Academic Press.

Wang, S.-K., S.-J. Lee, R.T. Lahey, Jr., and O.C. Jones, Jr. 1987. "3-D Turbulence Structure and Phase Distribution Measurements in Bubbly Two-Phase Flows," Int. J. Multiphase Flow 13(3).

Wilcox, D.C. 1998. Turbulence Modelling for CFD. La Cañada, Calif.: DCW Industries.

Woodling, R.A., and H.J. Morel-Seytoux. 1976. Multiphase fluid flow through porous media. Annu. Rev. Fluid Mech. 8:233-274.

Zaho, L., and K.S. Rezkallah. 1993. Gas-liquid flow patterns at microgravity conditions. Int. J. Multiph. Flow 19(5):751-763.

Zumbrunnen, D.A. 1998. Enhanced physical properties of composite materials produced by chaotic mixing. Pp. 689-690 in Proceedings of the National Science Foundation Design and Manufacturing Grantees Conference, January 5-8, Monterrey, Mexico. Arlington, Va.: National Science Foundation.

IV.D HEAT TRANSFER

Introduction

Many of the systems required for HEDS technologies involve single- and two-phase flow, heat and mass transfer (transport phenomena), including power generation and storage, propulsion, life support, thermal management, and in situ resource utilization. Because fluid flow and heat transfer are affected by reduced gravity or microgravity, they represent critical processes in efficient and reliable active power generation technologies. For example, in the absence of gravity, forced convection or cryogenic cooling of regulators, converters, control circuits, etc. (under "Power Management" in Figure III.A.2) will be required. More specifically, multiphase flow phenomena, which are highly gravity-dependent, are central to heat and mass transfer in many systems. Gas/liquid contacting for air purification plays an important role in chemical processes such as catalysis and in beneficiation techniques. Fundamental studies of such phenomena will contribute to process and system design for microgravity or fractional gravity environments.

Fluid mechanics and transport phenomena are inseparable, and both are significantly influenced by gravity (particularly for multiphase flow) and play essential roles in many processes that are important to HEDS mission-enabling technologies. For example, a difference in density caused by temperature and/or composition can produce buoyancy-driven flow, thus giving rise to convective heat transfer. Even for single-phase transport, a reduction of gravitational forces leads to the dominance of other forces normally obscured in terrestrial environments, such as surface tension effects (Ostrach, 1982). A number of different types of forces and fluid flows have been identified that can occur under microgravity conditions. For example, in such microgravity applications as two-phase flow with heat transfer (e.g., boiling) and thermocapillary migration of bubbles and droplets, thermocapillary flow is known to play an important role (Kamatoni, 1997). Extensive discussion of the microgravity issues pertaining to single-phase fluid flows as they relate to materials processing in space can be found in an earlier NRC report (1978), and a detailed account of the gravity effects on fluid flows, including identification of relevant scaling parameters, has been provided by Ostrach (1982). Within the last decade, the Committee on Microgravity Research has also reviewed the status of microgravity research (NRC, 1992, 1995); the review need not be repeated here.

Of the three modes of heat transfer—conduction, convection, and thermal radiation—conduction and radiation are not directly affected by gravity and will not be discussed, except in passing. Single- and multiphase fluid flow are strongly affected by gravity, but single-phase convective heat transfer can be more readily scaled and is, therefore, treated rather superficially. The discussion in the report will focus on convective heat transfer with phase change.

Heat conduction in solids and liquids is not influenced by gravity. Heat conduction in gases is also not affected by gravity, except that, indirectly, low gravity is also associated with low atmospheric pressure. This, in turn, results in low density and thermal conductivity of the gas and reduces the rate of heat transfer by conduction. Heat transfer in highly rarified gases is well understood (Eckert and Drake, 1972). In a free space (10^{-7} torr) environment, the molecular mean free path becomes long, and gaseous heat conduction across a gap becomes negligible. In turn, elastic and plastic deformations can become important and affect thermal contact conductance (resistance), but they have not been studied in microgravity environments. Lack of space experience may, however, cause unexpected problems. For example, nitrogen ice, imbedded in aluminum foam inside a dewar, has been seen to expand more than expected. The expansion of the nitrogen ice apparently caused two originally separated internal components inside the dewar to come in contact, providing a path for heat conduction and the increased evaporation of nitrogen, thereby shortening the life span of one near-infrared camera and multi-object spectrometer (NICMOS) on the Hubble Space Telescope (Harwood, 1997). This occurrence clearly demonstrated the peril of using materials (e.g., aluminum foam-nitrogen ice) in the construction of a high-tech thermos whose thermophysical properties are not understood and not well modeled, in this case shortening the life span of an important instrument.

Thermal radiation heat transfer is also not affected by gravity. However, the radiation characteristics of a surface can be modified over time by the space environment, and this needs to be considered in the design of the systems. For example, cryodeposits formed on thermal control surfaces can alter the thermal emittance, and ultraviolet radiation can modify the solar radiation characteristics of thermal control coatings. Solar panels on Mars are expected to accumulate dust and soil, and the performance of the photovoltaic cells would be degraded with time if the panels were not cleaned. Human habitats, cryogenic storage facilities, and other structures would also be affected by dust and their radiation characteristics modified. Effects of Martian soil and dust on the radiation characteristics of structural materials are not understood and should be studied.

Even though the phenomena of heat conduction and radiation are not influenced directly by gravity, heat transfer by combined conduction and radiation in porous insulations is relevant to HEDS technologies. High-efficiency thermal insulations are usually associated with the cryogenic storage of liquid H_2 to prevent heat penetration and boiloff; they can also be used to reduce heat losses from high-temperature (~1000 °C) fuel cells and can improve the temperature performance and control of furnaces by preventing leakage. To improve engineering design equations for low-pressure heat transfer in porous thermal insulations, void radiation, void gas conduction in intermediate and low pressures, and solid conduction need to be understood and modeled.

The phenomenological and technical concerns related to the thermal design of systems for HEDS applications are wide ranging and cannot be addressed fully in this account. Single-phase convective heat transfer is discussed only briefly below. The emphasis of the rest of the section is on convective heat transfer problems where gravity can greatly affect fluid motion, and for the sake of brevity, the discussion is focused on the following specific topics: (1) evaporation, (2) boiling, (3) condensation, (4) two-phase forced convection, (5) phase-change (melting and solidification) heat transfer, and (6) phase-change heat transfer in porous media.

Single-Phase Convection

Convective heat and mass transfer, i.e., convection of heat and mass, is always accompanied by fluid motion, and the term is used to describe heat transfer between a surface and a moving fluid or across an interface between two immiscible fluids in relative motion. Hence, convection is always influenced by the fluid motion and the state of the fluid. The literature on low-gravity fluid flows in bulk fluids has been reviewed by Ostrach (1982) and Myshkis et al. (1987) and will not be repeated here, though it should be noted that there have been some significant advances since the appearance of these reviews. State-of-the-art literature reviews of convective heat transfer could not be identified, even though the process is as complex as the accompanying fluid flow. Convective heat and mass transfer phenomena play important roles in many space-based technologies. Applications include subsystems for power generation, propulsion, life support, chemical and materials processing, and thermal management. For example, the weightless condition of space travel eliminates the familiar gravitational body force on liquids and gases, usually causing surface tension and contact angle (capillary effects) to dictate the static equilibrium shape of a liquid and vapor in a propellant tank. Yet, in this situation, gas-free propellant must still be delivered to thrusters, the center of mass must be predicted and controlled, fuel must be kept above freezing, the remaining fuel must be accurately gauged, and apparent anomalies from telemetry must be diagnosed.

Single-phase forced flow at sufficiently high velocities is not directly affected by gravity, but for intermediate velocities (say, a Reynolds number of 10^4 when based on the diameter of the tube as an element of a heat exchanger) where large temperature gradients exist between the fluid near the wall and the center, gravitational effects could modify the forced flow and heat transfer. The heat exchanger orientation, turbulence, and relaminarization could be additional factors. In addition, if temperature gradients exist within a fluid flowing at relatively low Reynolds numbers through a curved conduit, there may be separation into different density regions due to secondary flow effects. This effect will be even more pronounced for multiphase flows, where secondary flow can induce phase separation. Unfortunately, the interplay between gravity and secondary flows has not yet been thoroughly studied. In the absence of forced flow, the existence of temperature and concentration gradients in the presence of a gravitational field can lead to bulk flow of the fluid.

Single-phase external and internal natural convection heat transfer has received significant research attention during the last 75 years (Raithby and Hollands, 1998). Convective heat transfer scaling relations have been developed both experimentally and theoretically for a large number of natural (Jaluria, 1987) and mixed (Chen and Armaly, 1987) convection flow geometries, and these scaling relations can be used for thermal design purposes. For example, the external natural convection heat transfer coefficient for laminar and turbulent convection from a vertical plate scales as $g^{1/4}$ and $g^{1/3}$, respectively. However, the existing scaling relations have not been carefully compared against experimental data obtained under reduced gravity conditions. The scaling with gravity of thermal phenomena, i.e., the identification of boundaries at which the physics of phenomena changes with the gravity level, has been recommended by a recent NASA workshop. For example, there may be a need to predict (simulate) convective heat transfer from solar cell arrays, radiator panels, and habitable structures on Mars. Theoretical models are currently available for this task.

Recent advances in microfabrication techniques have led to the development of compact, microchannel chemical reactors in the United States, Europe, and Japan (Tuckerman and Pease, 1981; Ehrfeld, 1995; Wegeng et al., 1996; Tonkovich et al., 1998). Microchannel heat exchangers and chemical processing equipment have the advantages of very efficient heat/mass transfer, compactness, and high specific performance (i.e., productivity per unit volume). Such microchannel-based processing technologies have, as an example, potential application for in situ resource utilization because of their greatly reduced mass. Theoretically, the chemical processing rates in such microreactors should increase significantly owing to a decrease in the resistance to the species transport caused by a drastic reduction in the thickness of thermal and solutal boundary layers. Ideally, one would hope to achieve and to maintain sufficiently large reactor throughput by using parallel chemical processing in many small channels composing the reactor, but fouling of these microchannels is of concern.

It is clear that successful design of microchannel-based chemical reactors requires a fundamental understanding of the transport processes occurring on the microscale. For example, distinct oscillatory flow has been predicted in microchannels, when heat and mass transfer is accompanied by adsorption/desorption (Fedorov and

Viskanta, 1999). However, a similar physical situation has not been found in macrochannels, and the reasons for the difference are not known. There do not appear to be any direct effects of microgravity on flow and transport phenomena in microchannels, but miniaturization of the components reduces their sensitivity to gravity. An indirect effect of microgravity is that it may be difficult to control the purity of fluids in HEDS applications of the technology.

In extended operation, single-phase and phase-change heat exchange surfaces in terrestrial environments invariably experience performance degradation due to fouling. Passive means and active devices for enhancing single- and two-phase heat transfer and for mitigating fouling have recently been discussed (Sommerscales and Bergles, 1997). The influence of fouling caused by impurities or dust on heat-exchange performance (particularly for microchannels) when the heat exchanger is operated for extended periods of time should be studied in microgravity.

Research Issues

To understand low-Reynolds-number forced and mixed convection as well as natural convection, theoretical analyses and computational tools for single-phase heat transfer need to be developed and supported by parallel experimental efforts so that the models can be validated. The theoretical models should be able to predict not only microgravity fluid behavior but also the effect of body forces due to artificially imposed force fields. Induced forces such as centrifugal, surface tension, magnetic, electrostatic, and osmotic could be analytically evaluated as to their effectiveness in compensating for the absence of gravity field in various space subsystem applications. This would permit simulation of fractional, micro-, and variable gravity and thus would improve understanding and insight into fundamental phenomena. Theoretical/computational models validated against experimental data could then be applied with confidence to physical situations of interest.

To meet the future demand for growing active power generation, thermal management, and other systems for HEDS missions, passive and active heat transfer enhancement schemes suitable for microgravity conditions need to be assessed and developed for single-phase fluids to reduce the size and weight of the equipment. Examples of such methods include imposition of electric fields, variable gravity or rotation to enhance thin-film heat transfer, and liquid jet impingement cooling/heating of a surface, among others.

Microsystems may allow the size and mass of the needed equipment to be reduced. Of particular interest would be schemes that enhance heat transfer but do not increase pressure drop or pumping power proportionately.

Evaporation Heat Transfer

Evaporation is a surface (interface) mass transfer process that occurs when the liquid molecules near the surface experience collisions that increase their energy above that needed to overcome the surface binding energy. When the temperature of the vapor/gas mixture is lower than the interface temperature, evaporation may proceed only if the parent liquid is superheated and thus supplies the necessary heat (Lock, 1994). Evaporative cooling that occurs at the liquid surface as a result of evaporation is a phase-change heat-transfer process that is associated with the latent heat of vaporization of the liquid. The energy required to sustain the evaporation must come from the internal energy of the liquid, which then must experience a reduction in temperature (the cooling effect).

Evaporating drops, sprays, and liquid films are widely used industrially and in space devices such as space radiators for life support and waste heat management. For example, thin evaporating liquid films produce high heat transfer rates and are used in heat pipes, sweat coolers, grooved evaporators, and other enhanced heat-transfer-surface devices that depend on the formation of thin-film regions owing to capillary action (Bankoff, 1990; Faghri and Khrustalev, 1997). As a basic physical process, the evaporation of a thin liquid layer plays a key role in heat transfer. Some transport processes can be modeled using a classical continuum transport model with special modification to boundary conditions to account for nonequilibrium effects. Wayner (1998) discusses recent progress. Obviously, a number of high-efficiency heat-transfer devices are dependent on the heat-transfer characteristics of the evaporating thin layer. As the space applications of two-phase devices are developed, the importance of convection and heat transfer in evaporating thin liquid films will become more important.

The behavior of thin films under the action of various mechanical, thermal, or structural forces has recently been discussed in a comprehensive review (Oron et al., 1997). Theoretical advances in the evaporation/condensation of thin films were also discussed, including the effects of mass loss/gain, vapor thrust, capillarity, and thermocapillarity. Unsolved problems were identified, but the effects of gravity on thin film evaporation were not considered. Stability and dryout of thin liquid films are an important consideration since they can determine the allowable heat flux in forced-convection, subcooled boiling or concurrent annular flow. Stability regimes of thin liquid films in the presence of surface tension, van der Waals forces, hydration, and elastic strain interactions have been studied theoretically (Majumdar and Mezic, 1998). The study showed that the interplay among these highly nonlinear forces can result in a wide variety of regime maps, but the maps have not been validated experimentally. An interesting application of thin film evaporation phenomena is the electrostatic liquid-film radiator, an ultralight radiator proposed for electric power generation in space (Kim et al., 1994).

Macroscopic thin liquid films are relevant to some HEDS technologies. For example, a knowledge of micro- and macrolayer evaporation (e.g., at a bubble's base and at a bubble boundary) is necessary for predicting nucleate boiling heat transfer in microgravity when buoyancy is strongly reduced and transport processes are determined by the properties of the interface alone (Straub, 1994). The effects of thermocapillary and capillary pressure flow, evaporation, condensation, and coalescence mechanics need to be understood for predicting nucleate boiling, and future research needs have been discussed (Dhir, 1998). An understanding of convection driven by evaporation in thin liquid layers may also provide a basis for determining the nucleate boiling mechanisms in the macrolayer. For example, the thin liquid sublayer beneath the bubble is believed to be a key element for enhancement of heat transfer in thin (macro-layer thickness) liquid films. In a microgravity environment, the effects of surface tension forces, thermocapillary forces, and the disjoining pressure force and convection on the microlayer thickness, stability of the liquid film, and mechanisms leading to film rupture or dryout are expected to be different than in $1\ g_0$.

In spite of their relevance to a fundamental understanding of the heat transfer mechanisms in nucleate boiling and in two-phase flow in space applications, the thermal conditions taking place in thin liquid layers with a free surface have not been investigated under microgravity. Heat transfer processes in thin liquid layers (<1 mm) with a free surface, in which convection is driven by the surface tension, have not been studied, and nonequilibrium evaporation (or condensation) in weightless conditions across an interface between a viscous liquid and a viscous gas has not been characterized. In the past, almost all investigators concentrated on the convective instability in the liquid layer rather than heat transfer in the layer and evaporation from the free surface. Evaporation and latent heat are expected to play important roles in the onset of convection, and the convection itself in thin liquid layers will be influenced by surface tension under microgravity conditions.

With high heat fluxes, the heat and mass transfer rates may be limited by a critical mechanism forced by a physical process. In a gravity field these phenomena are partially masked, but they can be very significant in microgravity environment. Some examples include the following:

- Shear stresses at the liquid/vapor interface in an axially grooved condenser, which are due to vapor/liquid interaction combined with the effects of thermocapillarity, can cause the capillary structure to be flooded with liquid and the appearance of liquid recirculation zones. This will lead to unacceptable thermal resistance in the condenser.
- The existence of thick liquid films attached to the extended evaporating meniscus in a capillary tube can significantly change the existing methods of predicting dryout in capillary-driven devices.
- Shear stresses induced on the liquid/vapor interface by thermocapillarity and complex three dimensional vapor flow can cause the recession of the evaporating liquid meniscus in a groove and lead to an unstable mode of operation in the axially grooved evaporator.
- Rotating thin films offer enhanced evaporative heat transfer characteristics owing to the film thinning effects associated with rotational body forces.

Research Issues

Some research issues for surface-tension-driven single-phase free surface flows are identified in Section IV.B. Additional areas of appropriate research include (1) theoretical analyses and experiments to observe evaporation-driven fluid motion in very thin evaporating layers, (2) measurement of heat transfer coefficients in very thin liquid films undergoing evaporation for a range of Marangoni numbers, (3) characterization of the Marangoni instability phenomena in droplets undergoing radiative and/or evaporative cooling, and (4) determination of the induced convection on the droplet evaporation rate, which is required for properly designing droplet radiators.

Some unexpected critical phenomena have been encountered in such devices as long heat pipe evaporators, and high thermal resistance has been noted in loop heat pipe evaporators. Fundamental investigations under microgravity conditions are needed to better understand the thermal/fluid behavior in capillary structures during evaporation at high heat fluxes. Heat and mass transport may be restricted by the initiation of critical mechanisms forced by physical phenomena at high fluxes. In a gravity field, these phenomena may be masked but they can be very important in a microgravity environment.

Boiling Heat Transfer

Boiling is a phase change process in which vapor bubbles are formed in a superheated liquid layer adjacent to a heated surface or on a heated surface. Boiling can also be considered as an evaporation process that involves creation of discrete vapor/liquid interfaces on a heated surface. Single-component boiling, which is a formation of pure vapor from a superheated pure liquid, begins when the wall (surface) temperature exceeds the saturation temperature of the liquid. Boiling is known to be a very efficient mode of heat transfer. Knowledge of boiling heat transfer under microgravity conditions, with gravity levels varying from 1 g_0 to 10^{-6} g_0, is needed for the design of power generation (e.g., Rankine cycle), thermal management, fluid handling and control, on-orbit storage and supply systems for cryogenic propellants and life support fluids, and for cooling the electronic packages associated with various instrumentation and control systems. The subsystems affected by the phenomena include heat exchangers-boilers; on-orbit storage and supply systems for cryogenic propellants and life-support fluids; and systems for cooling electronic packages, instrument packages, and control systems.

During the past 50 years boiling has been the subject of much research because it is a very efficient and, technologically, a very important heat transfer process. Most of the research was on boiling under normal gravity conditions, and the process has received a lot of attention; however, controversies about the mechanisms controlling boiling persist, and a complete understanding of the critical heat flux (CHF) phenomena remains elusive. Numerous accounts of boiling are available, including that of Dhir (1998), which discusses boiling fundamentals and cites relevant references. In summary, there are two types of boiling heat transfer—pool and forced flow. Pool boiling refers to boiling under natural (buoyancy-driven) convection conditions, whereas in forced-flow boiling (whether internal or external), liquid flow over the heated surface is imposed by external means. The boiling heat transfer data are usually presented in the form of a boiling curve in which the wall heat flux is plotted against the difference between the surface temperature and the liquid saturation temperature. The nucleate, transition, and film boiling regimes as well as the critical (maximum) heat flux are identified from the boiling curve. CHF sets the upper limit of fully developed nucleate boiling for safe operation of equipment.

Pool Boiling

Gravity is one of a large number of system variables that influence the dependence of the nucleate boiling heat flux on the wall temperature (Siegel, 1967; Dhir, 1991, 1998). The magnitude and direction of the gravitational acceleration with respect to the heater surface influence the fluid dynamics and thermal boundary layers and, consequently, the bubble trajectory. Pool boiling heat transfer results obtained under reduced gravity conditions have been contradictory (Lee et al., 1997; Sitter et al., 1998). Some earlier nucleate pool boiling data obtained using terrestrial facilities (i.e., free-fall drop towers, parabolic aircraft flights, and sounding rockets) indicated that nucleate boiling heat transfer was insensitive to reductions in gravity (see Straub et al., 1990; see also Lee et al.,

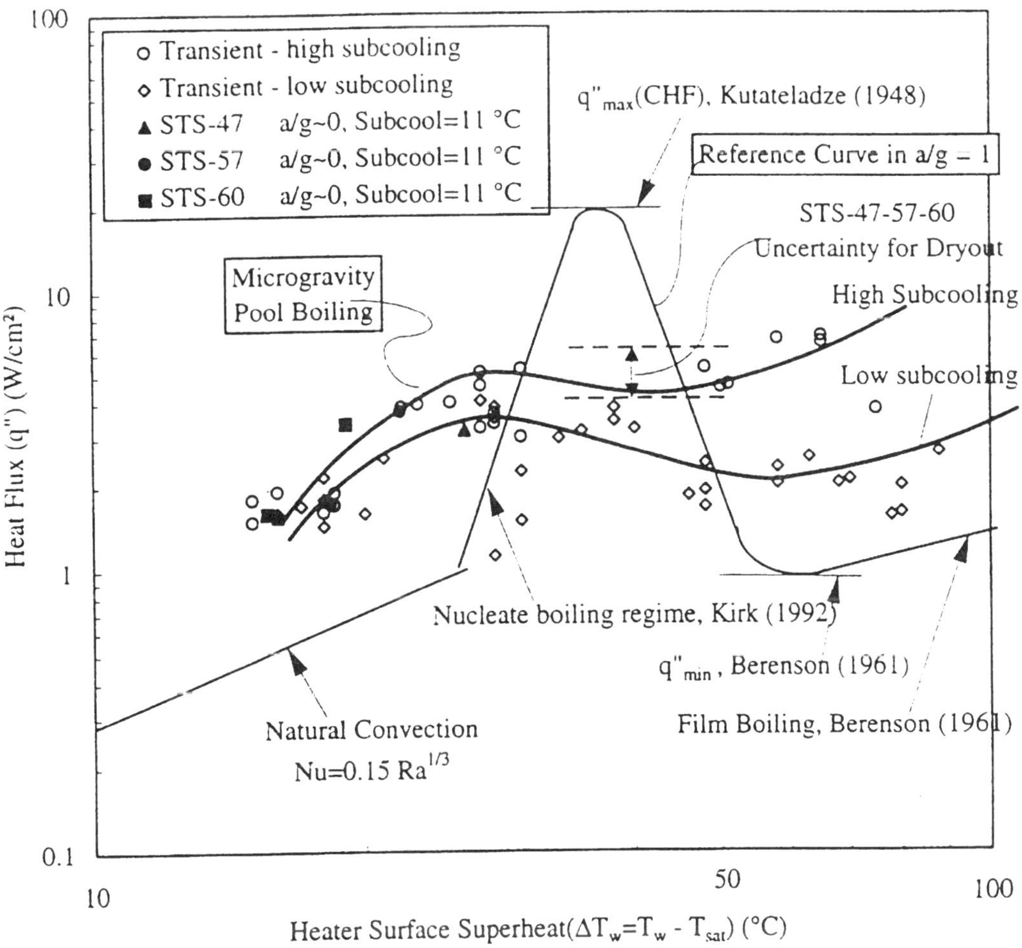

FIGURE IV.D.1 Approximate composite microgravity pool boiling curves for R-113 from steady and quasi-steady measurements made during shuttle flights STS-47, STS-57, and STS-60. SOURCE: Lee et al. (1997). Reprinted with permission from the American Institute of Aeronautics and Astronautics.

1997, for additional citations), whereas other data (Weinzierl and Straub, 1982; Merte et al., 1994) indicated enhancement of nucleate boiling heat transfer in microgravity. The difference in geometry of the heated surfaces, the variable quality of the gravity, and the short times available for conducting experiments contribute to the apparent contradictions in findings about the role of gravity in nucleate pool boiling heat transfer (Lee et al., 1997).

In recent experiments steady-state pool boiling of R-113 has been achieved, and a pool boiling curve has been generated (Lee et al., 1997, 1999). Analysis of the transient data has revealed that steady-state nucleate boiling heat transfer under microgravity conditions is enhanced relative to that in Earth gravity, whereas the CHF is considerably reduced. Approximate microgravity pool boiling curves for R-113 have been constructed using available data from quasi-steady measurements on STS-47, -57, and -60 and available correlations (Lee et al., 1997). Two curves, one for low and one for high levels of subcooling, are compared in Figure IV.D.1 with a reference curve for normal gravity. Pool boiling with fluids whose wetting characteristics are unlike those of R-113 and with surfaces different from the highly polished quartz used in these experiments are expected to produce different behavior than indicated in the figure.

Controversy persists about the dependence of CHF on gravity. According to the Kutateladze (Verkin and Kirichenko, 1976) and the Zuber (Dhir, 1998) theories, CHF should scale as $g^{1/4}$. However, for very low gravities, its functional dependence on gravity is weaker than is predicted from these theories. The reasons for the weaker dependence of CHF on gravity under microgravity conditions are not fully understood. On the other hand, Kirichenko and his coworkers took measurements at high pressures and observed a stronger dependence of CHF on gravity (i.e., ~ $g^{2/5}$) (Verkin and Kirichenko, 1976). This empirical finding over a range of pressures is consistent with the predictions of a simple theoretical model. The reasons for the disagreement between the two theories are not known.

The minimum (film boiling) heat flux, corresponding to complete dryout in microgravity, can also be anticipated to be much less than that in Earth gravity. Nucleate boiling heat transfer experiments were conducted by Ohta et al. (1998) under 10^{-4} g_0 conditions using a NASDA TR-1A No. 5 rocket. These investigators observed a marked heat transfer enhancement for ethanol when microlayers underneath the attached primary bubbles occupied a large part of the heating surface. However, the heat transfer deteriorated as time progressed owing to the extension of dry patches in the microlayers.

Experimental data from five space flights have recently been reported and analyzed (Merte et al., 1997; Lee et al., 1999). Measurements of parameters associated with bubble dynamics and heat transfer in pool boiling of fluorocarbon R-113 were made at about 10^{-4} g_0. Boiling characteristics, including vapor bubble dynamics associated with nucleation, bubble growth/collapse, and surface tension, were examined from a series of photographs, and heat transfer coefficients were determined from measured heater surface temperatures. Steady-state pool boiling was achieved and was attributed to surface tension effects. A large vapor bubble observed to hover near the heater surface produced a maximum 32 percent enhancement in heat transfer. A peculiar phenomenon was discovered that was referred to as vapor bubble migration, in which numerous tiny vapor bubbles nucleated and then moved toward a large bubble attached to the heater. An enhancement of about 30 percent in heat transfer was noted, but there was a significant decrease of the CHF in microgravity. Existing theoretical models are unable to predict CHF under pool boiling conditions. For example, Suzuki et al. (1999) reported that the CHF measured in microgravity experiments is two to four times as great as predicted by the existing theories.

It is well established that bubbles suspended in a fluid will move when subjected to a temperature gradient, owing to the action of the resulting interfacial tension gradient (Subramanian, 1992). Such motion, termed capillary migration, may have important implications for pool boiling heat transfer from a heated surface. The driving force for the bubble motion is the shear stress at the interface, which is a consequence of the temperature dependence of the surface tension. Recent results of spacecraft experiments on thermocapillary migration of bubbles have been reported (Balasubramanian et al., 1996), but the effect of the surface-tension-driven bubble motion on pool boiling heat transfer in microgravity has not been studied. Theoretical analyses (Subramanian, 1992) have identified two dimensionless parameters (i.e., Reynolds number and Marangoni number) that govern bubble motion in a liquid. In practical situations for space applications, it is expected that a wide range of values of these parameters will be encountered, depending on the thermophysical properties of the fluids involved and the temperature gradients imposed. Study of bubble motion during nucleate pool boiling is complicated by the fact that there may be mutual interference and distortion of the nearly spherical shape as small vapor bubbles coalesce to form larger ones, as observed by Lee et al. (1999).

Important questions remain to be addressed about the mechanisms and character of nucleation, dynamic behavior of the vapor bubbles, nucleate and transition boiling heat transfer, transition from nucleate to transition boiling, boiling stability, partial dryout behavior, drying/rewetting processes on a heated surface, and critical heat flux under microgravity conditions. The physics of bubble growth and detachment, bubble merger (coalescence) at and away from the heated surface, and vapor removal and the contributions of various mechanisms to the total heat transfer rate for pool and forced flow boiling under reduced gravity conditions are not fully understood. For example, the presence of thin liquid layers under vapor bubbles growing on heater surfaces is believed responsible for the enhancement of nucleate pool boiling observed in microgravity relative to that in Earth gravity. The formation of these large-scale thin layers is poorly understood and has not been adequately described for pool and flow boiling by a theoretical model. No correlations suitable for thermal design purposes for pool and low-velocity-flow boiling heat transfer exist. There is also a need to understand microlayer evaporation when its thickness is of the same order of magnitude as the root-mean-square value of surface roughness. The heat transfer relations, based on dominating buoyancy and hydrodynamic effects, are not applicable under low-gravity conditions.

In summary, the available results on pool boiling heat transfer under microgravity conditions are contradictory, and a mechanistic understanding of the various regimes of boiling is still lacking. Also, no rational basis exists for predicting pool boiling heat transfer, including CHF. Enhanced understanding of pool boiling requires the ability to model bubble nucleation, bubble dynamics, and heat transfer mechanisms. The mechanisms that need to be considered include microlayer evaporation at the bubble base, evaporation on the bubble boundary, transient conduction, thermocapillary convection, and convection induced by bubble motion. Some processes such as microlayer evaporation and flow induced by thermocapillary forces are greatly impacted by the Marangoni effect (see Section IV.B), but the magnitude of the effect cannot be quantified unless the operating conditions, such as gravity level, fluid, heated surface, and so forth, are specified.

Research Issues for Pool Boiling

To answer critical questions about pool boiling and CHF, research is needed on the fundamental mechanisms that govern bubble nucleation, growth, and interaction with the force field; bubble departure from the heater surface; and subsequent bubble transport in the superheated fluid. Bubble-bubble interactions, bubble rolling and sliding on macro/micro liquid layers, and bubble agglomeration under microgravity conditions are all in need of study. At higher heat fluxes, the transport mechanisms between the mushroom-shaped larger overhead bubbles and the frothing microlayer (very high density of small bubbles) on the heater surface must be established.

Nucleate boiling phenomena in thin and thick films need to be studied in the microgravity environment. Such phenomena are relevant to some types of heat pipes and space radiators; they differ significantly from those observed in pool boiling and remain undefined.

Active and passive schemes for enhancing pool boiling heat transfer (including fluid additives, surface coatings, and macro- and microsurface geometry modifications) are recommended for study. Such schemes have a potential of reducing the size and mass of the devices needed for thermal management.

Flow Boiling

Forced-convection boiling heat transfer and the pressure drop in uniform and nonuniform cross-section channels (conduits) under microgravity conditions must be measured. Potential flow boiling applications in HEDS missions include Rankine cycle power generation, heat pumps, life-support systems, and thermal management. Applications of two-phase technologies for these systems in space vehicles and lunar and Martian habitats promise to significantly increase thermal efficiency and to reduce the hardware mass to be launched. Two-phase systems can provide higher heat transfer rates at uniform temperatures under variations of heat load than single-phase systems.

As already discussed, distinct regimes of flow boiling have been identified in which the dominant heat transfer mechanism varies as the two-phase mixture progresses through a heated duct. Current research on the development of mechanistic models for nucleate boiling under microgravity conditions has been recently reviewed (Dhir and Hassan, 1998) and need not be repeated here. Over 100 references on both pool and forced-flow boiling are cited in that document. For example, the very important problem of forced-flow nucleate boiling under reduced gravity has received only very limited research attention, and there is no known international effort under way to develop mechanistic models for nucleate boiling or to obtain CHF data under low-velocity conditions in microgravity. Fundamental research needs are identified for developing a basic understanding of the mechanisms responsible for heat transfer and vapor removal from the vessel wall.

In spite of the fact that forced flow boiling is an attractive means of heat transfer in the microgravity space environment owing to its efficient transport of energy, only two series of aircraft trajectory experiments have been performed. Saito et al. (1994) used Japanese experimental aircraft to study the low-gravity, low-velocity (<6.7 cm/s) flow boiling of water on a rod heater placed in a square channel at about 0.01 g_0 for 20 s. The photographs reveal that under Earth gravity, small bubbles are detached from the heater rod surface, whereas under microgravity conditions, the bubbles hardly detach from the heater rod. The bubbles grow as a result of direct heating from the rod and/or coalesce to become much larger and surround the heater rod. Their measurements were limited to low nucleate boiling heat fluxes, and no data for CHF were taken. Tests of pressure drop and CHF aboard a NASA DC-9 aircraft were performed by Abdollachian et al. (1996). Unfortunately, there were some serious deficiencies in the experiments (i.e., an electrical tape was used to heat a glass tube) and insufficient details were provided in the report, so these test data cannot be used to thoroughly assess a model.

Subcooled forced-convection nucleate boiling experiments with R-113 under terrestrial gravity conditions have demonstrated that if buoyancy is significant relative to bulk liquid momentum, then a decrease in the buoyancy normal to and away from the heater surface enhances heat transfer (Kirk et al., 1995). In addition, it has been shown that the effect of the bulk flow velocity on heat transfer is dependent on surface orientation. Reference is made to recent accounts on flow boiling for the effects of, among others, subcooling, flow velocity, heater surface orientation, and internal vs. external flow, on flow boiling under the influence of gravity (Dhir, 1991, 1998; Brusstar and Merte, 1998; Hewitt, 1998). Recently, Brusstar and Merte (1998) concluded that the CHF for forced convection in microgravity is comparable to that for vertical upflow under normal gravity for flow velocities exceeding the buoyant terminal velocity of a vapor bubble in pool boiling. However, they argued that for flow velocities lower than this, the CHF in microgravity is expected to be much lower than in the presence of buoyancy, since the forces acting on the vapor are reduced substantially. Yamada and Fujii (1999) have reported on short (~10 s) drop tower two-phase flow and heat transfer experiments under microgravity conditions. They found that in microgravity forced flow boiling (of Fluorinert F-22) heat transfer is less than in Earth gravity and pressure drop greater.

Research Issues for Flow Boiling

Many questions are still wide open, and to resolve them would require both experimental and theoretical work on forced boiling heat transfer under microgravity conditions. A few specific topics have been identified for research, including innovative experiments to simulate microgravity conditions on Earth that would allow performing long-term two-phase flow and heat transfer experiments and experiments to obtain data needed for validating models and for developing design correlations. Experimental results obtained on board an aircraft are difficult to characterize owing to the often unknown influence of transient effects, and the question often arises as to whether true steady-state conditions are present to a great enough degree to draw valid conclusions.

Condensation Heat Transfer

Condensation in the context of heat transfer is a phase change process that occurs when a saturated vapor comes in contact with a surface at a lower temperature. Condensation processes require that the enthalpy of phase change be removed through the wall. Three distinct modes of condensation are possible: direct contact condensation, in which the vapor being condensed and the subcooled liquid that is condensing it are mixed, and film and dropwise condensation, in which the vapor and liquid are separated by a solid surface. Direct contact condensation produces very high rates of phase-change heat transfer (Marto, 1998), but care must be taken to avoid instabilities and condensation-induced loads. Film condensation occurs if the condensate film wets a wall and a complete film of liquid covers it. The film thickness will grow as it flows down, say, along a vertical wall under the action of gravity. Dropwise condensation occurs when the condensate does not wet the wall and instead, droplets of condensate nucleate at small pits and other surface imperfections and these droplets grow rapidly by direct vapor condensation upon them and by coalescence. Many factors influence condensation heat transfer (Marto, 1998); gravity is only one.

Film condensation heat transfer for laminar flow along a vertical wall is known to scale with $g^{1/4}$ and for turbulent flow with $g^{1/3}$ (Marto, 1998). Unfortunately, no film condensation experiments have been performed under low-gravity conditions, and it is uncertain if the scaling relations are valid for $g \rightarrow 0$. There has been relatively little research on the detailed mechanisms operative at the film interface between condensed liquid and its vapor under low-gravity conditions (Bankoff, 1994; Oron et al., 1997). Experiments and theoretical analyses are needed to determine the condensate film layer growth and stability, the development of interfacial disturbances, and the heat transfer rate. In shear-controlled condensers, interfacial stability, the condensation coefficient, and the effects of noncondensable gases on the condensation coefficient are important issues in the design of the components intended for operation under microgravity. Understanding of the phenomena and practical correlations are needed by designers of equipment for life support, thermal management, and power generation systems. An example of a life-support system is the carbon dioxide condenser/evaporator that is required for the closed-cycle regenerator.

A recent review of theoretical interphase mass transfer during condensation and evaporation has been prepared (Rose, 1999). The calculated heat transfer coefficients for dropwise condensation of steam have been compared with experimental data. The results are limited to normal gravity conditions, but the results do not reveal any direct dependence on gravity. In a low-gravity environment interfacial forces such as surface and thermocapillary forces can become significant and could affect the interfacial boundary conditions and dropwise condensation heat transfer. The fundamentally different stability characteristics of condensate film and how they differ from those of films of comparable scale in the absence of condensation is not fully understood. Consideration of the combined effects of reduced body force and thermocapillary forces suggests the existence of a convective pattern arising in the presence of condensation that can only be revealed under low-gravity conditions.

The fundamental fluid physics for condensate film growth, film instability, and the resulting interfacial motion under reduced gravity, and the corresponding implications for heat transfer have been little studied, are poorly understood, and deserve research attention. Recent work has shed light on the thermocapillary mechanisms driving fluid motion and instabilities in liquid layers with nonuniform temperature. The basically different instability and convective behavior of films in reduced gravity will likely lead to a very different relationship between temperature and wall heat flux (i.e., condensation curve) for both laminar and turbulent flow from that on Earth. For HEDS applications and systems, this topic is worthy of both theoretical and experimental research attention.

Research Issues

Theoretical and experimental research on interfacial transport phenomena involving phase transition under transient and steady-state conditions is suggested. Condensation is influenced by both gravity and interfacial forces. There is a need to identify flow and transport regimes as well as their boundaries by properly scaling the thermal phenomena with gravity. Boundaries, where the physics of the phenomena change with the gravity level, need to be understood and clearly delineated. Fundamental studies of condensation phenomena under reduced and microgravity conditions need to be undertaken. In parallel with experiments, detailed theoretical analyses should be carried out to develop an understanding of the fundamental fluid physics responsible for condensate film growth, film instability, and the resulting interfacial motion under reduced gravity, and the corresponding implications for forced-flow condensation heat transfer.

Two-Phase Forced Convection Heat Transfer

Gas/liquid mixtures occur in numerous situations relevant to space missions (Swanson et al., 1989). The application of two-phase flow and heat transfer technology in future power-generation and thermal management systems for space vehicles and for lunar and Martian habitats promises to significantly increase thermal efficiency and reduce the system mass that must be launched. As compared with single-phase systems, two-phase systems can provide larger heat transfer rates at uniform temperature under large variations of heat load. Studies in Earth gravity show that heat transfer rates and frictional pressure drop depend on how the phases are distributed in the duct (i.e., the flow regimes). Since gravity plays a significant role in the development of flow patterns, these flow regimes are expected to change in low gravity and microgravity conditions. Similarly, heat transfer and pressure drop will be altered with variations in gravity. Surface tension is expected to be a dominant force in determining two-phase flow patterns in gas/liquid mixtures in microgravity. However, inclusion of such a force in theoretical analysis of slug flow in a pipe, for example, can lead to unexpected results that are inconsistent with experimental data (Taitel and Witte, 1996).

Studies (Bousman et al., 1996; Jayawardena et al., 1997) have identified three distinct flow regimes at reduced gravity conditions: bubbly, slug, and annular, with transitions of bubbly-slug and slug-annular. Very recently, a critical literature survey on two-phase flow in reduced gravity was completed and a model was developed for predicting flow regimes in microgravity (Diev et al., 1998). The investigators have shown that four flow regimes are sufficient for characterizing two-phase flow in microgravity and for ground testing of thermal control systems: annular, bubbly, plug/slug, and stratified. The stratified flow regime cannot occur in microgravity, but its occurrence is very probable during ground testing.

In annular flow the liquid flows as a thin film along the tube wall and as droplets in a gas/vapor core. The flow regime is relatively simple, occurs over a wide range of gas and liquid flow rates, and has received the most research attention under reduced gravity conditions (Fore et al., 1996). Pressure drop, film thickness, and heat transfer were measured for annular gas/liquid mixtures at reduced gravity abroad NASA KC-135 aircraft. Air and two liquids, water and 50 percent aqueous glycerin, were used as fluids. Pressure drop measurements agree reasonably well with published correlations. Measured film thicknesses compare well with correlations derived from ground-based vertical annular flow data. Heat transfer coefficient data for each fluid have been compared with established empirical correlations. Hydrodynamic and heat transfer for two-phase slug flows in reduced gravity environment have been measured by the same investigators (Fore et al., 1997). The measured heat transfer coefficients at reduced gravity were found to be lower than predicted by normal-gravity correlations.

For annular flow the pressure drop data agree well with the well-known Lockhart-Martinelli correlation, whereas for the slug flow regime the data do not correlate well either with the Lockhart-Martinelli or the homogeneous flow correlations. The heat transfer results for the annular two-phase flow are mixed. For some fluid combinations (i.e., air/50 percent aqueous glycerin) the data follow reasonably well the established turbulent flow model, whereas for the air/water system the model overpredicts the Nusselt number. For slug flow, the measured heat transfer coefficients are lower at reduced gravity than predicted by normal-gravity correlations (Fore et al., 1997). This difference can be attributed to the lower turbulence levels that should exist in reduced gravity, although no turbulence measurements were made. In summary, very limited data exist for annular and slug flow regimes, and there is partial agreement between existing models and experimental data.

Liquid sprays are being used widely in many industrial, manufacturing, agriculture, and food production processes and other applications requiring rapid and effective cooling. To overcome the deleterious effects of microgravity on two-phase heat transfer, spray cooling has a potential for use in various thermal management and life support systems in the HEDS context. Experiments on and theoretical analysis of vaporizing droplets and sprays, and on impingement heat transfer to liquid sprays, relevant to HEDS technologies have been very sparse. Experiments with droplets impinging on high-temperature (from 425 to 567 K) surfaces were conducted at $10^{-4}\ g_0$ using a drop shaft (Tokura et al., 1995). Apparent heat fluxes increased with the collision velocity of the droplet up to about 10^7 W/m^2. The effect of gravity on spray cooling characteristics was investigated by means of parabolic flight maneuvers (Kato et al., 1995; Sone et al., 1995). Either water or CFC-113 was sprayed from a single nozzle onto a circular chromium-plated surface with wall superheating between 100 and 400 °C. In the experiments the gravity ranged from 2 g_0 to ~0.01 g_0. Spray cooling characteristics (i.e., heat fluxes vs. superheats) for both water and CFC-113 were measured at low and high volumetric spray fluxes. At high spray fluxes, gravity did not affect the heat transfer characteristics. Since the spray patterns for water and CFC-113 were different, the differences in the experimental findings cannot be attributed to the effects of gravity alone. The experimental results demonstrate that the effects of variable gravity on the two-phase heat transfer in spray cooling can be overcome.

Research Issues

Some progress has been made in understanding how microgravity affects two-phase flow heat transfer in simple (straight) ducts. Future studies would need to be directed toward bringing experimental and theoretical closure to simple ducts and focus on more complicated geometries that are expected to arise in practical systems in numerous HEDS technologies.

More experimental data under reduced and microgravity conditions would be needed to increase the confidence limits in the empirical constants and/or support the development of physically based models for forced-convection two-phase-flow heat transfer mechanisms. For the bubbly flow regime, no heat transfer measurements have been made under reduced or microgravity conditions. In such a flow regime, the capillary-induced migration of the bubbles in the presence of temperature gradients perpendicular to the walls of the conduit may have a significant effect on the phase distribution and convective heat transfer.

Additional research on vaporizing droplets, sprays, and spray cooling needs to be performed, because many gravity-related issues, such as the dynamics of evaporating drop deformation, breakup, and coalescence, are not understood in microgravity. The reasons for heat transfer enhancement in the low-heat-flux regime below CHF, the differences in the trends for the CHF for different fluids, the effect of spray patterns and droplet size distribution on heat transfer characteristics, and the effect of gravity on the heat transfer coefficient in the film boiling region: all these, and more, need to be addressed for successful application of the technology in space missions.

Solid/Liquid Phase-Change Heat Transfer

Solid/liquid phase-change heat transfer phenomena are relevant to HEDS mission-enabling technologies and subsystems. One concern is understanding how altered transport phenomena in microgravity influence the operation of the subsystems. Examples of subsystems or processes include freezing and thawing in stagnant fluid lines, radiators for rejecting waste heat and their start-up (melting) from the frozen or partially frozen state, latent heat thermal energy storage (LHTES) units, start-up (from the frozen state) of liquid cooled nuclear reactors, melting of nuclear reactor core under severe accident conditions, and freezing and start-up of heat pipes under off-design operating conditions, among others. LHTES is needed to ensure a constant heat supply for power generation during the shade period of the orbit. Void formation and void location can impact continuous delivery of heat, but the process has not been studied in microgravity and is not understood. During phase-change heat transfer, such as melting, freezing, and sublimation, there is usually an intrinsic change in density associated with the transition (Ostrach, 1982). The fluid motion produced by the difference in density between the two phases and buoyancy-induced flow under 1 g_0 conditions is known to affect the local heat transfer rate during melting and solidification and can greatly affect the solid/liquid interface and the solid/melt fraction (Viskanta, 1983).

Solar heat receivers, for example, employing phase-change materials (PCMs) have an advantage over sensible heat receivers: they require less mass because they possess higher energy storage densities. The effects of sedimentation of the denser phase and buoyancy due to expansion or contraction of the phase-change material can combine to induce flows in the melt of the phase-change material, affecting heat transfer. The flows induced by these forces can occur under normal gravity conditions, but their effects are usually masked by convective flow driven by buoyancy. In a low-gravity environment, buoyancy will be absent, but with void formation during solidification or melting, thermocapillary forces will come into play and will affect fluid motions in the melt. The combination of density difference and sedimentation-induced fluid motion and thermocapillary effects could influence the rate of melting/solidification in LHTES units and impact their thermal performance, but fluid motion in the melt has been ignored in predicting the performance of an LHTES unit (Hall et al., 1998). Some phase-change heat transfer processes are controlled or substantially affected by gravity. The effects of vibration on melting of an unfixed PCM under variable gravity conditions (Shirivanian et al., 1998) and of density change on unfixed rectangular phase-change material in a low-gravity environment (Asako and Faghri, 1999) have been studied theoretically, but experimental data do not exist for validating the predictions. The more critical problem of solidification heat transfer has not been studied either theoretically or experimentally.

Research Issues

To address the questions related to solid/liquid phase-change heat transfer under reduced or microgravity conditions, it will be necessary to obtain a comprehensive and detailed description of the solid/liquid phase change processes under microgravity conditions. Several critical issues that will have a great impact on the science and technology of melting and solidification could perhaps be studied on Earth instead of in space, by producing artificial low gravity. By imposing an electromagnetic force that opposes Earth's gravity it may be possible to simulate the oscillatory gravity environment of space and to damp natural convection-flow-induced sedimentation.

The formation of voids when the PCM used in solar active power generation systems freezes needs to be studied and resolved. Realistic computational models of LHTES capable of simulating relevant physical processes occurring in microgravity need to be developed and validated.

Phase-Change Heat Transfer in Porous Media

Capillary and porous structures are used widely in two-phase devices such as heat pipes, capillary pumped loops, and loop heat pipes to provide liquid transport and enhanced heat transfer during evaporation and condensation in spacecraft fluid and thermal management systems, but neither boiling nor condensation in porous structures has been studied under reduced-gravity or microgravity conditions (Khrustalev and Faghri, 1997). For example, capillary heterogeneity, induced by variation in permeability, has application in heat pipes operating in a microgravity environment. Phase change (melting or freezing) of the working fluid in the porous (wick) structure of the heat pipes under reduced gravity has not been studied and is not fully understood. Knowledge of the phenomena is necessary for starting up a frozen heat pipe or shutting down a high-temperature heat pipe under emergency conditions.

Evaporation of a liquid from porous (wick) structures or micropores under reduced gravity is relevant to the design and efficient operation of heat pipes, micro heat pipes, and capillary-driven devices such as capillary pumped loops (Ku, 1997; Faghri, 1999; Peterson et al., 1998). During startup of a heat pipe, the vapor flow is in rarified or free molecular flow regimes and lacks continuous flow characteristics. But the transient and often nonequilibrium evaporation processes are not completely understood. Evaporation of a liquid from liquid-vapor menisci attached to a heated, highly curved solid could affect vapor flow (shear) in microfilm and thick films but has not been studied in microgravity.

A concrete example of a two-phase heat transfer device that is employed by NASA for spacecraft instrument thermal control is the capillary pumped loop (CPL) (Ku, 1997). This device is capable of transporting large heat loads over great distances and with very small temperature differences across the system. At the heart of the device is an evaporator, which serves as both a heat-absorbing element and a fluid-circulating pump. A typical evaporator consists of a porous tubular wick that is force-fitted within an axially grooved aluminum tubing outer shell. The liquid flows axially inside the flow channel and radially through the wick to reach the heating surfaces. As the liquid is heated to the saturation temperature set by the reservoir, vapor bubbles form at the heating surfaces and migrate until vented into the grooves (channels). Surface tension prevents migration of vapor bubbles into the wick structure. At the same time, menisci are formed at the liquid/vapor interfaces resulting in capillary forces that circulate the fluid throughout the loop. There are still technical challenges facing CPLs and loop heat pipes (LHPs). One of the major issues surrounding LHP operation is the temperature hysteresis, i.e., the loop exhibits different operating temperatures under the seemingly same conditions, depending on whether the heat load is increasing or decreasing. The physics leading to hysteresis is not fully understood and has not been addressed in the open literature, and unsteady mathematical models to simulate the behavior of LHP operation are needed (Kaya et al., 1999).

Recent experiments on forced convection nucleate boiling in porous media-filled ducts revealed that when boiling occurs, three zones (liquid, liquid/vapor, and superheated vapor) develop (Miscevic et al., 1998). A characteristic boiling curve was determined; it revealed that boiling in a porous medium appears at very low wall superheats. The heat flux at small superheats can be considerably greater than for pool boiling in the absence of porous media, and the critical heat flux (CHF) can also be considerably greater than in the case of pool boiling. A mechanistic explanation of the observations is not yet available. Recent experiments on CHF enhancement by porous structures on heat-dissipating surfaces have indicated a significant (twofold) increase in the critical heat flux for one of the modulated porous layer coatings (Liter and Kaviany, 1998). This suggests that such coatings may find application in microgravity in cases where it is desirable to enhance heat dissipation from a surface by boiling of a liquid.

In spite of the fact that capillary-driven two-phase flow devices can operate in microgravity, a number of technical issues have not been addressed and require testing. Examples include start-up from the frozen state and shutdown, as well as off-design operation such as boiling in the wick due to overheating. The ability of the wick to act as a liquid pump depends on the thermal design of the heat pipe and on the operating conditions. Of critical importance because they affect the capacity and thermal performance of a given heat-pipe design are such factors as wick design and geometry, and interfacial stability (Faghri, 1995). For example, in ordinary heat pipes the liquid and vapor flow in opposite directions within the pipe during operation such as may occur at high heat flux limits, nucleate/film boiling in wick, freezing/thawing of working fluid, and other critical phenomena that will affect and/or limit heat pipe design and performance.

Research Issues

Among the phenomena relevant to many HEDS mission-enabling technologies are the forced and surface-tension-gradient-driven single- and two-phase flows, evaporation, condensation, and boiling heat transfer that occur in porous media under reduced-gravity or microgravity conditions. Fluid thermal behavior in capillary porous structures at high heat fluxes under microgravity conditions needs to be understood; several areas have been identified for research, including the following: (1) study of the obstruction of the liquid transport to the evaporator by the incipience of nucleate boiling, which can cause evaporator dryout in an axially grooved or other heat pipe; (2) determination of the increase in the overall thermal resistance due to the formation of a vapor zone in the porous structure as a result of boiling in the evaporator, which may finally lead to its dryout; (3) study of the vapor zone in the porous structure of the inverted-meniscus-type evaporator that may cause oscillations in CPL performance; and (4) determination of the conditions for nucleation in a wick structure when bubble nucleation is caused by hot spots on the evaporator wall. Vapor bubbles in the wick of a heat pipe are undesirable because they can obstruct liquid circulation driven by capillary action.

References

Abdollachian, D., J. Quintal, F. Barez, J. Zahm, and V. Lohr. 1996. Study of Critical Heat Flux and Two-Phase Pressure Drop Under Reduced Gravity. NASA CR-198516. Cleveland, Ohio: NASA Lewis Research Center.

Asako, Y., and M. Faghri. 1999. Effect of density change on melting of unfixed rectangular phase-change material under low-gravity environment. Pp. 57-63 in Proceedings of the ASME Heat Transfer Division—1998. R.A. Nelson, Jr., et al., eds. HTD-Vol. 361-3. New York: American Society of Mechanical Engineers.

Balasubramanian, R., C.E. Lacy, G. Wozniak, and R.S. Subramanian. 1996. Thermocapillary migration of bubbles and drops at moderate values of the Marangoni number in reduced gravity. Phys. Fluids 9(4):872-880.

Bankoff, S.G. 1990. Dynamics and stability of thin heated liquid films. J. Heat Transfer 112:538-546.

Bankoff, S.G. 1994. Significant questions in thin liquid film heat transfer. J. Heat Transfer 116:10-16.

Bousman, W.S., J. McQuillen, and L.C. Witte. 1996. Gas-liquid patterns in microgravity: Effects of tube diameter, liquid viscosity, and surface tension. Int. J. Multiph. Flow 22:1035-1053.

Brusstar, M.J., and H. Merte, Jr. 1998. An experimental and analytical approach to modeling the CHF for forced convection boiling in microgravity. Pp. 231-236 in Heat Transfer 1998: Proceedings of the 11th International Heat Transfer Conference 2. Singapore: Hemisphere Publishing.

Chen, T.S., and B.F. Armally. 1987. Mixed convection in external flow. Handbook of Single-Phase Convective Heat Transfer. S. Kakac, R.K. Shah, and W. Aung, eds. New York: John Wiley & Sons.

Dhir, V.K. 1991. Nucleate and transition boiling heat transfer under pool and external flow conditions. Int. J. Heat Fluid Flow 12:290-314.

Dhir, V.K. 1998. Boiling heat transfer. Annu. Rev. Fluid Mech. 30:365-401.

Dhir, V.K., and M.M. Hassan. 1998. Science Requirement Document for a Mechanistic Study of Nucleate Boiling Heat Transfer under Microgravity Conditions. Third draft, July 1998. Cleveland, Ohio: NASA Lewis Research Center.

Diev, M.D., A.I. Leontiev, and A.V. Shchetinin. 1998. A software for forecasting the flow patterns in microgravity. Pp. 475-479 in Proceedings of the ASME Heat Transfer Division—1998. R.A. Nelson, Jr., T. Chopin, and S.T. Thynell, eds. HTD-Vol. 361-5. New York: American Society of Mechanical Engineers.

Dullien, F.A.L. 1998. Capillary effects and multiphase flow in porous media. J. Porous Media 1:1-29.

Eckert, E.R.G., and R.M. Drake, Jr. 1972. Analysis of Heat and Mass Transfer. New York: McGraw-Hill.

Ehrfeld, W., ed. 1995. Microsystem Technology for Chemical and Biological Microreactors. DECHEMA Monograph, Vol. 132. Frankfurt am Main: DECHEMA.

Faghri, A. 1995. Heat Pipe Science and Technology. Washington, D.C.: Taylor and Francis.

Faghri, A. 1999. Recent advances in heat pipe analysis and simulation. Annual Review of Heat Transfer, Vol. 8. C.-L. Tien, ed. New York: Begell House.

Fahgri, A., and D. Khrustalev. 1997. Advances in modeling enhanced flat miniature heat pipes with capillary grooves. Enhanced Heat Transfer 4:99-109.

Fedorov, A., and R. Viskanta. 1999. Heat and mass transfer dynamics in the microchannel adsorption reactor. Microscale Thermophysical Engineering 3:101-139.

Fore, L.B., L.C. Witte, and J.B. McQuillen. 1996. Heat transfer to annular gas-liquid mixtures at reduced gravity. J. Thermophys. Heat Transfer 10:633-639.

Fore, L.B., L.C. Witte, and J.B. McQuillen. 1997. Heat transfer to two-phase slug flows under reduced-gravity conditions. Int. J. Multiph. Flow 23:301-311.

Hall, C.A., III, E.K. Glapke, J.N. Cannon, and T.W. Kerslake. 1998. Modeling cyclic phase change and energy storage in solar heat receivers. J. Thermophys. Heat Transfer 12:406-413.

Harwood, W. 1997. Hubble instrument faces shorter lifespan. Space News, March 31-April 6, p. 3.

Hewitt, G.F. 1998. Boiling. Chapter 15 in Handbook of Heat Transfer, 3rd Ed. W.M. Rohsenow, J.P. Hartnett, and Y.I. Cho, eds. New York: McGraw-Hill.

Jaluria, Y. 1987. Basics of natural convection. Handbook of Single-Phase Convective Heat Transfer. S. Kakac, R.K. Shah, and W. Aung, eds. New York: John Wiley & Sons.

Jayawardena, S.S., V. Balakotaiah, and L.C. Witte. 1997. Flow pattern transition maps for microgravity two-phase flows. AIChE J. 43:1637-1640.

Kamatoni, A. 1997. Surface tension driven convection in microgravity. Pp. 487-499 in Space Cooperation into the 21st Century (7th ISCOPS), Vol. 96, Advances in the Astronautical Sciences. P.M. Bainum, G.L. May, M. Nagamoto, et al., eds. San Diego, Calif.: American Astronautical Society.

Kato, M., Y. Abe, Y.H. Mori, and A. Nagashima. 1995. Spray cooling characteristics under reduced gravity. J. Thermophys. Heat Transfer 9:378-381.

Kaya, T., T.T. Hoang, J. Ku, and M.K. Cheung. 1999. Mathematical modeling of loop heat pipes. AIAA Paper No. 99-0477. Reston, Va.: American Institute of Aeronautics and Astronautics.

Khrustalev, D., and A. Faghri. 1997. Boiling heat transfer in the miniature axially-grooved discrete heat sources. Enhanced Heat Transfer 4:163-174.

Kim, H., S.G. Bankoff, and M.J. Miksis. 1994. The cylindrical electrostatic liquid film radiator for heat rejection in space. J. Heat Transfer 116:986-992.

Kirk, K.M., H. Merte, Jr., and R. Keller. 1995. Low-velocity subcooled flow boiling at various orientations. J. Heat Transfer 117:380-386.

Ku, J. 1997. Recent advances in capillary pumped loop technology. AIAA Paper No. 97-3870. Reston, Va.: American Institute of Aeronautics and Astronautics.

Lee, H.S., H. Merte, Jr., and F. Chiarmonte. 1997. Pool boiling curve in microgravity. J. Thermophys. Heat Transfer 11:216-222.

Lee, H.S., H. Merte, Jr., and F.P. Chiarmonte. 1999. Pool boiling phenomena in microgravity. Pp. 395-400 in Heat Transfer 1998: Proceedings of the 11th International Heat Transfer Conference, August 23-28, 1998, Kyongju, Korea, Vol. 2. J.S. Lee, ed. Singapore: Hemisphere Publishing.

Liter, S.G., and M. Kaviany. 1998. CHF enhancement by modulated porous-layer coating. Pp. 165-173 in Proceedings of the Heat Transfer Division—1998, Vol. 1. R.A. Nelson, Jr., K.S. Ball, and D. Kaminski, eds. New York: American Society of Mechanical Engineers.

Lock, G.S.H. 1994. Latent Heat Transfer. Oxford: Oxford University Press.

Majumdar, A., and I. Mezic. 1998. Stability regimes of thin liquid films. Microscale Thermophysical Engineering 2:203-213.

Marto, P.J. 1998. Condensation. Chapter 14 in Handbook of Heat Transfer, 3rd Ed. W.M. Rohsenow, J.P. Hartnett, and Y.I. Cho, eds. New York: McGraw-Hill.

Merte, H., Jr., H.S. Lee, and J.S. Ervin. 1994. Transient nucleate pool boiling in microgravity—some initial results. Microgravity Science and Technology VII(2):173-180.

Merte, H., Jr., H.S. Lee, and F.P. Chiarmonte. 1997. Nucleation, bubble growth, and heat transfer in extended microgravity pool boiling. Pp. 354-365 in Proceedings of the Joint Xth European and VIth Russian Symposium on Physical Sciences in Microgravity, Vol 1. V.S. Avduyevsky and V.I. Polezhaev, eds. Noordwijk, Netherlands: European Space Agency.

Miscevic, M., L. Tadrist, J. Pantaloni, and R. Yu. 1998. Forced convection boiling inside a duct filled with a sintered fibrous medium. J. Porous Media 1:135-146.

Myshkis, A.D., V.G. Babskii, N.D. Kopachevskii, L.A. Slobozhanin, and A.D. Tyuptsov. 1987. Low-Gravity Fluid Mechanics. Berlin: Springer-Verlag.

National Research Council (NRC). 1978. Material Processing in Space. Washington, D.C.: National Academy of Sciences.

NRC, Space Studies Board. 1992. Toward a Microgravity Research Strategy. Washington, D.C.: National Academy Press.

NRC, Space Studies Board. 1995. Microgravity Research Opportunities for the 1990s. Washington, D.C.: National Academy Press.

Ohta, H., M. Kawaji, H. Azuma, K. Kawasaki, S. Okada, S. Yoda, and T. Nakamura. 1998. Microgravity pool boiling on a transparent heating surface (4th Report Experiments by TR-1A Rocket). Pp. 443-444 in Proceedings of the 35th National Heat Transfer Symposium of Japan, Nagoya. Tokyo: Heat Transfer Society of Japan.

Oron, A., S.H. Davis, and S.G. Bankoff. 1997. Long-scale evolution of thin liquid films. Rev. Mod. Phys. 69:931-980.

Ostrach, S. 1982. Low-gravity flows. Annu. Rev. Fluid Mech. 14:313-345.

Peterson, G.G., L.W. Swanson, and F.M. Gerner. 1998. Micro heat pipes. Pp. 295-337 in Microscale Energy Transport. C.-L. Tien, A. Majumdar, and F.M. Gerner, eds. Washington, D.C.: Taylor and Francis.

Raithby, G.D., and K.G.T. Hollands. 1998. Natural convection. Chapter 4 in Handbook of Heat Transfer, 3rd Ed. W.M. Rohsenow, J.P. Hartnett, and Y.I. Cho, eds. New York: McGraw-Hill.

Rose, J.W. 1999. Interphase matter transfer, the condensation coefficient and dropwise condensation. Pp. 89-104 in Heat Transfer 1998: Proceedings of the 11th International Heat Transfer Conference, August 23-28, 1998, Kyongju, Korea, Vol. 1. J.S. Lee, ed. Singapore: Hemisphere Publishing.

Saito, M., N. Yamaoka, K. Miyazaki, M. Kinoshita, and Y. Abe. 1994. Boiling two phase flow under microgravity. Nucl. Eng. Des. 146:451-461.

Shirvanian, A., M. Faghri, Z. Zhang, and Y. Asako. 1998. Numerical solution of the effect of vibration on melting of unfixed rectangular phase-change material under variable-gravity environment. Numerical Heat Transfer, Part A 34:257-278.

Siegel, R. 1967. Effects of reduced gravity on heat transfer. Pp. 143-228 in Advances in Heat Transfer, Vol. 4. J.P. Hartnett and T.F. Irvine, Jr., eds. New York: Academic Press.

Sitter, J.S., T.J. Snyder, J.N. Chung, and P.L. Marston. 1998. Terrestrial and microgravity pool boiling heat transfer from a wire in an acoustic field. Int. J. Heat Mass Transfer 41:2143-2155.

Sommerscales, E.F.C., and A.E. Bergles. 1997. Enhancement of heat transfer and fouling mitigation. Pp. 197-253 in Advances in Heat Transfer, Vol. 30. J.P. Hartnett, T.F. Irvine, Jr., Y.I. Cho, and G.A. Green, eds. San Diego, Calif.: Academic Press.

Sone, K., N. Sone, T. Oka, Y. Abe, Y.H. Mori, and A. Nagashima. 1995. Spray cooling under reduced gravity: Heat transfer characteristics over a wide range of wall-superheating. Pp. 543-544 in Proceedings of the 32nd National Heat Transfer Symposium of Japan, 1995, Yamaguchi, Japan, Vol. II. Tokyo: Heat Transfer Society of Japan.

Straub, J., M. Zell, and B. Vogel. 1990. Pool boiling in a reduced gravity field. Pp. 91-112 in Proceedings of the 9th International Heat Transfer Conference, Vol. 1. G. Hestroni et al., eds. Washington, D.C.: Hemisphere Publishing.

Straub, J. 1994. The role of surface tension for two phase heat transfer in the absence of gravity. Exp. Thermal Fluid Science 9:253-273.

Subramanian, R.S. 1992. The motion of bubbles and drops in reduced gravity. Pp. 1-42 in Transport Processes in Bubbles, Drops and Particles. R.P. Chabra and D. Dekee, eds. New York: Hemisphere Publishing.

Suzuki, K., H. Kawamura, Y. Koyama, and Y. Aoyaina. 1999. Experiments on subcooled water in microgravity (observation of bubble behavior and burnout). On CD-ROM in Proceedings of the 5th ASME/JSME Joint Thermal Engineering Conference, March 15-19, 1999, San Diego, California. New York: American Society of Mechanical Engineers.

Swanson, T.D., A. Juhasz, W.R. Long, and L. Ottenstein, eds. 1989. Workshop on Two-Phase Fluid Behavior in a Space Environment. NASA report CP-3043. Washington, D.C.: NASA Goddard Space Flight Center.

Taitel, Y., and L. Witte. 1996. The role of surface tension in microgravity slug flow. Chem. Eng. Sci. 51:695-700.

Tokura, I., Y. Hanaoka, and H. Saito. 1995. Droplet impingement on a heat transfer surface in microgravity. Pp. 549-550 in Proceedings of 32nd National Heat Transfer Symposium of Japan, 1995, Yamaguchi, Japan, Vol. II. Tokyo: Heat Transfer Society of Japan.

Tonkovich, A.L.Y., D.M. Jimenez, J.L. Zilka, M.J. LaMont, Y. Wang, and R.S. Wegeng. 1998. Microchannel chemical reactors for fuel processing. Technical Report. Richland, Wash.: Pacific Northwest National Laboratory.

Tuckerman, D.B., and R.F.W. Pease. 1981. High-performance heat sinking for VLSI. IEEE Electron Device Lett. EDL-2:126-129.

Verkin, B.I., and Y.A. Kirichenko. 1976. Heat transfer under reduced gravity conditions. Acta Astronautica 3:471-480.

Viskanta, R. 1983. Phase-change heat transfer. Pp. 153-222 in Solar Heat Storage: Latent Heat Materials, Vol. I. G.A. Lane, ed. Boca Raton, Fla.: CRC Press.

Wayner, P.C., Jr. 1998. Interfacial forces and phase change in thin liquid films. Pp. 187-224 in Microscale Energy Transport. C.-L. Tien, A. Majumdar, and F.M. Gerner, eds. New York: Taylor and Francis.

Wegeng, R.S., C.J. Call, and M.K. Drost. 1996. Chemical system miniaturization. Proceedings of the AIChE Spring National Meeting, New Orleans, La., February 25-29. New York: American Institute of Chemical Engineers.

Weinzierl, A., and J. Straub. 1982. Nucleate pool boiling in microgravity environment. Pp. 21-27 in Proceedings of the Seventh International Heat Transfer Conference, Vol. 4. U. Grigull et al., eds. Washington, D.C.: Hemisphere Publishing.

Yamada, H., and T. Fujii. 1999. Convective heat transfer of the two-phase flow under microgravity. Proceedings of the 5th ASME/JSME Joint Thermal Engineering Conference, March 15-19, San Diego, Calif. Paper No. AJTE99-6416. New York: American Society of Mechanical Engineers.

IV.E SOLIDIFICATION

Solidification is a phase change initiated by the nucleation of one or more crystallites in the liquid and proceeding by the growth of these nuclei as latent heat is removed from the solidification front. The impingement of two growing nuclei of different orientation results in a grain (crystal) boundary in the solid. Nucleation occurs by a statistical fluctuation and is either homogeneous in the liquid or heterogeneous on foreign particles or surfaces. In a pure substance, the liquid and solid coexist in equilibrium at a single melting/freezing temperature T_M (for fixed pressure); it is generally possible, however, for the liquid to persist in an undercooled state below T_M provided care is taken to suppress heterogeneous nucleation. In a multicomponent system, there is a range of temperatures at each of which solid and liquid of different composition coexist in equilibrium.

Nucleation is the ultimate amplification process in which an event on an atomic scale is amplified by growth to the macroscopic scale. Indeed, this is the reason for the use of cloud and bubble chambers to reveal atomic/nuclear events via nucleation in a metastable phase. In a solidification context, heterogeneous nucleation is sensitive to minute concentrations of foreign particles or the detailed condition of container surfaces. The driving forces for both nucleation and growth (freezing) increases with undercooling, but the kinetic response to the driving forces decreases with undercooling. Rapid cooling generally favors nucleation relative to growth and results in a fine-grained material.

For growth or solidification to proceed, the latent heat released at the solid/liquid interface must be transported away. In multicomponent systems, components must also be transported to or from the interface to adjust any composition differences between the liquid and solid. This energy and mass transport that accompanies solidification influences the shape evolution of the moving interface and the associated mode of solidification. The influence arises from two usually competing effects: the so-called point effect of diffusion or heat flow, in which local projections of the solid into the liquid facilitate the transport of energy and matter to/from the interface, and the capillary effect, which favors a minimum interface area (more precisely, free energy); the scale and extent of the projections are determined by the balance between these two effects. For example, the point effect of heat flow accounts for the breakdown of a planar into a dendritic shape of a liquid/solid interface advancing into a pure undercooled liquid. In general, the two effects lead to local kinetic equations that determine the time-dependent differential geometry of the evolving interface. Mathematically, the problems described above belong to a class of free-boundary problems in which the configuration of the boundary (liquid/solid interface), on which conditions of temperature and composition fields hold, is not specified in advance but emerges as part of the solution.

The configuration of the advancing liquid/solid interface affects the microstructure of the solid left in its wake. For example, in the dendritic mode, solute may be rejected from the dendritic tips and arms; it then accumulates in the liquid interstices between these features. The result is a segregation pattern in the solid that reflects the dendritic configurations. It is this coupling between microstructure of the frozen solid and the morphological details of the evolving interface shape as related to the temperature and composition fields in the liquid that makes the evolving interface morphology of practical importance.

Solidification occurs in many of the enabling technologies for HEDS, as is indicated in Table III.G.1. Specific subsystems where solidification (freezing) occurs include latent heat-of-fusion thermal energy storage systems and off-design freezing of liquid lines, space radiators, and heat pipes. In addition to the obvious cases of traditional casting methods and unidirectional solidification (Larson and Pirich, 1982; Coriell and McFadden, 1990), it occurs in joining methods such as soldering and welding, in liquid-phase sintering (German et al., 1995), in direct manufacturing processes that depend on rapid solidification of metal powder deposited layer by layer and melted by laser scanning, and in crystal growth from the melt (Hurle et al., 1987; Alexander and Rosenberger, 1990).

The effect of gravity levels on solidification stems from the magnitude of buoyancy-induced convection in the liquid. This affects the distribution of temperature and composition in the liquid near the liquid/solid interface and accounts for the segregation pattern in the solid, as explained above. In addition, buoyancy-driven convection affects the distribution of foreign particles and gas bubbles in the liquid and hence exerts an influence on nucleation rate and on porosity and inclusions in the solid. For example, castings made in microgravity are observed to have more porosity than corresponding castings at 1 *g*, perhaps because gas bubbles do not rise to the surface in microgravity (Abbaschian et al., 1996). It should be emphasized that convection may be driven by nonbuoyancy forces such as the Marangoni effect at a fluid/fluid interface. Discussions of gravity effects in solidification are given by Tabeling (1995), Hurle (1995), and Curreri and Stefanescu (1988).

Research on solidification has focused on fundamental scientific questions using both theoretical and experimental methods. Research on nucleation has focused more on control than on understanding. Suppression of nucleation allows large undercooling and the preparation of potentially interesting metastable solid materials. Nucleation control has been studied experimentally by the use of containerless solidification (Herlach et al., 1993; Shong et al., 1987; Hofmeister et al., 1987; Naumann and Elleman, 1986).

Theoretical research on growth has focused on the following generic question: Given some interface configuration with specified initial conditions of temperature and composition throughout the phases and specified conditions on an external boundary (e.g., a closed system), how will the system evolve? To make the problem complete, boundary conditions for the temperature and composition fields on the moving interface must be specified. These are often taken to be given by the phase diagram adjusted for the effects of capillarity (i.e., surface tension and local curvature), the so-called local equilibrium boundary conditions. To make the problem tractable, capillary effects are often assumed to be isotropic and convection is neglected.

With all the preceding assumptions, the problem becomes a well-posed standard free-boundary problem (Antar and Nuotio-Antar, 1993). Several computational methods have been developed to obtain solutions, but they do not apply to all ranges of parameters of interest (Wang and Sekerka, 1996). Therefore, the problem is often further simplified by developing steady-state solutions involving the simple geometries of an advancing planar or cellular interface or a growing dendrite tip (Billia and Trivedi, 1993), which are then analyzed for stability of shape. One usually finds that a range of steady state stable solutions are possible; the problem then becomes one of finding an additional principle (e.g., marginal stability) that selects the operating point of the system (see Wang and Sekerka, 1996; Langer, 1980).

Experimental tests of theory are based on the control of experimental conditions to realize as much as possible the assumptions underlying the theories. Thus, controlled experiments in microgravity are carried out to minimize convection, and materials are chosen that conform closely to such assumptions as isotropy. The elegant work of Glicksman et al. (1987), Glicksman et al. (1995a-c), Abbaschian (1996), and Bassler et al. (1995) exemplifies these experiments.

Additional theoretical and experimental work has explored the effect of crystal anisotropy, including the appearance of facets, the breakdown of the local equilibrium assumption at high growth speeds and its replacement by kinetic assumptions, and the effects of convection (Sekerka, 1986; Coriell and McFadden, 1993). In some cases, the appropriate physics is not yet fully understood.

Research Issues

Although considerable progress has been made in understanding the principles that determine pattern formation (e.g., cells, dendrites, spatial variation of composition) during solidification, it is still not possible to predict the microstructure of a casting except under very special controlled conditions. Research to address this lack of understanding would include (1) extension of the theoretical, computer modeling, and experimental work supported by NASA to advance the predictive capabilities of solidification theory by including the many complications of crystalline anisotropy, interface kinetics, convection, and the effects of suspended particles and bubble formation and (2) research at the technical level of casting and other practical solidification processes under different gravity conditions to determine the effect of gravity on such casting parameters as grain size, porosity, inclusion distributions, and segregation. The research at the technical level is needed to supplement the theoretical models, which do not yet have sufficient predictive capabilities.

References

Abbaschian, R. 1996. In-situ monitoring of crystal growth using MEPHISTO. Pp. 45-87 in Second United States Microgravity Payload: One Year Report. P.A. Curreri and D.E. McCauley, eds. Huntsville, Ala.: NASA Marshall Space Flight Center.

Alexander, J.I.D., and F. Rosenberger. 1990. Bridgeman crystal growth in low gravity: A scaling analysis. P. 87 in Low-Gravity Fluid Dynamics and Transport Phenomena. J.N. Koster and R.L. Sani, eds. New York: American Institute of Aeronautics and Astronautics.

Antar, B.N., and V.S. Nuotio-Antar. 1993. Materials processing. Fundamentals of Low Gravity Fluid Dynamics and Heat Transfer. Boca Raton, Fla.: CRC Press.

Bassler, B.T., W.H. Hofmeister, and R.J. Bayuzick. 1995. Examination of solidification velocity determination in bulk undercooled nickel. Materials Research Society Fall Meeting, Boston, Mass. Warrendale, Penn.: Materials Research Society.
Billia, B., and R. Trivedi. 1993. Pattern formation in crystal growth. Pp. 899-1073 in Handbook of Crystal Growth, Vol. IB. D.T.J. Hurle, ed. Amsterdam: North-Holland.
Coriell, S.R., and G.B. McFadden. 1990. Instability during directional solidification: Gravitational effects. P. 369 in Low-Gravity Fluid Dynamics and Transport Phenomena. J.N. Koster and R.L. Sani, eds. New York: American Institute of Aeronautics and Astronautics.
Coriell, S.R., and G.B. McFadden. 1993. Morphological stability. P. 785 in Handbook of Crystal Growth, Vol. 1B. D.T.J. Hurle, ed. New York: Elsevier.
Curreri, P.A., and D.M. Stefanescu. 1988. Low-Gravity Effects During Solidification. P. 147 in Metals Handbook, 9th Ed., Vol. 15. Metals Park, Ohio.: ASM International.
German, R.M., R.G. Iacocca, J.L. Johnson, Y. Liu, and A. Upadhyaya. 1995. Liquid-phase sintering under microgravity conditions. J. Met. 47(8):46-48.
Glicksman, M.E., and S.P. Marsh. 1993. The dendrite. Pp. 1075-1122 in Handbook of Crystal Growth, Vol. IB. D.T.J. Hurle, ed. Amsterdam: North-Holland.
Glicksman, M.E., E. Winsa, R.C. Hahn, T.A. Lograsso, R. Rubinstein, and M.E. Sellick. 1987. Isothermal dendritic growth. P. 37 in Materials Processing in the Reduced Gravity Environment of Space. R.H. Doremus and P.C. Nordine, eds. Materials Research Society (MRS) Symposia Proceedings, Vol. 87. Warrendale, Pa.: Materials Research Society.
Glicksman, M.E., M.B. Koss, L.T. Bushnell, and J.C. LaCombe. 1995a. The isothermal dendritic growth experiment: Implications for theory. P. 663 in Modeling of Casting, Welding, and Advanced Solidification Processes VII. M. Cross and J. Campbell, eds. Warrendale, Pa.: Minerals, Metals, and Materials Society.
Glicksman, M.E., M.B. Koss, L.T. Bushnell, J.C. LaCombe, and E.A. Winsa. 1995b. Dendritic growth of succinontrile in terrestrial and microgravity conditions as a test of theory. ISIJ International 35(6):1216.
Glicksman, M.E., M.B. Koss, L.T. Bushnell, J.C. LaCombe, and E.A. Winsa. 1995c. Dendritic growth in terrestrial and microgravity conditions. P. 13 in Fractal Aspects of Materials: Proceedings of Materials Research Society (MRS) Fall 1994 Symposium, Vol. 367. F. Family, P. Meakin, B. Sapoval, and R. Wool, eds. Warrendale, Pa.: Materials Research Society.
Herlach, D.M., R.F. Cochrane, I. Egry, H.J.Fecht, and A.L. Greer. 1993. Containerless processing in the study of metallic melts and their solidification. Int. Mat. Rev. 38:273-347.
Hofmeister, W., M.B. Robinson, and R.J. Bayuzick. 1987. Undercooling of bulk high temperature metals in the 100 meter drop tube. P. 149 in Materials Processing in the Reduced Gravity Environment of Space: Proceedings of the Materials Research Society (MRS) Symposium, Vol. 87. R.H. Doremus and P.C. Nordine, eds. Warrendale, Pa.: Materials Research Society.
Hurle, D.T.J. 1995. Crystallization processes. European Low-Gravity Physical Sciences in Retrospect and in Prospect. ELGRA Report. Paris: European Low Gravity Research Association (ELGRA).
Hurle, D.T.J., G. Müller, and R. Nitsche. 1987. Crystal growth from the melt. P. 313 in Fluid Sciences and Materials Science in Space. H.U. Walter, ed. New York: Springer-Verlag.
Langer, J.D. 1980. Instabilities and pattern formation in crystal growth. Rev. Mod. Phys. 52:1-28
Larson, D.J., and R.G. Pirich. 1982. Influence of gravity driven convection on the directional solidification of Bi/MnBi eutectic composites. P. 523 in Materials Processing in the Reduced Gravity Environment of Space: Proceedings of the Materials Research Society (MRS) Symposium. G.E. Rindone, ed. Warrendale, Pa.: Materials Research Society.
Naumann, R.J., and D.D. Elleman. 1986. Containerless processing technology. P. 294 in Material Science in Space. B. Feuerbacher, H. Hamacher, and R.J. Naumann, eds. New York: Springer-Verlag.
Sekerka, R.F. 1986. Phase interfaces: Morphological stability. P. 3486 in Encyclopedia of Materials Science and Engineering. M.B. Bever, ed. New York: Pergamon.
Shong, D.S., J.A. Graves, Y. Ujiie, and J.H. Perepezko. 1987. Containerless processing of undercooled melts. P. 17 in Materials Processing in the Reduced Gravity Environment of Space: Proceedings of the Materials Research Society (MRS) Symposium, Vol. 87. R.H. Doremus and P.C. Nordine, eds. Warrendale, Pa.: Materials Research Society.
Tabeling, P. 1995. Solidification and nucleation. European Low-Gravity Physical Sciences in Retrospect and in Prospect. ELGRA Report. Paris: European Low Gravity Research Association (ELGRA).
Wang, S.-L., and R.F. Sekerka. 1996. Computation of the dendritic operating state at large supercoolings by the phase field model. Phys. Rev. E 53(4):3760.

IV.F CHEMICAL TRANSFORMATION

Combustion

Behavior of Combustion Phenomena in Microgravity

Gravitational acceleration influences combustion phenomena because of the large density differences that appear as a consequence of the large temperature differences that result from the exothermic chemical reactions that characterize combustion processes. Density changes of nearly a factor of ten in combustion gases are not uncommon.

That buoyancy forces are important in many combustion phenomena on Earth is evident through examination of the Grashof number, the ratio of buoyancy to viscous forces. For Earth gravity and a density ratio of about 10, this force ratio is not small for physical scales of the size of about 0.1 m or more. Because the Grashof number increases as the cube of the physical scale, the influence of buoyancy increases rapidly compared to viscous effects as the scale approaches the usual laboratory scales of experimental investigation. Consequently, "quiescent" combustion experiments in earthbound laboratories are nearly impossible to conduct unless some element of free fall is present. Drop towers, for example, are earthbound laboratories in which quiescent experiments can be conducted, but they are limited to experimental times of approximately 10 s and less, generally between 2 and 5 s.

For forced flows to overwhelm buoyancy and eliminate its influence, i.e., for the Reynolds number to be much larger than the square of the Grashof number, forced-flow velocities on the order of a meter per second or more are needed to suppress the influence of buoyancy. Slow-flow combustion phenomena are therefore difficult to investigate on Earth without interference from buoyancy.

While some combustion phenomena are not influenced by buoyancy, several important ones are: mixture flammability, instability, gas diffusion flames, droplet combustion, particle cloud combustion, smoldering, and flame spread (Law, 1990; Sacksteder, 1990; NRC, 1995). The importance of these phenomena to HEDS and the associated reduced-gravity environments stems either from their role in fire safety for space travel and the habitation of distant planets or from their use in processes such as materials production or construction during space missions. Each of these combustion phenomena is addressed separately below. Because those aspects of the phenomena that are of HEDS interest are interrelated, the research recommendations are grouped together at the end of the section rather than listed under each phenomenon.

Mixture Flammability

Whether a premixed mixture of fuel and oxidizer is flammable following a sufficient input of ignition energy is a question of importance to fire safety. Mixtures exhibit both lean and rich flammability limits on Earth, with the limits for upward-propagating flames being wider than those for downward-propagating flames. For downward-propagating flames the burnt gases are above the unburnt gas, while for upward-propagating flames the opposite is true, which leads to a curvature in flame shape as the burnt gases tend to rise under buoyancy into the unburnt gases to enhance flammability. This curvature produces flame stretch, which influences the flame temperature and hence flame propagation. Depending on the magnitude and sign of the product of the strain rate with the transit time through the flame and the Lewis number (the ratio of the thermal diffusivity to the mass diffusivity of the less abundant reactant), flame stretch can widen or narrow the flammability limits (Law, 1990). At reduced gravity and reduced flame stretch, it is possible for the limits to be outside the normal-gravity limits (Law, 1990; Sacksteder, 1990).

At one time, because flammability limits were thought to exist only as a result of the influence of gravity, it was thought that there would be no flammability limits at zero gravity. However, flammability limits at reduced gravity have been found, and they are hypothesized to exist because of the relatively enhanced influence of radiative losses at reduced gravity and/or the effects of chemical kinetics; as such, they would be fundamental limits (Law, 1990).

Near-limit premixed flames at reduced gravity exhibit unusual behavior not observed at normal gravity. For Lewis numbers less than unity, spherically expanding flames propagate and then extinguish. The extinction occurs as a result of enhanced radiative loss and the reduced effects of flame stretch as the flame radius increases. Under certain circumstances, "stationary" spherical flames, or flame balls, have been observed (Ronney et al., 1998). Because these flame balls, which require radiative heat losses for their stability, are "convectionless," they cannot exist at normal gravity, where there is convective flow through buoyancy (Law, 1990; Sacksteder, 1990).

Flame Instabilities

Flame front instability, which results in flames whose surfaces are not smooth but instead contain cellular structures, is due to effects associated with heat and mass diffusional processes and hydrodynamic effects that give rise to the curved shape of upward-propagating flames. Buoyancy is stabilizing for downward propagation, and its removal makes the remaining effects dominant, thus allowing for the near-limit phenomena observed in reduced gravity such as the flame balls mentioned above (Law, 1990).

Gas Diffusion Flames

In gas jet diffusion flames, in which the flame is formed from a jet of gaseous fuel issuing from a burner tube into an oxidizer, buoyancy is generally important. The flame stabilizes as a premixed flame near the burner rim, where the flow speed and flame speed match. Because buoyancy affects the flow speed near the burner rim, removal of buoyancy affects the stabilization mechanism. Additionally, laminar flames are longer and wider in reduced gravity than in normal gravity and generate more soot. Radiation losses increase, and flame temperatures decrease (Sacksteder, 1990).

Droplet Combustion

Diffusion flames surrounding liquid fuel droplets become spherical in the absence of gravity, mirroring the configuration employed in classical droplet-burning theory in which the square of the droplet diameter decreases nearly linearly with time and the flame diameter decreases as the droplet diameter decreases. At reduced gravity, however, unsteady effects are observed in which burning rates and flame diameters initially increase slowly with time. Additionally, soot production is enhanced, and a soot shell may form at the location where the thermophoretic transport of soot back toward the droplet surface is balanced by the outward drag on the soot particles from the outward fuel flow (Law, 1990; Nayagam et al., 1998). At normal gravity, the soot mantle is swept away from the lower, windward side of a droplet and consumed in the upper reaches of the flame plume above the droplet.

Cloud Combustion

Arrays or clouds of droplets or combustible particles may exhibit different flame propagation characteristics at normal and reduced gravity. With settling at normal gravity, upward-propagating flames propagate through an initially richer mixture while downward-propagating flames propagate through an initially leaner mixture. Clouds that may not be flammable because of such settling (they are either initially too rich or too lean) may sustain propagation when uniformly dispersed under reduced gravity (Sacksteder, 1990).

Smoldering

Smoldering combustion, i.e., the slow surface oxidation of a combustible solid, has practical ramifications for safety. In normal gravity, buoyancy enhances oxygen transport to and product removal from the reacting surface. In reduced gravity, this transport mechanism is absent. Results to date show that carbon monoxide production in smoldering combustion is enhanced substantially in reduced gravity (NRC, 1995; Stocker et al., 1996), but the prevalence of this effect is unknown.

Flame Spread

Flame spreading over solid and liquid surfaces has direct implications for fire safety and materials selection. Generally, it is classified as either opposed flow or concurrent flow (with respect to the oxidizer flow). Upward spread in normal gravity is concurrent and tends to be acceleratory. Opposed flow spread tends to allow steady spread to develop. In the absence of a wind, flame spread at reduced gravity is of the opposed flow type, with the opposing flow with respect to the flame equal to the flame propagation speed (West et al., 1996).

In opposed-flow flame spread, the spread rate is determined by the upstream transfer to the unburnt fuel of the heat needed to vaporize it and how this heat transfer is affected by chemical kinetics. Flame extinction occurs at high velocities as a result of kinetic effects and flame blowoff. Flame extinction has also been found to occur at reduced gravity in quiescent environments as a result of radiation loss effects that reduce the flame temperature and propagation speed (this effect is suppressed in normal gravity because of the necessary presence of the induced flow). The reduced speed makes it difficult for oxygen to be transported to the flame, and extinction occurs (Altenkirch et al., 1998).

For thin fuels, the limiting oxygen concentration below which propagation will not occur is higher in reduced gravity than in normal gravity. However, a low-speed opposing flow enhances flammability for thin fuels at reduced gravity, such that the limiting oxygen concentration is below that for normal-gravity downward spread (Law, 1990; Olson, 1991). Consequently, there appears to be a minimum oxygen concentration below which spread does not occur, approximately 15 percent for cellulosic materials (Olson, 1991). The limiting concentration is higher in normal gravity, which indicates that flammability at reduced gravity may be greater, although flame spread rates are lower, under certain circumstances, than at normal gravity.

For spread over thick fuels, there appears to be no steady state at reduced gravity. The increased in-depth conduction needed to raise the temperature of the heated layer in the solid causes the flame to spread more slowly than for thin fuels, which enhances heat loss by flame radiation. The enhanced radiation causes a reduction in flame size such that the flame shrinks into a region of continually decreasing oxygen concentration. Eventually, the decreased oxygen transport results in flame extinction. Apparently, the higher spread rates for thin fuels prevent this phenomenon, and so thin fuels exhibit steady spread (Altenkirch et al., 1998).

For liquid fuels, surface tension gradients cause hot liquid fuel to be drawn out from under the flame and brought in front of it to establish the spread rate. For shallow pools, in which buoyancy would be absent even at normal gravity, when the flame spreads at a uniform rate, normal and reduced gravity give the same result. Numerical modeling predicts that pulsation will occur in microgravity with forced convection (Schiller and Sirignano, 1996), although experimental results seem not to indicate such pulsation (Ross and Miller, 1996, 1998). A tentative explanation for this discrepancy is that modeling is two-dimensional while the scale of the experiments implies three-dimensionality, and expansion normal to the propagating flame is thought to be responsible for dampening the pulsation.

Implications of the Behavior of Combustion Phenomena in Microgravity for Spacecraft Design and Operations

Differences in combustion phenomena at normal and reduced gravity have implications for fire safety, spacecraft materials selection and utilization, especially interior materials, environment selection, interior environment exchange and ventilation, fire detection and suppression, propulsion, and (potentially) manufacturing/materials synthesis that relies on the maintenance of exothermic chemical reactions (NASA 1992a,b). Flammability of materials is an issue of signal importance to spacecraft fire safety. Because spacecraft inhabitants are virtually captive within the spacecraft, it is imperative that construction materials be selected so that the threat of fire is minimized, and care should be taken to ensure that the spacecraft breathable environment also minimizes fire risk. Methods of installation, for example, that preclude electrical overheating, smoldering, and flaming, are necessary to minimize fire risk.

While the respiratory system responds to the partial pressure of oxygen, fires respond to the concentration of oxygen. Consequently, judicious selection of breathable environments, whether they be in the spacecraft per se or within inhabitant's space suits, should maintain suitable partial pressures of oxygen while minimizing, insofar as possible, oxygen concentration.

Detection systems designed with the assumption of buoyancy in mind (e.g., smoke detectors) need rethinking. Technologies different from those usually used on Earth (e.g., radiation sensors) or different applications of existing technologies (e.g., the use of forced ventilation for environment throughput, as opposed to reliance on buoyancy-driven flow for smoke detectors, as is done on Earth) may be necessary. Fire suppression systems must not produce end products toxic to humans (e.g., Halon extinguishers), and they must not produce situations that could pose additional fire safety concerns (e.g., liquid invasion of electrical systems). Fire suppression requires transport of the suppressant to the fire, and this transport often is affected by reduced gravity. It is, for example, beneficial to deliver water or carbon dioxide to the base of a fire, a process that can be aided by buoyant inflow at normal gravity.

While it is unlikely that any combustion process in combustion-based propulsion systems will be affected per se by reduced gravity, because of the relatively high velocities developed compared to buoyancy-generated velocities, care should be taken to ensure that the overall system is not adversely affected by reduced gravity without some compensation for that environment. For example, fluid transport of combustibles is affected by reduced gravity, and that should be taken into account in propulsion systems design.

There may be a potential for combustion to be used in necessary manufacturing processes. Consequently, depending on the process, reduced gravity may play a role, e.g., in the transfer of heat from flames for processing and in direct, high-temperature materials synthesis.

Affected Technologies

As discussed in Chapter III, the technologies affected in the presence of reduced gravity that relate to combustion include fire detection and suppression technologies, as mentioned above; electrical/electronic packaging to minimize the potential for overheating and smoldering combustion; ventilation control in the presence of a fire; and fluid distribution systems associated with propulsion. In addition, there is the potential, though as yet undetermined, influence on technology for materials synthesis during exothermic reaction.

Research Issues

The main issue surrounding combustion as it relates to HEDS activities is safety, which has implications for materials selection, environment selection, fire detection, fire management, and fire suppression. Further work on materials and environment selection is needed, but the emphasis should be on robust fire detection and suppression in reduced gravity.

As discussed previously, a number of areas of reduced-gravity combustion research are particularly relevant to improving fire detection and suppression in a reduced gravity environment, including flammability and flame behavior (such as flame instabilities and dynamics under different gravity conditions); diffusion-flame structure and behavior for gaseous, liquid, and solid fuels, especially the production of soot and toxic products in such flames and conditions necessary for their extinction; smoldering rates at reduced gravity, with special emphasis on conditions for initiation and termination of smoldering and on products of smoldering combustion; and, finally, flame spread phenomena for various types of fuels. Besides safety-related issues such as these, there is a need, albeit a less pressing one, for reduced-gravity research on materials synthesis and materials processing through combustion.

Without practical, agreed-upon means for detection and suppression, fire looms as a potential cause of mission failure. Fluid distribution in propulsion systems and the potential for materials synthesis are combustion areas in which implications for successful HEDS activity may reside, and which also deserve attention, though certainly not to the same extent as fire safety.

Pyrolysis

As described in Chapter III, pyrolysis is the mechanism by which chemical transformations are brought about by application of heat. Only a small number of the many pyrolysis processes that could be of interest for HEDS are described in Chapter III. The large temperature changes in practical pyrolysis imply appreciable gravity effects by virtue of the associated changes in density. It may therefore be expected that reduced-gravity issues arise in pyrolysis. The effects would involve influences of gravity on the transport of reactants and products to and from regions of pyrolysis, and so pyrolysis processes that may be of use, such as oxygen-production processes, will need study to determine these effects.

Many pyrolysis processes that occur during combustion are discussed in the preceding section. For example, soot production in diffusion flames involves pyrolysis of the gaseous fuels in fuel-rich regions of the flow. These pyrolysis processes are of importance in fire protection aboard spacecraft, for example.

Research Issues

There are many different kinds of pyrolysis processes, with reactants and products in different phases and with very different temperature thresholds and time scales. For this reason, microgravity studies of pyrolysis would need to be pursued on individual bases, separate for each process. One potentially relevant process is the high-temperature recovery of oxygen from silica in lunar regolith. Another is soot production from hydrocarbon fuels. A third is gaseous fuel production from cellulosic fuels. Each process proceeds at successively lower temperatures. Many other pyrolysis processes with potential relevance to HEDS can be identified, but each would need to be evaluated separately with respect to its HEDS relevance and its sensitivity to microgravity. However, it can be noted that general categories of processes affected by gravity include gas production from solids and gas-phase chemical transformations in a flow field.

Solution Chemistry

While the interaction of individual molecules in a solution is not expected to be directly affected by gravity levels (except possibly in the case of very large molecules such as proteins), there are numerous ways in which the effects of gravity on bulk fluid flow might either inhibit or enhance the efficiency of the chemical reactions carried out on a HEDS mission. Incomplete mixing of reactants is possibly the chief area of concern for solution reactions in reduced gravity. On Earth, density-driven convection, particularly in reactions requiring heating, is relied upon as the default method of mixing reactant solutions. While this driving force would be absent or reduced in low gravity, complete mixing could still be accomplished in most cases by the use of mechanical stirrers and by paying careful attention to the design of reaction chambers.

On the other hand, for a multiphase mixture of immiscible phases with different densities, the surface area at which a reaction could take place might be greatly enhanced in microgravity. After mixing, density differences would rapidly separate such a solution into layers on Earth, whereas in reduced gravity the dispersed droplets, with their greater surface area, could remain suspended for a longer period of time, allowing reactants in the two phases to interact at a higher rate. Surface tension would still be present, however, as a driving force for coalescence of the phases and thus a reduction in the surface area of reaction.

A better understanding of phase distribution and separation in low gravity is cited as a need in a number of sections in this report and so is not discussed in detail here, except to note that such an understanding might also be applied to enhancing the efficiencies of some types of chemical reactions, such as those involving immiscible phases, in low gravity.

References

Altenkirch, R.A., L. Tang, K. Sacksteder, S. Bhattacharjee, and M.A. Delichatsios. 1998. Inherently unsteady flame spread to extinction over thick fuels in microgravity. Pp. 2515-2524 in Twenty-Seventh Symposium (International) on Combustion. Pittsburgh: Combustion Institute.

Law, C.K. 1990. Combustion in microgravity: Opportunities, challenges, and progress. AIAA-90-0120. New York: American Institute of Aeronautics and Astronautics.

National Aeronautics and Space Administration (NASA). 1992a. 1991 Integrated Technology Plan for the Civil Space Program. NASA TM-107988. Washington, D.C.: NASA.

National Aeronautics and Space Administration (NASA). 1992b. Review of NASA's Integrated Technology Plan for the Civil Space Program. NASA TM-107966. Washington, D.C.: NASA.

National Research Council (NRC), Space Studies Board. 1995. Microgravity Research Opportunities for the 1990s. Washington, D.C.: National Academy Press.

Nayagam, V., J.B. Haggard, Jr., R.O. Colantonio, A.J. Marchese, F.L. Dryer, B.L. Zhang, and F.A. Williams. 1998. Microgravity *n*-heptane droplet combustion in oxygen-helium mixtures at atmospheric pressure. AIAA J. 36(8):1369-1534.

Olson, S.L. 1991. Mechanisms of microgravity flame spread over a thin solid fuel: Oxygen and opposed flow effects. Combust. Sci. Technol. 76(4-6):233-249.

Ronney, P.D., M.-S. Wu, H.G. Pearlman, and K.J. Weiland. 1998. Experimental study of flame balls in space: Preliminary results from STS-83. AIAA J. 36(8):1361-1368.

Ross, H.D., and F.J. Miller. 1996. Detailed experiments of flame spread across deep butanol pools. Pp. 1327-1334 in Proceedings of the Twenty-Sixth Symposium (International) on Combustion. Pitttsburgh: Combustion Institute.

Ross, H.D., and F.J. Miller. 1998. Flame spread across liquid pools with very low-speed opposed or concurrent airflow. Pp. 2723-2729 in Proceedings of the Twenty-Seventh Symposium (International) on Combustion. Pitttsburgh: Combustion Institute.

Sacksteder, K.R. 1990. The implications of experimentally controlled gravitational accelerations for combustion science. Pp. 1589-1596 in the Proceedings of the Twenty-Third Symposium (International) on Combustion. Pittsburgh: Combustion Institute.

Schiller, D.N, and W.A. Sirignano. 1996. Opposed-flow flame spread across n-propanol pools. Pp. 1319-1325 in Proceedings of the Twenty-Sixth Symposium (International) on Combustion. Pittsburgh: Combustion Institute.

Stocker, D.P., S.L. Olson, D.L. Urban, J.L. Torero, D.C. Walther, and A.C. Fernandez-Pello. 1996. Small-scale smoldering combustion experiments in microgravity. Pp. 1361-1368 in Proceedings of the Twenty-Sixth Symposium (International) on Combustion. Pittsburgh: Combustion Institute.

West, J., L. Tang, R.A. Altenkirch, S. Bhattacharjee, K. Sacksteder, and M.A. Delichatsios. 1996. Quiescent flame spread over thick fuels in microgravity. Proceedings of the Twenty-Sixth Symposium (International) on Combustion. Pittsburgh: Combustion Institute.

IV.G BEHAVIOR OF GRANULAR MATERIALS

Lunar and Martian Regolith

Characterizing and understanding the behavior of granular materials in different gravitational fields are of critical importance to the HEDS program. In order to carry out tasks ranging from determination of the energy requirements for excavation operations and estimation of the load-carrying capacity of extraterrestrial surfaces, through the handling of granular materials in conveying systems and the prediction of local surface properties of planetary bodies under various gravitational conditions, the influence of gravity on the behavior of granular materials must be better understood. Jaeger, Nagel, and Behringer (1996), in their useful overview of many aspects of granular material behavior, discuss the conditions under which these material systems behave like solids, liquids, or gases, depending on environmental conditions. Granular materials of interest in this discussion are soils, regoliths, and other similar resources that can exhibit cohesion and are arrangements of rigid particles in frictional contact. Because gravity contributes to the normal stress force of interaction between particle surfaces and frictional forces are typically proportional to the normal (hence gravitational) forces, the elastic behavior of granular soils, which has both axial and radial force components, is strongly dependent on gravity. In fact, when granular particle assemblies are produced in terrestrial environments with particle density levels that are sufficient to maintain continuous contact between adjacent particles, gravity creates internal stress distributions that are highly nonhomogeneous and anisotropic. Many experiments have shown that the internal stress distributions produced even in highly simplified (identical particle) granular assemblies are distinctly different when test conditions are repeated using the same apparatus. Great care thus will be required to isolate gravitational effects from effects produced by random assembly variations. It is also extremely difficult to conduct terrestrial experiments on the behavior of granular systems (other than angle-of-repose experiments) that are not controlled by container boundaries.

The ability to construct stable structures and roadways that can carry loaded vehicles on planetary surfaces (geotechnical engineering) is a critical element in designing equipment for testing, processing, and transporting these soils. The soil properties depend strongly on the shape of the individual particles, which can vary from very angular to well rounded. Specifically, it is known that owing to the thermal and mechanical bombardment events undergone by the Moon, its surface materials include nearly spherical glass particles and very irregular fine particles. In contrast, because of the apparent hydrologic epoch that characterized some portion of Mars's geologic history and because Mars continues to possess an atmosphere that sustains aeolian erosion, it is likely that Mars regolith is composed of less-abrasive granular materials and therefore has a mechanical behavior different from that of the lunar regolith. It is already known that the lunar surface is characterized by the extreme density of its soil only a few centimeters beneath its surface (Carrier and Mitchell, 1990), compaction that exceeds that produced by even heavy compaction equipment on Earth. Hence, even though it will be extremely difficult to excavate material or even drill holes in the lunar surface without employing such techniques as vibratory excavation (Klosky, 1997), undisturbed lunar soil should be able to carry heavy loads.

The behavior of dust in reduced gravity is also an important consideration. While the composition, particle shape, and atmospheric loading of Mars dust is neither fully known nor understood at this time, their potential impact on future HEDS missions must be considered. Lunar dust, whose characteristics were briefly described above, presents a different set of problems. It is considered to be one of the most difficult design constraints for lunar base construction since the lunar fines are abrasive minerals that, in the desiccated lunar environment, are electrostatically sticky and adhere very tenaciously to most surfaces. Furthermore, there are still questions about what types of attractive forces cause the dust to cling to surfaces and therefore about the cleaning methods that can be used to manage lunar dust (Perko, 1998).

The soil's void ratio (the ratio of void to solids), characteristic particle dimensions, and relative density (the ratio of soil density to that of solid soil without voids) all bear directly on the soil strength. The presence of smooth inclusions even at a level of a few percent can cause a significant decrease in soil strength (Klosky, 1997). Granular materials under self-weight are used terrestrially to construct structures such as dams, road embankments, mine waste dumps, and ore stockpiles. The edges of such piles cannot be steeper than the angle of repose. If more material is added and the maximum stability angle is exceeded, avalanches occur. Well-established failure criteria have been discussed in the literature, but the dominant design criteria for frictional, granular material continue to be Mohr-Coulomb (M-C) failure measurements (Wood, 1990; Craig, 1992). These measurements of granular shearing usually focus on the mean properties of the system and use techniques such as conventional triaxial compression (Klosky, 1997). Shear stress is applied to a sample under normal load, and the grains respond elastically up to the yield point, where shear displacement occurs. When this failure occurs, the measured shear force decreases. Simultaneous measurement of shear stress and normal stress define M-C failure envelopes. A linear connection through the maxima of these plots has a slope that can be interpreted as the friction angle. The intercept where the normal stress is zero gives c, the apparent cohesion (the shear stress necessary to overcome the cohesion), which is the force that enables the soil to cling together in opposition to the forces tending to separate it into parts. Cohesion implies a surface-surface particle interaction such that as cohesion decreases, particles move more readily and pack more densely under a given set of stress conditions. The interaction forces between grain particles include hard-body interactions, friction, and inelasticity, as characterized by a coefficient of restitution less than unity. The dissipative nature of these interactions causes even an energetic (possessing random motion) collection of grains to coalesce into a dense, compact state. Such a compact state is usually very inhomogeneous with regard to the forces acting on individual grains, and large fluctuations in the local forces have been observed (Behringer et al., 1999), with recent studies beginning to provide new insight into their characteristics.

Research Issues

The continued measurement and characterization of lunar and Martian regolith would have a high priority in any attempt to establish the groundwork needed for the exploration activities that would support the development of extraterrestrial base stations. While important characteristics of the mean properties of the regolith are measured by such techniques as Mohr-Coulomb failure criteria, other aspects are much less well understood: the kinetics of Coulomb friction, the internal variables and energy fluctuations, and the effects of agitation on particle size separation.

Kinetics of Granular Flow

The spontaneous flow of granular materials out of the base of a silo or hopper requires the presence of a driving force such as gravity. At reduced gravity, such as lunar or Martian gravity, the flow will likely change, but in ways that are not now obvious; for instance, unlike a true fluid, the hydrostatic pressure is apparently insensitive to the depth of the granular material in containers. The contact forces between the grains, and the static friction with the sides of the container, allow the sand in the hourglass to flow through the orifice at a nearly constant rate. Thus, when granular material is held in a silo, no height-dependent static pressure head occurs, as it would with a liquid. The pressure reaches a maximum value independent of height, and these flows will require further investigation of gravitational effects (Jaeger et al., 1996). Because many factors contribute to the soil properties, the analysis of these flows is not trivial. Gravity could be replaced by using another driving force such as electrical or magnetic. The former would require imparting a charge to the particles and then subjecting them to an electric field. If the particles were ferromagnetic, such as the agglutinates in an Apollo 11 sample, then a magnet could drive the flow (Agosto, 1985)

While the Mohr-Coulomb failure criteria for granular material has permitted the measurement of important mean properties of granular material subjected to shear, the measurements are in a three-dimensional system, so that the normal stress is not separated from the shear. It would be useful to complement this information with direct measurements of the forces and displacements of individual grains in a shearing experiment. Studies of three-dimensional systems show that in systems with 10 to 100 grains, fluctuations in stress are enormous, often more than an order of magnitude greater than the mean stress; this factor is usually ignored in modeling, impairing the ability to predict the state of such granular materials (Miller et al., 1996). Stresses developed in static piles of cohesionless granular material have received considerable attention (Savage, 1997). Dense, slow flows and rapid, gaslike flows (Schäfer et al., 1996) are useful idealizations for the development of models (Jaeger et al., 1996), and real systems often display both flows simultaneously in different spatial domains. As discussed above, particles coalesce into a dense, packed state, so that to maintain granular material flow at low density, energy must be continuously supplied, for instance, by shaking. To achieve a flow of dense material, enough shear stress must be applied to exceed some yield point, where grains begin to slide past each other. Applied stress (including gravity) is therefore an important parameter, but other important effects include convection, size separation, and mixing.

Using a two-dimensional system, where the effects of gravity are effectively removed, some recent experiments have shown that the packing density of grains subjected to shear undergo a novel kind of phase transition at a critical packing density (Behringer et al., 1999; Veje et al., 1999). In these experiments the grains are simulated by the use of photoelastic disks that are birefringent under stress/strain so that cross polarizers show light/dark regions. The axially emitted light shows bright and dark bands in the polariscope, and from this observation, it is possible to determine the applied shear forces. The disks, which are on a smooth slippery sheet, are confined at their inner and outer radii by roughened surfaces that apply shear stresses when rotated. The width of the characteristic shear band that forms near the inner wheel depends slightly on the packing fraction and is about six disks in radial direction. Virtually all azimuthal motion of the disks occurs in this band, and the remaining outer disks remain nearly frozen. The disks in the shear band dilate, compacting the disks in the outer region of the experiment. Rearrangements of this type can also occur at dense packings and influence the statistical and mean properties of the flow over relatively long times. In studies of shear from the inner radius, it is observed that there is a critical packing fraction of the grains, Γ_c, of 0.77, at which point the system undergoes a change from complete slipping, where the disks can remain indefinitely at rest without shear, to a state of nonslipping dynamics, with closely packed grains subject to some shear stress at all times. This transition has some resemblance to a phase transition. Just above Γ_c, fluctuations are temporally intermittent, and the resulting stress chains tend to be long. As Γ is increased further, the system becomes more homogeneous, because strong contacts now deform enough to allow other contacts to take up the load. The strong dependence of the dynamic behavior on grain packing density gives new insight into the properties of granular materials.

When granular materials are agitated in Earth gravity, size segregation of grains can occur by several mechanisms, causing preferential filling of the space beneath large particles by smaller particles (Jenkins and Louge, 1997). Gravity imparts buoyant forces that separate grains of different sizes. These forces compete with gradients in concentration and energy and can result in convection cells in which particles with different properties separate (Knight et al., 1993). In the absence of gravity, only the simpler balance between gradients in concentration and in fluctuation energy remains; important studies in this area are in progress (Jenkins and Louge, 1997). It would appear that vibrating containers can be designed for operation in microgravity that will be capable of separating particles by their size (Rosato et al., 1987; Williams, 1976). An area receiving considerable attention in the physics community is the stirring of granular materials by means of paddles to provide mixing without segregation; the design principles have been discussed by Khakhar et al. (1997) and Ottino (1990). While these systems would likely be used in pressurized environments other than the vacuum of space, it is important to note that the process for separating particles by size has not been validated in environments that are at pressures below 10 torr (Pak et al., 1995).

Research Issues

Modeling and predicting the behaviour of granular materials is an important activity that has been the subject of renewed interest. The classical descriptions of dense granular material use static arrays of grains with typically undetermined Coulomb friction forces. Recent studies, described above, show that local forces have large fluctuations and are very sensitive to small perturbations in packing density, and such effects must be included in statistical descriptions and models of granular materials. The modeling of granular materials under applied stress is important to a number of HEDS activities, such as construction, surface transport, and materials processing at low pressure. Studies that separate or counter the effects of gravity while examining the effects of shearing on granular behavior in three dimensions would help in understanding the phase transition observed at a critical packing density of granular material.

References

Agosto, W.N. 1985. Electrostatic concentration of lunar soil minerals. Pp. 453-464 in Lunar Bases and Space Activities of the 21st Century. W.W. Mendell, ed. Houston: Lunar and Planetary Institute.

Behringer, R.P., D. Howell, L. Kondic, S. Tennakoon, and C. Veje. 1999. Predictability and granular materials. Physica D: Nonlinear Phenomena 130(1-2):1-17.

Carrier, W.D., III, and J.K. Mitchell. 1990. Geotechnical engineering on the moon. Pp. 51-58 in de Mello Volume: A Tribute to Prof. Dr. Victor F.B. de Mello. E. Blucher, ed. São Paulo, Brazil: Editora Edgard Blücher.

Craig, R.F. 1992. Soil Mechanics. London: Chapman and Hall.

Jaeger, H.M., S.R. Nagel, and R.P. Behringer. 1996. RMP colloquium: Granular solids, liquids, and gases. Rev. Mod. Phys. 68:1259.

Jenkins, J.T., and M.Y. Louge. 1997. Pp. 539-542 in Powder and Grains 97: Proceedings of the 3rd International Conference. R.P. Behringer and J.T. Jenkins, eds. Brookfield, Vt.: A.A. Balkema.

Khakhar, D.V., J.J. McCarthy, and J.M. Ottino. 1997. Radial segregation of granular mixtures in rotating cylinders. Phys. Fluids 9:3600-3614.

Klosky, J.L. 1997. Behaviour of composite granular materials and vibratory helical anchors. Ph.D. dissertation. University of Colorado.

Knight, J.B., H.M. Jaeger, and S.R. Nagel. 1993. Phys. Rev. Lett. 70:3728.

Miller, B., C. O'Hern, and R.P. Behringer. 1996. Stress fluctuations and continuously sheared granular materials. Phys. Rev. Lett. 77:3110.

Ottino, J.M. 1990. The Kinematics of Mixing: Stretching, Chaos and Transport. Cambridge: Cambridge University Press.

Pak, H.K., E. van Doorn, and R.P. Behringer. 1995. Phys. Rev. Lett. 74:4643.

Perko, H.A. 1998. Surface cleanliness-based dust adhesion model. Pp. 495-505 in Space 98: Proceedings of the Sixth International Conference on Engineering, Construction, and Operations in Space. R.G. Galloway and S. Lokaj, eds. Reston, Va.: American Society of Civil Engineers.

Rosato, A., K.J. Shandburg, F. Prinz, and R.H. Swendson. 1987. Why brazil nuts are on top: Size segregation of particulate matter by shaking. Phys. Rev. Lett. 58:1038-1040.

Savage, S.B. 1997. Pp. 185-194 in Powder and Grains 97: Proceedings of the 3rd International Conference. R.P. Behringer and J.T. Jenkins, eds. Brookfield, Vt.: A.A. Balkema.

Schäfer, J.J., J.S. Dippel, and D.E. Wolf. 1996. Force schemes in simulations of granular materials. J. Physique I(6):1751-1776.

Veje, C.T., W. Daniel, W. Howell, and R.P. Behringer. 1999. Kinematics of a two-dimensional granular Couette experiment at the transition to shearing. Phys. Rev. E. 59(1):739-745.

Williams, J.C. 1976. The segregation of particulate materials: A review. Powder Tech. 15:245-254.

Wood, D.M. 1990. Soil Behavior and Critical State Soil Mechanics. Cambridge: Cambridge University Press.

V

Other Concerns

V.A INDIRECT EFFECTS OF REDUCED GRAVITY ON DESIGN

It has been pointed out that reduced gravity can have indirect effects on systems and components by setting design requirements that differ from the corresponding requirements on Earth. Often, these indirect effects have to do with reduced structural forces or loads under reduced gravity. Another such indirect effect concerns the products of wear and decay, which are presumably less easily collected and managed in reduced gravity than in Earth gravity. There can be no confidence in the success and safety of long-duration crewed missions unless such indirect effects are well identified, understood, and managed. The following paragraphs address some of these issues.

Piping Systems

Pipe flows of fluids or particulates often contain a number of phases distributed in various ways that are not necessarily uniform or steady and that are greatly influenced by the presence or absence of gravity. Pipes are not necessarily straight; they incorporate bends, elbows, and other fittings. Therefore, any nonuniform density of flowing fluids or particulates will cause fluctuating forces on the piping structure. If the piping system (always more elaborate and massive than supposed in the original system conception) is designed to minimize the mass delivered to space, it will presumably be very flexible and will therefore respond to fluctuating forces in extreme and possibly destructive ways. Even in robust terrestrial systems, the familiar phenomena of "surge" and "liquid hammer" are always of concern to the designer (Streeter and Wylie, 1985). Either single- or multiphase flow through a piping system can cause potentially undesirable flow-induced loads aboard a spacecraft. Moreover, the potential for flow induced vibration is great.

The structural dynamics of piping systems in microgravity is an interesting subject that has received little attention to date. Fluid networks are generally controlled by valves, and valve actions are obviously important for the dynamics of such systems, initiating or aggravating unwanted pressure waves. Cavitation is sometimes involved in such behavior. It may be that gradual, slow-acting valves will be wanted for space systems. Certainly, magnetically controlled valves could be helpful and would also eliminate the need for seals.

Since pipe bends and fittings impart transverse accelerations to the fluids passing through them, phase mixing or phase separation might be achieved by suitable design of the piping layout. This idea will be discussed later, in connection with artificial gravity.

One may conclude that careful analysis should be made of the system dynamics of entire fluid networks for HEDS applications in reduced gravity, including structural effects as well as the multiphase fluid flow phenomena treated in Section IV.C.

Bearings

It has been pointed out that many HEDS systems and components will have rotating machinery, and these will require bearings to maintain shaft positions and to bear loads under various speed requirements. A variety of bearing types can be used, depending on the nature of the loads to be borne, and these loads obviously will depend on the gravity level or its equivalent.

By way of example, one may recall that cooled rotating-element bearings have been used in fuel pumps on the Shuttle, but they require replacement after each mission because of rapid wear. These will presumably be replaced in the future by film bearings designed to be lighter, more efficient, and more durable. Especially if bearings are to last for years, as HEDS will require, film bearings will be preferred, running at very high speeds with no contact wear (San Andres, 1996).

Cryogenic fluids such as hydrogen or oxygen, if already available on the spacecraft for propulsion or other purposes, are suitable for the films as they generally have a low viscosity. Cryogenic film bearings have been extensively studied for Shuttle applications (San Andres, 1995, 1996). One concern about such bearings is the stability of the film. The cryogenic liquid will often be supplied near its critical point and so would be subject to bubble formation, flashing, or cavitation, both because heat generated in the bearing may raise the film's temperature and because pressures may drop sufficiently to cause phase change. Should such phase change occur, then gravity may affect the bearing film itself and will certainly affect the processes of collection and recirculation of the film liquid.

The foregoing concerns emphasize the importance of maintaining thermal balance between a bearing and its surroundings, by effectively collecting and dispersing the heat generated by friction. Analysis and experiments that look at the behavior of bearings with different proposed designs need to be carried out in environments that realistically simulate microgravity. Furthermore, bearings will clearly be important collection sites for the heat-transmission network of a spacecraft, which we know is subject to microgravity concerns. It should be noted that cryoturbopumps have not yet been tested under realistic space conditions.

Another stability concern arises because of the response of the film to load. When a film bearing provides support for a rotating shaft, the manner in which the film works is not affected by gravity (unless the film is a multiphase fluid, as discussed above). However, for journal bearings, the bearing loads are conventionally transverse to the horizontal shaft and are due to weight (gravity). Unloaded bearings, that is, with no transverse force, can be dynamically unstable (Pincus and Sternlicht, 1961). If gravity is absent, the load is gone, the film behavior is changed, and the liquid film can be expected to suffer severe stability problems. Thus, the dynamic behavior of film journal bearings in microgravity or variable gravity will need study.

In certain applications, magnetic bearings may offer a means of preventing surface friction and erosion, and their use certainly merits study, on the assumption that HEDS missions will have ample onboard or station power. Magnetic bearings would seem to be especially useful in microgravity, where their purpose would be more to position machine elements than to bear a large load.

Seals

Seals are needed whenever it is necessary to prevent or control flow through a region where machine elements slide with respect to each other. In a typical film bearing, fluid is introduced but then must leave, perhaps along the shaft, perhaps through a concentric "damper seal." In terrestrial applications, film fluid may be collected by gravity, reprocessed, and then returned to the bearing. In microgravity, where this kind of collection is impossible, the solution might be to "flood" the bearing, encapsulating the entire machine in film fluid. Seals, like bearings, are subject to wear and failure.

HEDS technology will also require devices intended to rotate or slip relative to each other, but only intermittently (robots, for instance). Therefore, starting friction in joints can be a problem. If there is a need to lubricate or to protect such surfaces in space, the joints may need to be encapsulated in a seal.

In the interest of simplicity and the avoidance of occasions for wear, fatigue, erosion, and failure during long HEDS missions, bearings, seals, and valves should be avoided when possible. That would mean eliminating shafted devices in favor of more passive elements that might adequately serve the same function. An example would be the production of rotation by eccentric fluid jets rather than by shaft rotation, for the purpose of liquid/vapor separation. This is discussed in Section B of this chapter. While some shafted pumps and turbines will probably still be needed for high-performance systems, surely the list of troublesome elements can be greatly reduced by clever use of fluid mechanics and electromagnetics. Of course, when purely fluid mechanisms are substituted for solid-surface motions, a degree of control over kinematics is lost, and stability difficulties can be expected as a consequence.

Robots and Articulated Structures

For HEDS missions, robots will presumably be used to assist the crew, and they should be designed with an endurance measured in years. As mechanical devices, they will experience wear, decay, and fatigue, and provision will need to be made to collect the waste products of wear and decay generated over long periods of time. How robots should be designed to include long-lasting joint bearings and appropriate dynamic controls are important topics for HEDS, requiring microgravity research.

The extreme structural flexibility of low-mass components of space robots that will function over large volumes of space means that movements will be complex and of large amplitudes. The structural damping of oscillations is typically weak, and a substantial fraction of the time allotted for an operation can be spent waiting for unwanted vibrations to subside (Longman, 1994). In fact, robot movements can cause rotational displacement, or even the tumbling of a spacecraft (Longman, 1994). In microgravity, joints of articulated structures may fail to position members precisely, and joint hysteresis may become a factor in the dynamic behavior of such structures.

Tanks and Antennas

Antennas for communication, and also solar collectors and space radiators, will need to be very large and precisely positioned for distant HEDS missions of large scale. Also, tanks for storage of liquids and gases will no doubt be very large. As is always true for space applications, these devices will be designed to have the lowest mass feasible. Such large tanks and antennas may therefore be expected to have complex and technically important dynamical behavior in microgravity, just as will the elaborate piping systems postulated for many power- or environmental-control systems. The larger, simpler, and thinner these "plates and shells," then the greater the number of modes, and their amplitudes, of the oscillations that can be amplified following excitation by the various vibrations and transient forces inherent in space operations. Such oscillations, even if not immediately catastrophic, could cause fatigue failure if maintained over the time of a HEDS mission.

It is well understood that structural design and fabrication are quite different in the absence of gravity from design and fabrication on Earth (Swaim et al., 1994), and it is not necessary to review this topic in this report. However, the dynamic behavior of mechanical devices in microgravity certainly merits ongoing, device-specific research for HEDS. It must be borne in mind that HEDS missions will typically impose a significant range of gravity levels, and all spacecraft elements will be expected to function in variable gravity.

Summary of Concerns Prompted by Considering the Indirect Effects of Reduced Gravity

The following concerns arise in connection with the indirect effects of reduced gravity:

- Phase change in bearing films;
- Thermal balance in film bearings;
- Stability of bearing films in microgravity applications;
- Magnetic bearings and valves (whether they are useful);
- Flooding, encapsulation of lubricated bearings in microgravity;
- Starting friction of bearings and joints, including "cold welding" effects;
- Joint hysteresis;
- Endurance, wear, and decay;
- Avoidance of bearings and the like in the interest of simplicity and reliability;
- Dynamics of plates and shells (antennas and tanks);
- Surge and liquid hammer in flexible piping as a result of flow-control operations;
- Structural dynamics of piping networks and similar components;
- Dynamics of robotic and other structures; and
- Device performance in variable gravity.

References

Longman, R.W. 1994. Tutorial overview of the dynamics and control of satellite-mounted robots. Pp. 237-258 in Teleoperation and Robotics in Space: AIAA Progress in Astronautics and Aeronautics, Vol. 161. S.B. Skaar and C.F. Ruoff, eds. New York: American Institute of Aeronautics and Astronautics.

Pincus, O., and B. Sternlicht. 1961. Theory of Hydrodynamic Lubrication. New York: McGraw-Hill.

San Andres, L. 1995. Thermohydrodynamic analysis of fluid film bearings for cryogenic applications. J. Propulsion Power 11(5):964-972.

San Andres, L. 1996. Thermohydrodynamic Analysis of Cryogenic Liquid Turbulent Flow Fluid Film Bearings. Cleveland, Ohio: NASA Lewis Research Center.

Streeter, V.L., and E.B. Wylie. 1985. Pp. 521-543 in Fluid Mechanics, 8th Ed. New York: McGraw-Hill.

Swaim, P.M., et al. 1994. Use of manipulators in assembly of space station freedom. Pp. 443-473 in Teleoperation and Robotics in Space: AIAA Progress in Astronautics and Aeronautics, Vol. 161. S.B. Skaar and C.F. Ruoff, eds. New York: American Institute of Aeronautics and Astronautics.

V.B MICROGRAVITY COUNTERMEASURES

Previous chapters of this report show that low gravity generally complicates the design and operation of systems for HEDS. Low gravity is also harmful to biological systems, leading to the crew's loss of bone and other tissue. A recent report (NRC, 1998) reviews the relevant results and discusses the various mechanisms that may underlie such effects. Such mechanisms are not yet fully understood, and that 1998 report urged research to "provide mechanistic insights into the development of effective countermeasures for preventing bone and muscle deterioration during and after spaceflight" (p. 232). Thus, while microgravity poses a wealth of interesting research questions, it would seem to be a matter of basic interest for HEDS that NASA learn how to supply appropriate degrees of artificial gravity as a microgravity "countermeasure."

In the past the potential importance of artificial gravity was emphasized (NRC, 1987), but subsequent planning documents (NASA, 1991) have given this topic a low priority, and concepts for creating artificial gravity for spacecraft have been, and remain, hypothetical and theoretical.

During a typical HEDS mission, the gravity level will vary from zero to near-Earth values; this variability of gravity is itself a problem, because a design suited to one gravity level may be quite inappropriate for other levels that are encountered in the course of the mission. Artificial gravity could eliminate this variability by making possible the continuous adjustment of effective gravity level during the course of a mission.

Concerns about gravity level will not be equal for all parts of a system; some components will be unaffected, while for others, especially those involving multiphase fluids, gravity level will be vitally important. Therefore, artificial gravity might be thought important for an entire spacecraft or perhaps only for certain critical components. Even for such systems as unmanned, low-thrust "freighters," artificial gravity may be useful.

Earlier, it was pointed out that for most processes, scale and gravity are grouped together; in effect, a sufficiently small-scale device will be operating in a microgravity regime whatever the actual level of gravity. Thus, the effects of varying gravity can in principle be avoided by scale reduction, if it can be accepted that the device operates entirely in the microgravity regime. However, the limitations of small scale and of microgravity operation would generally not be desired, and artificial gravity would generally be the appropriate "countermeasure" for reduced and variable gravity.

In principle, artificial gravity can be supplied by any means that introduces a "body force," and magnetic fields could do that. The most familiar and, presumably, reliable method is to impose rotation in order to provide a centrifugal force that mimics gravity, subject to certain limitations (to be discussed). There are many ways that rotation can be imposed. The mechanical rotation of a spacecraft or component is the most obvious, but introduction of swirl by fluid-mechanical means is an equally important, if more subtle, method, which may be especially important for dealing with components. The following paragraphs describe some possible methods of supplying rotation at various scales—from the whole spacecraft to the level of the components. It should also be realized that on a transient or local basis, any means that produces acceleration will mimic gravity. Possibilities among this more extended class of gravity substitution are also discussed.

Spacecraft Rotation

Artificial gravity can be supplied by imparting rotation to a spacecraft ,and low-thrust, station-keeping kinds of rockets can easily do this. A rotating spacecraft might look like a dumbbell, or twirler's baton, with a very long, low-mass straight section connecting two or more major masses of the spacecraft (Figure V.B.1).

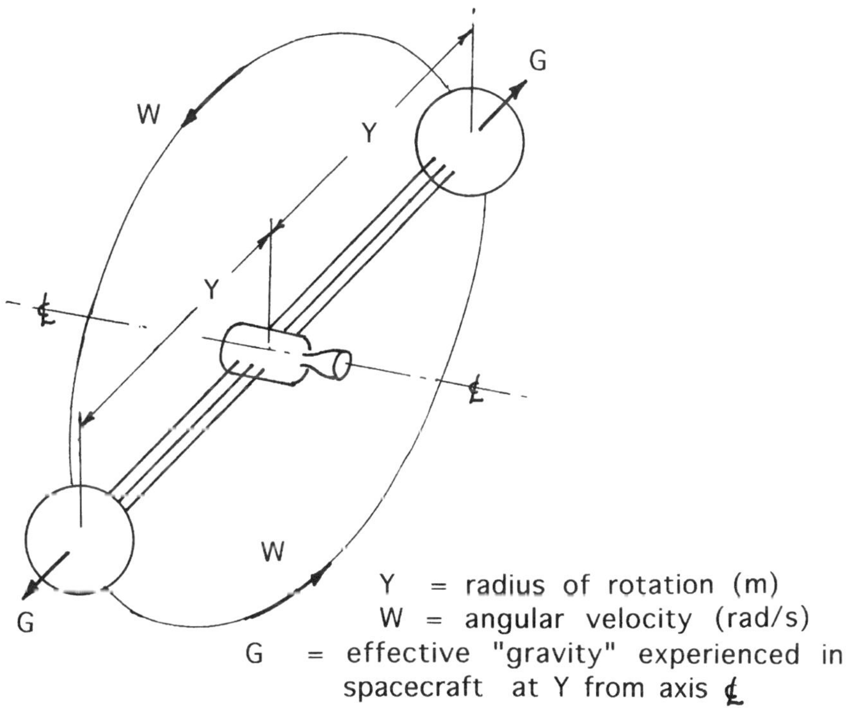

FIGURE V.B.1 Schematic of spacecraft rotation about axis.

It has been proposed that the connecting section might be a tether (in principle, a long flexible cable) (Hoyt and Forward, 1995). There may well be other structural devices of equal or greater usefulness. How long should the radius of rotation be? The longer the connecting section, the lower the angular velocity required to achieve a given effective gravity G at the ends where the masses are. For example, if two equal main masses are separated by 1,800 m, then G = g_0 (Earth gravity) will be achieved with a rotation rate of only about 1 rpm. For 500 m, that same rotation rate will provide G = 0.28 g_0, and for 200 m, the same G = 0.28 g_0 will require about 1.6 rpm rotation.

The significance of angular velocity is perhaps made clear by considering motion experienced in a plane normal to an axis of general rotation at an angular velocity W. The motions are supposed to occur in a region of the plane centered at a radial distance Y from the assumed center of general rotation. In such a case, the effective gravity provided by centrifugal force is

$$G = W^2Y.$$

The centrifugal force (G) is the desired artificial gravity effect—the reason for applying rotation. There are unwelcome side effects of rotation, however: the Coriolis force and a gravity gradient. The gravity gradient is the less troublesome of these side effects; if the radius of rotation (Y) is hundreds of meters, displacements of a few meters cause little change in the effective gravity.

Coriolis force is a greater problem; it is the force that needs to be supplied if a velocity is to be experienced as rectilinear motion in the rotating plane (Goldstein, 1959), and it is proportional to both that velocity and to the angular velocity of rotation (W). This Coriolis effect, if strong, could presumably be disorienting and even nauseating for an astronaut trying to move in a rotating frame, because it entails a force perpendicular to his intended movement. For physical systems and components, Coriolis force would represent a complication, but not necessarily a harmful one. For the cited example of separation distance 500 m (Y = 250 m) and G = 0.28 g_0, the Coriolis force is less than 0.1 G if motion in the rotating frame is at a velocity of less than 1.3 m/s. For the same G and the same velocity of motion, if the radius is longer (say, Y = 900 m), the Coriolis term will be only half as important. Of course, an increase of rotational velocity (W) will, in itself, also diminish the relative importance of the Coriolis effect (which is proportional to W) in comparison with the centrifugal force, which is proportional to W^2.

In other words, Coriolis effects associated with movements at the ends of the connecting section (separation member) would be higher for shorter separations and quite substantial for separation distances of the order of 100 m (which might be typical of separations used for radiation protection of crews in nuclear-powered spacecraft). One may then ask, How serious are the consequences of high Coriolis effects, both for crew and for technical systems? Thus a research topic would be the study of Coriolis effects and reduced-gravity effects in combination. Similarly, shorter separations imply greater gravity gradients within the spacecraft, the significance of which also requires evaluation.

Because an adjustable level of effective gravity achievable by spacecraft rotation could be beneficial for both crew health and component performance, microgravity research should take account of the following:

- Physical effects should be studied as functions of gravity level. That is, it should be recognized that gravity level is not necessarily "given" but might be "designed." For example, above what gravity level will a two-phase heat transfer loop be preferred to a single-phase device? Clearly, the catalog of regime-change zones discussed in Section IV.A, to be achieved through the research recommended in this report, will be essential to guide the design of any artificial-gravity system, of whatever scale.
- Prediction of component behavior and subsequent optimization of the system would both become more complex, because gravity would be a new free parameter—a design option. The engineering virtues of simplicity, reliability, and safety could be pursued as functions of gravity.
- Coriolis effects should be studied in combination with reduced gravity. For example, in ordinary fluid mechanics, Coriolis forces cause secondary flows. Such effects could combine in interesting ways with the effects of reduced gravity. The effects of gravity gradients in reduced gravity should also be studied.

Spacecraft rotation has not yet been developed or designed in practical terms; neither the difficulties nor the costs have been explored. Certain practical issues are obvious: Any tether system with its attachments must be strong enough to bear a large centrifugal load, and the system must be secure from destruction by meteorite impact. The rotation rate must be well controlled, including starting and stopping. Also, it is clear that the rotation of a spacecraft would raise concerns about the orientation of antennas for communications or perhaps of arrays for the collection of solar or laser energy.

Liquid/Vapor Separators

Any power system that uses a Rankine cycle requires a condenser and means of ensuring that, after the condenser, there is no carry-under of vapor with the condensate (see Section III.B). On Earth, vapor carry-under is prevented by gravitational separation in the hot well of the condenser, but this technique is ineffective in space. Moreover, noncondensable gases may infiltrate the system, and they also must be separated and removed.

NASA has envisioned that artificial gravity in the form of centrifugal acceleration can be used to accomplish liquid/vapor separation downstream of a condenser. The approach used may be the direct one of mechanical rotation (the so-called rotary fluid management device), or a more indirect, passive one that induces swirl by tangential flow injection. These two approaches are discussed below, and possible extensions of the passive approach are mentioned as well.

Rotary Fluid Management Device

The rotary fluid management device (RFMD) developed by Sundstrand, and more recently at the Johnson Space Center, is discussed in pages 98-100 of Brown and Alario (1990) and shown schematically in Figure V.B.2. It is essentially a rotating drum, intended to concentrate liquid at the outer surface. The complexity of the sketched system is associated with the recirculation to make the ultimate separation more complete and to provide thermal

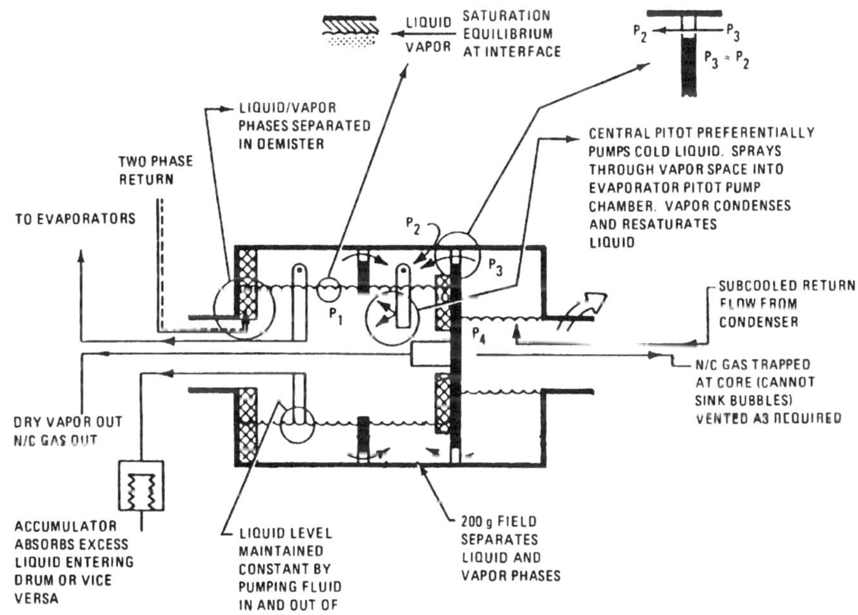

FIGURE V.B.2 Rotary fluid management device.

balance, since heat is released in condensation. The rotational speed of the drum required to provide on the order of 100 g_0 is quite modest. Assuming the drum radius is about 0.1 m, rotation would be at 1,000 rpm, and the tip speed would be about 10 m/s.

Apart from problems of operational stability, this device, which depends on forced rotation, requires a power supply, an electric motor or turbine, and bearings to support the shaft. The problems of bearings in microgravity are discussed in Section V.A. If device simplicity is required for maintenance-free reliability, it would seem that this approach will have to be much refined if it is to be a part of long-range HEDS missions.

Free Vortex Separator

The RFMD just described is considered an active device, in that rotation is imposed by mechanically spinning a tank body about a shaft. In NASA's concept of a free vortex separator (FVS), the tank is stationary and rotation is produced by eccentric injection of the mixture to be separated (Shoemaker and Schrage, 1997) (Figure V.B.3). A free vortex, or cyclone, is thereby established in a stationary container, and the device's operation is thought of as indirect, or passive.

This method of establishing rotation requires little power and is simple; it needs no shaft, bearings, or motor hardware. On the other hand, the free-vortex method lacks the definite kinematic control that the rotating tank of the RFMD provides and so is more subject to instability or breakdown of the flow pattern or failure of the vortex to form as desired.

The FVS has been studied at NASA's Glenn Research Center in the laboratory and on the center's DC-9 aircraft. Results usually showed a strong, stable gas core on the axis (like a bathtub drain), into which injected bubbles are swept, as illustrated in Figure V.B.3. Under some circumstances, however, the vortex did not form or was unstable. At low through-flow or for extremely fine droplet size, the performance was not satisfactory. While work on the project has been very successful and is continually yielding design improvements to extend the operating range of the device, it is not yet clear that the FVS in its present form will be applicable to the power-cycle duty for which the RFMD is being developed.

Perhaps more-radical changes of concept are needed. For example, injection at a much larger radius, followed by a funnel-like convergence to something like the present tank diameter, might yield stronger free vortices, with little size or weight penalty. In any event, the principle of passive rotation by means of free-vortex arrangements should be aggressively pursued, exploring as wide a variety of potential configurations and applications as ideas permit.

Oscillation as a Microgravity Countermeasure

Rotation can provide a steady and quite uniform substitute for gravity. However, other means, unsteady or nonuniform, might also be useful. Oscillations can obviously provide accelerations that intermittently mimic gravity (circular motion can be represented as the superposition of two linear oscillations that are orthogonal and 90 degrees out of phase.) Oscillations of interest could be very slow or very fast; acoustic vibrations could be useful. The oscillation parameters (frequency, amplitude, wave form) would depend on the purpose to be served and the nature of the medium.

In microgravity, the purpose might be to separate phases, as in the case of the RFMD, or to mix phases. The reversal of force direction implicit in oscillations suggests that oscillations would be especially useful when mixing is the purpose, perhaps to enhance heat transfer or increase combustion rates. Ultrasonic vibration is a well-known mixing technique. Conversely, oscillations of a mixture can concentrate heavier particles at anti-nodes, depending on design parameters (panning for gold is an example). Possibilities for mixing or unmixing of phases by oscillations or wave processes should be studied for application to HEDS systems, and it should be noted that machinery complications such as the need for bearings can be avoided this way, although fatigue and erosion could still be limiting problems.

Regarding the biological problem of bone loss in microgravity, the National Research Council (1998) has suggested the possible significance of unsteady or pulsating mechanisms for changes of bone mass. Analysis

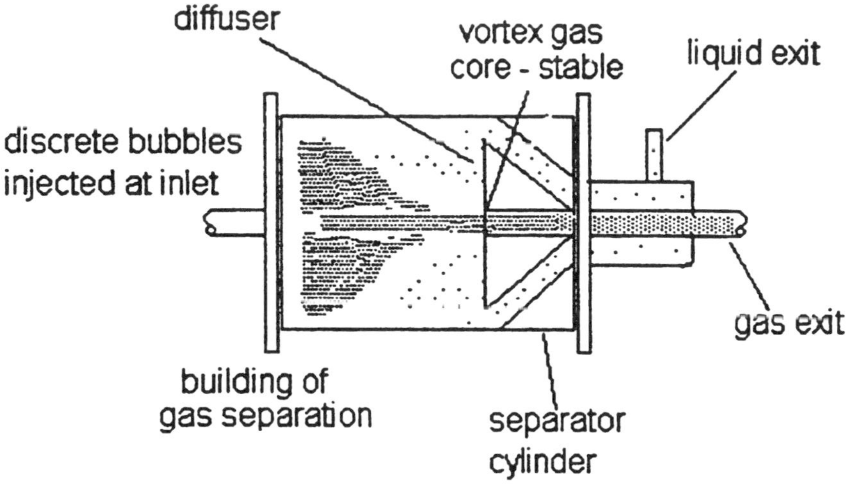

FIGURE V.B.3 Free vortex separator showing bubble distribution (model 5). SOURCE: Shoemaker and Schrage (1997). Courtesy of NASA.

reported in Weinbaum et al. (1994) suggests that the favorable effect of mechanical loading of bone is proportional both to strain amplitude due to load and to frequency of load application, for frequencies below about 25 Hz. If this is correct, then low forces imposed at rather high frequency might be just as beneficial as a much higher load statically imposed. Conceivably, then, an unsteady or oscillatory system could provide microgravity countermeasures for the benefit of both technical devices and crew.

Flow Deflection as a Microgravity Countermeasure

Sinuous steady flow can also be a means of phase management. A natural example is river meander, which increases with time by continuous transport of soil from the outer bank of a river bend to the inner bank, as a result of secondary (viscous) flow in the boundary layer of the river bottom (Callander, 1978). Another example is wing icing, caused by supercooled water droplets failing to be deflected along with the air flow as they approach the wing leading edge; in effect, while the air "misses" the wing, the more massive water droplets hit and freeze. This, of course, is a bulk effect, not a viscous boundary-layer one.

Piping and fitting layouts in spacecraft could be designed to take advantage of such effects, with due account taken of the interplay between bulk and boundary-layer phenomena, in order to manage phase distribution in such systems, as mentioned in Section III.B. NASA has studied phase separation in step-diffusers,[1] furnishing one example of this class of possibilities.

Research on Countermeasures

Various research needs and suggestions for the provision of artificial gravity for HEDS appear in the foregoing paragraphs. They underscore the importance of a theme that has appeared throughout this report, namely that knowledge of physical behaviors as functions of gravity level must be developed through research. This theme

[1]Singh, B., Lewis Research Center. Multiphase Flow and Phase Change in Space Power Systems. Presentation to the Committee on Microgravity Research, October 14, 1997, at NASA Lewis Research Center, Cleveland.

extends to such performance qualities as simplicity, reliability, and safety and is exemplified by the need to study Coriolis effects in combination with reduced gravity.

The principle of passive rotation by means of free-vortex arrangements should be aggressively pursued, and swirl generation at large radii merits consideration. More generally, flow deflection should be studied as a general microgravity countermeasure for piping and fittings. Possibilities for mixing or unmixing of phases by oscillations or wave processes should also be studied.

The importance of microgravity countermeasures for systems and components of a spacecraft has been shown, especially for the management of multiphase fluids, and a variety of potential means have been identified, ranging from spacecraft rotation to localized vibrations. For HEDS, it is clear that the human body is also in need of microgravity countermeasures, and a range of means will probably be feasible for both systems and crews. It seems logical that microgravity scientists will be able to help in the conception and development of systems that protect bone and muscle by means acceptable to astronauts.

Indeed, economy and simplicity demand that microgravity countermeasures for all these technical devices and systems be coordinated, along with those for biological purposes, through liaison activities in the various technical and biological communities. It would be extremely wasteful to develop and deploy redundant measures to counter the effects of microgravity on physical systems and crew members, when a single countermeasure might effectively serve both purposes. Developments of microgravity countermeasures to maintain both the technical and the human components of a spacecraft and its long-duration HEDS mission, should be guided by results obtained in microgravity research. It should also be made clear to the HEDS community how microgravity research can explain the need for, and the value of, artificial gravity, especially for long missions; in other words, the microgravity research community should provide the motivation for NASA to solve perceived problems, using spacecraft rotation or other appropriate and feasible means. NASA should consider planning in-space research on partial gravity using rotational or other means to achieve desired gravity levels for HEDS systems or their components. Research on ways to counter the effects of reduced or variable gravity on HEDS technology should encourage a wide variety of ideas, even if doing so delays engineering development of intended mission hardware, because new ideas will be the basis for devices of the future. And a trip to Mars or beyond is indeed in mankind's future; it will surely require engineering developments well beyond those now already under way or planned.

References

Brown, R., and J. Alario. 1990. Space station mechanical pumped loop. Pp. 83-130 in Thermal-Hydraulics for Space Power, Propulsion and Thermal Management System Design: AIAA Progress in Astronautics and Aeronautics, Vol. 122. S.B. Skaar and C.F. Ruoff, eds. New York: American Institute of Aeronautics and Astronautics.

Callander, R.A. 1978. River meandering. Annu. Rev. Fluid Mech. 10:129-158.

Goldstein, H. 1959. Classical Mechanics. Reading, Mass.: Addison-Wesley.

Hoyt, R.P., and R.L. Forward. 1995. Failsafe multistrand tether SEDS technology. Fourth International Conference on Tethers in Space, April 1995, Washington, D.C.

National Aeronautics and Space Administration (NASA). 1991. Integrated Technology Plan for the Civil Space Program: Strategic Plan. Washington, D.C.: NASA.

National Research Council (NRC), Aeronautics and Space Engineering Board. 1987. Space Technology to Meet Future Needs. Washington, D.C.: National Academy Press.

NRC, Space Studies Board. 1998. A Strategy for Research in Space Biology and Medicine in the New Century. Washington, D.C.: National Academy Press.

Shoemaker, J.M., and D.S. Schrage. 1997. Microgravity fluid separation physics: Experimental and analytical results. AIAA Paper 97-0886. New York: American Institute of Aeronautics and Astronautics.

Weinbaum, S., S.C. Cowin, and Y. Zeng. 1994. A model for the excitation of osteocytes by mechanical loading-induced bone fluid shear stresses. J. Biomech. 27:339-360.

V.C PREDICTIVE MODELS, RELIABILITY, AND PROBABILISTIC RISK ASSESSMENT

Earlier in this report, the importance for HEDS of system reliability was emphasized. To address reliability with appropriate rigor and completeness, probabilistic risk assessment (PRA) techniques would be very helpful. PRA is a proven methodology that has been widely and successfully used in the aircraft and nuclear industries. It employs fault and event trees to determine the relative likelihoods of the failure modes of specific designs (Green and Bourne, 1972; Shooman, 1968). Moreover, it highlights weaknesses in designs and components and can therefore be used to promote design reliability, simplicity, redundancy, and maintainability.

However, successful use of PRA depends on having physically based predictive models of system and component behavior. Such models are currently lacking for reduced-gravity effects and must be developed on the basis of NASA-supported research. In particular, a better understanding of scaling-law boundaries and multiphase flow and heat transfer phenomena is needed.

There is, clearly, a logical path leading from microgravity research to modeling, to risk assessment, and, finally, to reliability. This is a path that, if followed, should provide an integrated, consistent design process for HEDS systems and components.

References

Green, A.E., and A.J. Bourne. 1972. Reliability Theory. New York: Wiley-Interscience.
Shooman, M.L. 1968. Probabilistic Reliability: An Engineering Approach. New York: McGraw-Hill.

VI

Summary of Recommended Research on Fundamental Phenomena

BASIS FOR RECOMMENDATIONS

The discussions in the preceding chapters are intended to expose the wide range of questions that exist about the effects of reduced gravity on the roles of the various phenomena operating in HEDS technologies. In deciding which of those phenomena should be recommended subjects of NASA research, the committee considered such factors as the potential of a given phenomenon to affect a wide range of HEDS technologies, the potential importance of the affected technologies, and the potential magnitude of reduced gravity's effect. The research topics were then integrated to formulate specific recommendations. In this chapter, the recommendations for specific fundamental research are listed, and in Chapter VII, more general programmatic recommendations for managing the research are made. Topics are listed by number in the perceived order of importance within each discipline, and a recommendation's overall priority across all disciplines is indicated by the notation "higher," "medium," or "lower." These priorities are assigned with the understanding that setting priorities for research is an uncertain process that risks prejudging the results of the recommended research. The recommendations are based on the supposition that systems and processes important for HEDS must be of assured, predictable, and reliable performance. Before such assurances can be given, a much better understanding of the effects of gravity must be gained through research. Especially, it is clear that systems depending on multiphase phenomena will in turn depend strongly on gravity level, in ways that are not now well understood. It will also be essential for NASA to develop a description, across a wide range of physical phenomena and the corresponding dimensionless parameters, of the manner in which changes in gravity level produce fundamental changes in system performance.

It also seems clear that it will not be possible to arrive at confident design conclusions through a costly process of repeated tests in space of the many important systems and components. Rather, the approach should be to develop physically based computational models that are then fully assessed against relevant microgravity data. These models could be used to optimize design and to direct and support the scale-up of those space experiments used to evaluate subsystems and systems that cannot be tested at full scale in space. Such models would also provide the necessary basis for establishing desired reliability and safety levels for HEDS systems.

The foregoing approach obviously depends on a well-planned analytical and experimental research program, carried out on Earth and in space, to quantify the effects of gravity level. Efficient, multivariate experimental test matrices will need to be developed to sample the unknown behavior, so as to model it as efficiently as possible and at reasonable cost. The committee urges that NASA support for, and direction of, such research be continued and intensified, as it is essential to the success of the HEDS enterprise.

RECOMMENDED RESEARCH

Surface or Interfacial Phenomena

In microgravity, surface-tension-related phenomena may dominate fluid behavior. In particular, the handling and storage of cryogenic fluids may entail the use of vanes, wicks, and screens that rely on capillarity to control the location of the fluid. Heat pipes and capillary pumped loops also depend on surface tension effects. Condensation, evaporation, boiling, and sublimation are influenced by both gravity and interfacial forces. Marangoni effects lead to convection, which plays an important role in two-phase heat transfer, provides stirring of welding pools, and can be used to promote phase separation. The committee believes that research in the following areas should advance understanding of surface phenomena.

1. Ongoing work on the physical basis of wetting, including the hysteresis effect, the dynamics of wetting, and the correct description of wetting below the scale of the correlation length of the wetting fluid, should be extended. Empirical and fundamental knowledge of the material combinations and conditions for good wetting and wetting agents is also needed. **[Priority—Higher]**
2. Capillary-driven flows and transport regimes that occur in evaporation and condensation heat transfer need further work, as does the determination of how the flow regime boundaries scale with gravity. Extension of the current work on Marangoni convection is urgently needed, including the complications introduced by the geometrical configurations in the various multiphase flows. Dynamical work on the oscillations of liquid drops or bubbles and on the resonances between the applied forces or accelerations (e.g., *g*-jitter) and capillary modes of motion of a mass of liquid in a container, including the so-called sloshing problems and unstable modes, needs to be extended. **[Priority—Higher]**
3. Experimental determination of the parameters that enter into the Marangoni and other relevant dimensionless numbers needs to be greatly extended, including the investigation of tensioactive agents that influence the magnitude and sign of the effect. Thermal and concentration gradients need to be taken into account in assessing the merits of designs where Marangoni flow is possible. **[Priority—Medium]**
4. Classical work on static equilibrium capillary shapes, which minimize the area of the surface of a mass of liquid (in the absence of gravity) subject to boundary conditions of a fixed volume and given contact angles at the perimeter of the surface, needs to be extended. Multiplicity and stability of solutions should be investigated. **[Priority—Lower]**

Multiphase Flow and Heat Transfer

NASA engineers have often avoided the use of multiphase systems and processes in spacecraft and satellites because they lack a basic understanding of multiphase flow and heat transfer phenomena in reduced-gravity environments. This has prevented the deployment of efficient and high-power-density active systems that otherwise might have been used for NASA's missions. Nevertheless, it appears that the reliable operation of numerous phase-change systems for propulsion, power production, and life support will be required during HEDS long-duration missions. Thus, multiphase flow and heat transfer technology is considered to be a critical technology for HEDS. Indeed, it is expected to be mission-enabling technology that, if properly developed, could lead to revolutionary changes in spacecraft hardware. The proposed research is intended to give NASA the ability to better understand the performance of multiphase systems in space and to provide the basis for the development of accurate computational capabilities. It is recognized that there will be a continuing need for experimental microgravity data and appropriate empirical correlations, since some physical phenomena and HEDS design issues go beyond our current, and anticipated near-term, computational capabilities. Nevertheless, the primary objective of the proposed research is the development of a reliable, physically based, multidimensional two-fluid model for the computational fluid dynamics (CFD) analysis of multiphase flow and heat transfer phenomena of importance to the HEDS program. To this end, experimental and analytical programs are required that will do the following:

1. Determine the key experimental variables and parameters on which CFD models in microgravity will depend, and conduct appropriate experiments to sample flow behavior efficiently and sufficiently well to identify all important multiphase phenomena that occur in microgravity. **[Priority—Higher]**
2. Perform the experimental and analytical investigations needed to identify the mechanisms governing the effect of gravity on multiphase phenomena, and develop flow-regime-specific models for the various interfacial and wall transfers (i.e., mass, momentum, and energy). These models should be formulated so that they can be used in a multidimensional two-fluid model, and they should include surface-tension-induced forces, the various axial and lateral forces on the flowing phases, and flow-regime-specific interfacial constitutive laws. This knowledge should allow the development of physically based models for the prediction of both adiabatic and diabatic flow regimes and flow regime transition in reduced and microgravity environments. The resultant two-fluid model should be suitable for use in multidimensional CFD solvers, and these computational models should be assessed against phase distribution and separation data taken at various gravity levels. **[Priority—Higher]**
3. Develop flow-regime-specific multidimensional models for multiphase turbulence. These models should also be suitable for use in multidimensional CFD solvers. **[Priority—Higher]**
4. Assess the effect of gravity on forced convective boiling and two-phased forced convective heat transfer and pressure drop. In particular, detailed measurements (i.e., data that can support the development of a multidimensional CFD model) are needed for the entire forced convection boiling curve, including the ebullition cycle (i.e., nucleate boiling), critical heat flux (CHF), and transition/film boiling. Active and passive heat transfer enhancement schemes should also be studied. **[Priority—Higher]**
5. Assess the effect of gravity on convective condensation heat transfer. In particular, detailed data that support the development of a multidimensional CFD model, or other models, should be taken in apparatus undergoing direct contact condensation and film-wise condensation. **[Priority—Medium]**
6. Study active and passive single-phase and two-phase heat transfer enhancement and pressure drop reduction schemes in order to increase efficiency and decrease the mass and volume of the equipment. **[Priority—Medium]**
7. Study the effect of gravity on the flow regimes, thermal limits, and stability of adiabatic two-phase and boiling flows in porous media. Pore-level fluid mechanics needs to be better understood if we are to predict the development of complex flow structures and regimes during boiling heat transfer and CHF in porous media. **[Priority—Medium]**

Multiphase System Dynamics

Multiphase systems may also exhibit global closed loop power/propulsion system instabilities, and these instabilities can cause serious operational problems and/or severe damage. The effect of gravity on these instabilities is currently not well understood, and the following research is needed if reliable phase change systems are to be developed and used for HEDS missions:

1. Detailed stability data on boiling and condensing systems is required for Earth gravity, g_0 (baseline), and for fractional, variable, and microgravity environments. These experiments should focus on static and dynamic instabilities in phase-change systems, and the data should be used to assess the predictive capabilities of various analytical models (including two-fluid CFD models) for the linear stability thresholds and the various nonlinear instability phenomena (e.g., limit cycles, chaos) that may occur in multiphase systems. **[Priority—Higher]**

Fire Phenomena

Assuring an acceptable level of fire safety in spacecraft and extraterrestrial environments is a critical aspect of the HEDS enterprise. To develop the knowledge needed for design of appropriate fire-detection and fire-suppression systems and for determining suitable fire-safety procedures, further research on combustion in reduced gravity is recommended:

1. The great majority of combustible materials available to fuel an unwanted fire during a HEDS mission are solids. High priority is given to experimental, theoretical, and computational studies of flame spread over surfaces of solid materials in microgravity and fractional gravity. These studies should focus on generic materials, both cellulose and synthetic polymers, and should include ignition requirements, flame-spread rates, flame structure and production of gaseous fuel from solid-fuel pyrolysis. **[Priority—Higher]**
2. Gravity effects in smoldering is another high-priority topic due to its relationship to electrical cable fires, for example. Specific studies for this topic should concern the initiation and termination of smoldering, propagation rates of smoldering fronts, and the production of hazardous or flammable products from smoldering, including conditions for transition from smoldering to flaming combustion. **[Priority—Higher]**
3. Systems for fire suppression operate best by delivering the suppressant to the base of the flame, where the gaseous fuels are generated. The mechanisms of suppressant flow and transport in fire situations at reduced gravity are different from those at normal gravity. There is therefore a need to improve understanding of suppressant transport at reduced gravity for gaseous, liquid, and solid suppressants. Because the carbon dioxide employed in the International Space Station is not likely to be the optimum suppressant for planetary operations, there is also a need for studying possible new suppressants, such as replacements for Halon, for HEDS applications. The need for an understanding of fire suppression is secondary to that for flame spread and smoldering, identified above, in the sense that the combustion process needs to be understood before addressing approaches to suppression. **[Priority—Medium]**
4. Diffusion-flame structure of gaseous, liquid, and especially solid fuels as affected by gravity levels, especially in relationship to the production of soot and toxic products in diffusion flames and the establishment of conditions for and mechanisms of diffusion flame extinction, are areas of research with quite high priority. Knowledge of the hazardous products of fires is extremely important and requires a knowledge of combustion behavior, to be developed in the recommended studies of flame spread and smoldering. Understanding of extinction is needed in connection with fire suppression. **[Priority—Medium]**
5. Combustion topics of importance but of somewhat lower priority to HEDS are the flammability and flame behavior of gaseous combustible mixtures, sprays, and dust clouds, especially the instability and dynamical behavior of such flames. These results would bear on the types of fire histories that may occur and on means for fire detection and suppression. These studies are less relevant to HEDS than those identified above because most of the combustible materials of concern are anticipated to be solids. **[Priority—Lower]**

Granular Materials

The activities of site preparation, habitat construction, mining activities, and the installation of transportation and energy distribution networks on the Moon and Mars depend on the ability to penetrate and excavate soils using the same types of heavy construction equipment and material manipulation techniques as are used on Earth. Current knowledge of lunar soil behavior indicates that such a transfer of systems and technologies is both impractical and ill-advised:

1. The average properties of regolith are measured by such techniques as the Mohr-Coulomb failure criteria, but less well established are the detailed interactions of granular material as a function of applied stress (frictional or *g*-level). Research on granular material under applied stress is needed to examine separately the effects of gravity and shearing on granular behavior and, using both modeling and experimental studies, to obtain a detailed description of granular material flow and behavior that accounts for sample history, internal variables, energy

fluctuations between particles, the effects of agitation, and particle size and shape, especially for operation at low pressure. **[Priority—Higher]**

2. In addition to the higher-priority research associated with geotechnical engineering on the Moon and Mars, another important aim of research in this area should be to gain a fundamental understanding of the behavior of dust in spacecraft and extraterrestrial environments. While difficulties are created by the fact that the transport processes associated with the initiation and sustenance of dust storms, though obviously influenced by gravity, are not yet understood, nevertheless, a predictive capability that permits calculation and control of dust transport and deposition will need to be developed. An understanding must be achieved of the cohesion and adhesion mechanisms that control dust attachment, where the attraction mechanism appears to be electrostatic. The behavior of dust in the vacuum environment of the Moon is also a serious problem for long-term system operations. **[Priority—Higher]**

Solidification and Melting

In the processes of welding, liquid-phase sintering, casting, and containerless melting/freezing of special materials, solidification from the liquid plays a central role. Gravity levels affect solidification through their effect on sedimentation of nuclei and convection in the liquid ahead of the advancing interface. These processes in turn determine the microstructure of the freezing solid. When a solid is heated above the melting point, liquid forms at the surface, and in the absence of gravity-induced convection, heat transfer and the propagation of the liquid/solid interface are expected to be greatly modified. Melting has been less studied than solidification but may be important in the operation of heat pipes and other two-phase devices.

The predictive capabilities of present theories of solidification are adequate only in relatively simple cases. Therefore, more work in solidification research is needed along the following lines:

1. The effect of gravity on the nucleation of solid from the melt via the distribution of nuclei and bubbles needs to be clarified. More comprehensive computer models are needed of the effect of gravity on the time-dependent evolution of a solidification front and the concomitant microstructure formation (e.g., cells, dendrites, spatial variation of composition), including the effects of crystalline anisotropy, interface kinetics, and convection, as well as suspended particles, and bubble formation. Also, there should be research at the technical level of casting and other practical solidification processes at different gravity levels to determine the effect of gravity on such casting parameters as grain size, porosity, inclusion distributions, and segregation. **[Priority—Lower]**

OTHER CONCERNS

Reduced-Gravity Countermeasures

Because the effects of reduced or variable gravity are generally a troublesome complication of system design for HEDS and have harmful consequences for human health, research should be undertaken on means to counter such effects. Such means would probably be mechanical in nature, involving rotation or vibration, and could range in scale from the whole spacecraft down to small, critical components.

If rotation were utilized, then research would be needed to evaluate the collateral Coriolis and gravity-gradient effects on hardware devices (and humans) as they experience solid or fluid motions in an artificial gravity environment. The purpose of such research would be to provide a rational basis for designing booms, tethers, and control methods for maintaining appropriate spacecraft rotation or, alternatively, for the solid or fluid rotation of spacecraft components separately. More generally, applied research looking toward economic and effective artificial gravity should emphasize applications common to both physical and biological systems.

Since no practical, large-scale artificial-gravity system has been developed, it is obvious that the many structural and system problems that arise would have to be studied from a design standpoint before feasible concepts and costs could be established.

Research on and development of reduced-gravity countermeasures must obviously proceed hand-in-hand with the microgravity research recommended elsewhere in this report, because microgravity research will establish the target gravity levels desired for various components and systems. In turn, the specific benefits of an artificial gravity system must be understood and weighed against the penalties (e.g., weight and cost) so that design trade-offs can be made. In other words, it is to be expected that artificial gravity will be part of integrated system designs for HEDS. **[Priority—Higher]**

Indirect Effects of Reduced Gravity

Indirect effects can set design requirements for components that are different from those in Earth gravity. That is, a component may operate according to phenomena unaffected by gravity level but may still behave differently in reduced gravity. Or the component may have been designed to take account of fractional gravity and therefore may be different from the corresponding components designed for Earth gravity.

Structural dynamics will affect the performance and integrity of piping and tankage subsystems that have been designed to be light and flexible to take advantage of the lower load-bearing requirements in low gravity. Research is needed to relate structural effects to the excitations that can arise from such causes as intermittent multiphase flows. Similarly, structural dynamics will affect the performance and integrity of large-surface structures such as space radiators, solar collectors, antennas, structural panels, tethers for various purposes, and robots of various types. Research is needed to relate the mechanical operations—such as positioning—of these devices to structural phenomena such as lightly damped vibration and buckling. Seemingly mundane components such as piping, valves, and bearings must be tailored to reduced- and variable-gravity environments. For example, products of wear and decay are presumably less easily managed in microgravity. Such concerns will be among the essential details of a central issue of supreme importance for HEDS, namely the effect of reduced and variable gravity on system reliability and safety.

The overall priorities of these topics are difficult to assign, since they would depend on the importance assigned to avoiding the various technical or biological consequences of reduced gravity enumerated in this report.

VII

Programmatic Recommendations

The design of successful HEDS systems and components, and decisions concerning alternative designs, will necessarily be done by the application of computational analysis codes that have been carefully developed and validated and that are based on physically correct mathematical models. This is true of all modern system engineering practice and will be especially true for the elaborate systems needed by HEDS. These codes will be needed at all scales, including those of the smallest components, but in many cases there is insufficient information available for the development of reliable computational models. Moreover, the data are often costly to obtain at the gravity levels of interest. For this reason, special attention must be directed toward the design of the experimental test matrices required for model (i.e., code) validation.

All these things being true, these codes and their underlying mechanistic models, even more so than the codes for aircraft engine analysis and design, for example, must be physically rather than empirically based. This is because of the need for fidelity in variable-gravity environments and because they must be robust and flexible enough to credibly evaluate innovative concepts.

NASA research must supply the information needed to build these models and codes, especially that needed to delineate the ranges of the applicable scaling laws. To ensure progress in developing the desired computational analysis and design capabilities, such research must clearly in part be directed research based on an evolving understanding of the programmatic needs. This will be a significant task, requiring patience, commitment, and careful peer review. In this context, NASA should investigate the possibility of consulting the U.S. Department of Energy-Naval Reactors program for help in designing research programs that can develop multivariate, physically based, multiphase computational models.

In the following paragraphs, suggestions are made concerning goals, research planning, and activities of NASA in support of gravity-related research for HEDS. A number of these recommendations were contained in the phase I report (NRC, 1997) and are reflected again here, in this more extensive study. It was thought then, and is still believed, that in view of the long time scale needed for the evolution of basic scientific concepts into practical applications, the suggested research programs will require a sustained commitment on the part of NASA to understand gravity-related issues.

A RESEARCH APPROACH FOR THE DEVELOPMENT OF MULTIPHASE FLOW AND HEAT TRANSFER TECHNOLOGY

The purpose of this section is to recommend that NASA give consideration to adopting an integrated approach for the conduct of research on multiphase flow and heat transfer technology for HEDS applications. As pointed out in earlier discussions of multiphase and heat transfer flow issues, the reliable operation of multiphase systems and processes in reduced-gravity environments is likely to be required if the HEDS program is to be successful. In particular, phase-change systems are likely to be necessary for power production, propulsion, and life support. Unfortunately, relatively little is currently known about the effect of gravity on multiphase systems and processes. Nevertheless, it is known that the impact can be large (e.g., the flow regimes will be strongly affected).

It should be stressed that the empirical approaches that have been used for the design of multiphase systems on Earth will not be reliable for space applications. Moreover, it will not be possible to parametrically test in space the required full-scale multiphase systems and subsystems until a suitable design is achieved. Rather, the approach recommended in this report is that a reliable, physically based analytical model be developed and qualified against appropriate terrestrial and microgravity data. The resultant computational model could then be used to analyze and optimize final designs and to compare the relative merits of various alternate designs (e.g., Rankine vs. Brayton cycle power plants). Significantly, these analytical models could be used to develop new design features and to analyze, and support, separate effects experiments and the scale-up of those systems and subsystems that cannot be tested at full scale in space. Indeed, it is vital that NASA have the capability to scale up data from small-scale tests if the HEDS research and development program is to be timely and cost-effective.

As is described in Chapter IV, the most detailed and advanced mechanistic models of this type that are currently available for the analysis of multiphase flow and heat transfer are multidimensional, multifield, two-fluid computational fluid dynamics (CFD) models. This type of model has been developed by the U.S. DOE-Naval Reactors program and used by the Navy for numerous applications on Earth. Similar models are being developed in France (for the Atomic Energy Commission of France) under the FASTNET program (Bestion et al., 1999), which is focused on nuclear reactor applications. Fortunately, it appears that these models can be extended to applications involving reduced-gravity environments (Alajbegovic et al., 1999; Lahey and Drew, 2000).

In order to carry out the various activities that are needed, as described above, to develop a reliable multidimensional, multifield, two-fluid CFD model for HEDS use, a directed program of experimental and analytical research will be required. The range of specific terrestrial and microgravity data needed to successfully develop and test the model is unlikely to be obtained in a reasonable time frame without a coordinated, focused effort on the part of NASA. In particular, careful attention must be paid to understanding and modeling such important phenomena as flow regimes (including flow-regime transitions), phase distribution, phase separation, multiphase turbulence, Marangoni forces, boiling/condensing heat transfer, multiphase pressure drop, static and dynamic instabilities, and condensation-induced loads in reduced-gravity environments. As part of this effort, the dimensionless parameters that characterize the relevant fundamental multiphase phenomena should also be cataloged and the various distinct operational regimes should be identified.

COORDINATION OF RESEARCH AND DESIGN

Considerable difficulty in the generation of this report arose from NASA's intra-agency organization, in the sense that outstanding research efforts in some centers were either poorly communicated to groups at other centers or duplicated the efforts of those other centers. The committee felt that such difficulties were symptomatic of a counterproductive "territoriality" that has been allowed by NASA senior management to develop and even to accelerate. Such organizational roadblocks will obviously be detrimental to the implementation of many of this committee's central research recommendations. The NASA office responsible for microgravity research (presumably the Microgravity Research Division) should diligently inform NASA at large about the issues of reduced gravity that are foreseen for space hardware design, so that these issues enter design thinking at the conceptual stage rather than as afterthoughts. Conversely, the microgravity research community should be kept apprised of design issues whose resolution requires the understanding that would be gained through phenomenonological research. The peer review process is crucial for ensuring the quality of supported research, but that does not preclude NASA from taking a leadership role in encouraging research on programmatic needs.

NASA should encourage the blending of conceptual design and phenomenological research. Often, phenomena depend strongly on boundary conditions or on the conceptual design of devices, and in such cases it becomes crucial to involve basic researchers in the development process in a fruitful way. For example, the subject of fundamental aerodynamics has been developed in close association with aircraft wing design. To accomplish this melding of research and design is not easy; it requires attentive and sophisticated management.

Equally, communication and coordination among basic researchers and system designers and users will be essential for HEDS success and should be actively encouraged. For example, problems of bearing and seal integrity in low or variable gravity suggest a need for novel bearing concepts, which no doubt would raise significant fundamental research issues. Also, new concepts for flow control devices to replace conventional valves should be encouraged and pursued through research, in order to provide long-term reliability and to assure the desired control of dynamic effects in distribution systems.

An especially significant example of the need to connect component development with research concerns swirling-flow devices intended to provide the centrifugal force needed for phase separations. While the basic principle of desired operation is understood, device-specific issues of great subtlety will determine success or failure. Therefore, a variety of configurational concepts should be explored, computationally to the extent feasible. Also, novel concepts for phase separation and mixing by wave processes, oscillations, or flow deflection should be explored through appropriate research activities.

It is further recommended that the process of gathering and exchanging information relevant to research directions in support of HEDS missions be strengthened by holding workshops and study groups attended by both mission technologists and microgravity scientists. The August 1997 workshop at NASA's Lewis Research Center in Cleveland was just such an endeavor; experts presented a well-organized, ambitious program on the various activities that are focused on Martian exploration. The program used carefully prepared presentations and workshop discussion panels to exchange very useful information. It was a learning experience for the participants and brought them up to date in a way that could not have been easily done by other means. The committee recommends that workshops of this kind be an ongoing feature of NASA's gravity-related research program. In particular, workshops should be held, at least biannually, at which technology issues are presented to the scientific community and relevant microgravity research is presented to the engineering community. Such workshops should culminate in two types of recommendations: the first type would involve assessment of the applicability of current research results to the technological issues pertaining to current plans for space exploration; the second type would focus on definition of future directions for microgravity research in support of the plans for space exploration. The presentations on both science and technology, along with the recommendations developed, should be published as a workshop proceedings. The recommendations for the directions of microgravity research should be reflected in NASA Research Announcements.

It should be emphasized that the goals of HEDS require the development of complex systems that depend on knowledge from traditionally distinct fields such as biology, fluid physics, and materials science. Such systems are needed to address life support needs and in situ resource recovery efforts. Therefore, it is suggested, as it was in the phase I report (NRC, 1997), that NASA might profitably initiate and support selected interdisciplinary research projects.

The problem of human bone loss in reduced gravity is clearly an interdisciplinary one and one that must be of great concern to the microgravity research community. While the dimensions of this problem are uncertain, its undeniable importance justifies encouragement and support of appropriate research by the microgravity and other relevant divisions in NASA. Specific research on the physical effects of spacecraft rotation is recommended in Chapter VI. Beyond that, novel concepts short of spacecraft rotation that might hold promise for countering the effects of reduced gravity simultaneously for humans and technical devices should be encouraged and pursued. Furthermore, fluid-mechanical expertise in the microgravity research community may help to explain fundamental bone-modeling processes, or at least to interpret such processes in relation to proposed countermeasures.

MICROGRAVITY RESEARCH AND THE INTERNATIONAL SPACE STATION

It is expected that the International Space Station will provide a unique platform for conducting long-duration microgravity scientific research and assessing the efficiency and long-term suitability of many of the technical systems important to HEDS. Unlike previous space shuttle experiments, which by necessity were confined to component-level tests and typically operated for only a few days, the semipermanent operating status of the International Space Station will permit testing of integrated hardware systems over time intervals that are long enough to address critical HEDS system operation requirements. By incorporating hardware systems that are designed for extended, autonomous operation, critical HEDS technology issues can be addressed, including: (1) conduct of multiphase flow and heat transfer experiments in support of computational model development; (2) identification of fundamental channel or system instabilities that are either produced by or modified by microgravity and the subsequent establishment of appropriate system modifications or control strategies to avoid or control those instabilities; (3) development of instrumentation and control systems that can tolerate gradual system degradation and yet respond automatically and appropriately to operational upsets, distinguishing between instrumentation degradation or failures and actual system failures, while maximizing safety, survivability, and long-term hardware reliability; and (4) development of the necessary autonomous operational principles that will be required to automatically shut down and "safe" complicated processing hardware units during serious system upsets, to serve purposes similar to what is accomplished by methodologies employed currently for spacecraft computer systems that experience unexpected upsets when they are far from Earth. An example of this type of opportunity is the development and extended operation of an autonomous water electrolysis system for the International Space Station that can convert wastewater into oxygen for life support (and hydrogen exhaust). Development of that system as an advanced, redundant life-support system for the International Space Station will mean the simultaneous evolution of a key ISRU[1] processing module that can be used for primary life support on crewed missions to Mars and for producing oxygen and hydrogen feed gas (for the production of other chemicals) on the surface of Mars or in the polar regions of the Moon, and for similar HEDS water electrolysis opportunities on other microgravity surfaces such as asteroids and icy comet cores. The extended operation of systems of this type in the microgravity environment of the International Space Station is a unique opportunity to subject future HEDS technology to realistic tests prior to actual deployment. The foregoing comments appeared in the phase I report (NRC, 1997) and are reiterated here with the added thought that the Space Station will be an especially important facility for the multiphase flow research program urged earlier in this chapter.

PEER REVIEW FOR REDUCED-GRAVITY RESEARCH

The NASA Research Announcement and peer review mechanisms have been of great benefit to the productivity and quality of NASA's gravity-related research, as pointed out in the phase I report (NRC, 1997). These mechanisms should be maintained as the areas of science and engineering affecting HEDS technologies are further developed, ensuring the pursuit of truly scientific objectives while enhancing the HEDS enterprise.

REFERENCES

Alajbegovic, A., D.A. Drew, and R.T. Lahey, Jr. 1999. An analysis of phase distribution and turbulence in dispersed particle/liquid flows. Chem. Eng. Commun. 174:85-133.

Bestion, D., J.-P. Clement, J.-M. Delhaye, P. Dumaz, J. Garnier, D. Grand, E. Herview, D. Lebaigue, H. Lemonnier, C. Lhuillier, J.-R. Pages, I. Toumi, and M. Villand. 1999. FASTNET: A proposal for a ten-year effort in thermal-hydraulic research. Multiphase Science and Technology 11:79-145.

Lahey, R.T., Jr., and D.A. Drew. 2000. The analysis of two-phase flow and heat transfer using a multidimensional, four-field two-fluid model. Proceedings of the Ninth International Topical Meeting on Nuclear Reactor Thermal Hydraulics (NURETH-9), October 3-8, 1999. Nuclear Engineering and Design, in press.

National Research Council (NRC), Space Studies Board. 1997. An Initial Review of Microgravity Research in Support of Human Exploration and Development in Space. Washington, D.C.: National Academy Press.

[1]In situ resource utilization.

Appendixcs

A

Statement of Task

COMMITTEE ON MICROGRAVITY RESEARCH

Microgravity Research in Support of Human Exploration and Development Objectives

Background A major NASA goal continues to be the exploration of space by humans, a fact reflected by the inclusion of the Human Exploration and Development of Space (HEDS) enterprise in the current organizational structure of NASA. There are numerous technological barriers that must be overcome in order for long-term exploration, development, and habitation of space by humans to occur, and many of these technical issues can be resolved only through a deeper understanding of the behavior of fluids and materials in a reduced gravity environment. NASA's Microgravity Science and Applications Division (MSAD) now resides within the HEDS enterprise, and there is a strong need to explore opportunities for microgravity research to support human exploration objectives through fundamental research.

Plan At the request of the MSAD, the Committee on Microgravity Research will undertake an assessment of scientific and related technological issues facing NASA's Human Exploration and Development of Space endeavor. The committee will look specifically at mission enabling and enhancing technologies which, for development, require an improved understanding of fluid and material behavior in a reduced gravity environment. These might range from construction assembly techniques such as welding in space, to chemical processing of extraterrestrially derived fuels and oxygen. The committee will identify opportunities that exist for microgravity research to contribute to the understanding of fundamental science questions underlying exploration technologies and make recommendations for some areas of directed research.

Some of the topics the committee will consider in developing its recommendations are the following:

- Fluids and materials behavior in reduced gravity;
- Technological barriers, related to low-gravity fluid and material behavior, that must be resolved in order for efficient human exploration of space to occur;
- The maturity of the science base underlying key technologies that have been identified for a space station, a lunar base, and interplanetary missions; and

- Areas of research that MSAD can pursue in order to help develop an understanding of the fundamental science underlying these technical issues.

Study Plan The study will be carried out in two phases. A preliminary (phase I) report will address current program balance and near-term scientific and technological issues. The final (phase II) report will address long-term exploration goals.

B

Symbols

c_p	Heat capacity, or specific heat, J/kg-K
D	Diffusion coefficient, m^2/s
D_H	Hydraulic diameter, m
Fr	Froude number
g	Acceleration due to gravity (not necessarily Earth's), m/s^2
g_0	Gravity level at Earth's surface, m/s^2
G	Artificial gravity level produced by rotation, m/s^2
j	Inlet velocity, m/s
k-ε	Turbulent kinetic energy (k) and dissipation (ε)
L	Characteristic length scale or distance between boundaries, m
n	Numerical exponent (greater than unity), dimensionless
N_{SUB}	Nondimensional inlet subcooling
p	Pressure in fluid, N/m^2
q''_{max}	Critical heat flux, W/m^2
r	Radial position, m
R	Radius of pipe, m
T_{sat}	Saturation temperature, K
T_W	Wall temperature
$\overline{u'}$	Axial velocity fluctuation, m/s
$\overline{v'}$	Lateral velocity fluctuation, m/s
$\overline{u'v'}$	Reynolds stress (divided by density), m^2/s^2
W	Angular velocity of general rotation (e.g., of spacecraft about a distant axis), radians/s
w	Flow rate, kg/s
x	Axial dimension, m
Y	Radial distance (e.g., of spacecraft) from axis of general rotation, or lateral position, m
y	Lateral dimension, m

α_k	Volume fraction of phase k
Δ	An overall difference of a property (e.g., $\Delta\rho$), measured or given at a certain position
ε	Amplitude of oscillation, m
Λ	Friction number, fL/D_H
ν	Kinematic viscosity of fluid, m^2/s
Γ_1	Single-phase pressure-drop-to-inlet-velocity ratio, $N\text{-}s/m^4$
Π_1	Two-phase pressure-drop-to-inlet-velocity ratio, $N\text{-}s/m^4$
ρ	Mass density of fluid medium, kg/m^3
σ	Surface tension, N/m
τ–ε	Reynolds stress (τ) and turbulent dissipation (ε)
ω	Angular velocity, radians/s

C

Glossary

agglutinates aggregates of pieces

albedo the fraction of incident light that reflects diffusely from a surface

aphelia the most distant points in solar orbits

beneficiation process of transforming raw materials into more concentrated or beneficial forms

bioreactor any system that is engineered to use biological agents for the transformation of one form of matter into another. This would include engineered composters, waste recovery systems, air purification systems using biological agents, and fermentors used for producing antibiotics or other biological products.

bioregenerative system system that makes use of metabolic processes in the conversion of waste to useful gases

Brayton cycle a dynamic scheme for power generation featuring high conversion efficiency and a single-phase working fluid but with the drawback of relatively low heat-rejection temperatures, requiring relatively large and massive radiators

capillarity the behavior of fluids under the influence of surface tension, such as the movement of water in a very small diameter tube, induced by the water's ability to wet the walls of the tube

cohesion the force that enables soil grains to cling together in opposition to the forces tending to separate soil into parts

comminution reduction of a mass to minute particles or fine powder

computational fluid dynamics (CFD) a numerical approach using high-speed computers for evaluating the conservation equations that describe single- and multiphase flows

Coriolis force in a rotating system, a force proportional to the angular velocity and the velocity of any motion as measured in the rotating system

cryogenics the study of the production of very low temperatures and their effects

dead comets comet cores that no longer experience strong particle field interactions with the Sun

duracrust a thin agglomerated soil layer that is observed at some locations on Mars

dynamic equilibrium the balancing of forces on bodies in motion

ecological system the ensemble sustaining a life-support system, with special emphasis on the quality, nature, and analysis of interacting systems

ejecta rocks and other solid debris thrown off the surface of a planetary body

electrolysis the production of chemical changes in a chemical compound by causing its oppositely charged constituents, or ions, to move in opposite directions under a potential difference

enabling technology a technology permitting the extension of human activity

Eulerian formulation a mathematical model written in a fixed coordinate system

excursive instabilities two-phase excursions from one operating state to another

film bearing A bearing in which forces are transmitted through thin but continuous fluid films (gas or liquid) that separate solid machine components

fines the silty sand regolith on the Moon with a median grain size of only 0.1 mm

fractional gravity gravity levels lower than Earth's but typical of those found on the surfaces of planets and their major satellites

friction a force that opposes the sliding of one surface over another and whose magnitude is proportional to the normal force between the surfaces

***g*-jitter (gravity-jitter)** inertia effect due to oscillatory accelerations arising from crew motions, machinery, rocket firings, and so on, occurring in spacecraft

gravity a body force per unit mass experienced as a result of mutual attraction with all other bodies, independent of electromagnetic or other forces

heat exchanger device that facilitates the transfer of heat from a hot source to a cold sink

heat pipe A container of two-phase fluid used to transfer heat efficiently

heat sink A reservoir to absorb thermal energy

homeostasis maintenance of constant values, for example for temperature, humidity, and gases in the atmosphere and other critical environmental parameters

hydrostatic pressure the equilibrium pressure acting at the base of a fluid which is proportional to the fluid's depth and gravity

ilmenite $FeTiO_3$, an oxygen-rich mineral found at moderate, but variable, concentrations on the Moon

in situ resource utilization the production of useful materials from resources that are acquired and processed away from Earth's surface

interfacial phenomena behavior, specially associated with boundaries between different phases, including those between similar phases of different materials

Lagrangian formulation a mathematical model written in a coordinate system that is moving with the fluid

Ledinegg instability an excursive instability in a boiling loop which may cause the flow to go to a new value, which in turn may result in the onset of critical heat flux

liquid-phase sintering densification at the sintering temperature using a matrix of solid particles and a viscous liquid

lunar magma molten lunar rock

Marangoni effect liquid convection caused by surface tension gradients at the free surface of a liquid or at the interface between two liquids

Marangoni flow movement of fluids under the influence of the Marangoni force

Marangoni force a force exerted by the surface tension gradient

multiphase used to describe any process involving a mixture of phases; a glass of ice water is a multiphase system

near-net shape refers to shaped object that is in nearly the final desired shape and therefore requires only minimal machining

normal stress the component of stress normal to a plane inscribed in a substance, tending to cause compressive or tensile strain in that normal plane, N/m^2

penetrator a device used to penetrate beneath the surface of a planetary body

phase a homogeneous and physically distinct state of aggregation of a substance, i.e., solid, liquid, or vapor phase

phase separation separation of a mixture of phases into individual component phases

physically based model a model of system behavior based on fundamental physical principles (e.g., thermodynamic laws) and the appropriate physical mechanisms (e.g., heat transfer, capillary flow), as opposed to an empirical model based primarily on experimental measurements and that incorporates only a limited theoretical understanding of the system

physicochemical system system that makes use of a combination of physical and chemical processes

power-to-mass ratio a characteristic of propulsion and other systems connoting the yield of power per unit of system mass, not including propellant mass

pultrusion a process for producing continuous fibers for advanced composites that involves pulling reinforcements through tanks (of resins), a preformer, and a die

radiation anything propagated as rays, waves, or a stream of particles, but especially light and other electromagnetic waves or the emission from radioactive substances

Rankine cycle a dynamic method of power generation that uses separate boilers and condensers with two-phase (liquid/vapor) mixtures with high conversion efficiencies and high heat-rejection temperatures, allowing reduced radiator mass and areas

reduced gravity gravity levels that are less than 1 g_0

regolith surface rock, especially used to describe the lunar surface soil

shear stress the component of stress parallel to a plane inscribed in a substance, tending to cause a shearing or slippage strain in that parallel plane, N/m^2

slurry a mixture of a liquid and insoluble solids

specific impulse thrust per unit mass of flow rate produced by burning rocket propellant at the rate of one pound per second, in seconds

static equilibrium the balancing of forces on bodies at rest

substoichiometric in chemical reactions, the participation of one species in less-than-exact chemical equivalence with other reacting species

surface tension attractive forces at a liquid surface that cause the surface to contract as far as possible

technical readiness level the maturity of a system, ranging from level 1 (a basic principle observed and reported) to level 8 (a design qualified for spaceflight)

touch-and-go exploration round-trip space missions of the Apollo type, which are characterized by relatively short surface stays

triple point the point on a pressure-temperature diagram representing the condition at which the solid, liquid, and vapor phases can exist together in equilibrium

Van der Waals force attractive intermolecular force arising from induced dipole moment

viscosity the property of fluids by virtue of which they offer a resistance to flow

volatiles molecules that can exist in a gaseous phase on Earth's surface

volatilization conversion to a gaseous or vapor phase

water gas historical designation of a gaseous fuel produced by reacting steam with very hot solid carbon

D

Acronyms

AFC	alkaline fuel cell
AMTEC	alkali metal thermal-to-electric conversion
BASE	beta-alumina solid electrolyte
CELSS	closed ecological life support system
CFD	computational fluid dynamics
CHF	critical heat flux (the limit of nucleate boiling heat transfer)
CHX	condensing heat exchanger
CMC	ceramic matrix composite
CMGR	Committee on Microgravity Research
CPL	capillary pumped loop
DIP	dynamic isotope power
DMD	direct metal deposition
DNS	direct numerical simulation
DOD	Department of Defense
DOE	Department of Energy
DWO	density-wave oscillation
EVA	extravehicular activity
EW	equivalent weight
FIV	flow-induced vibration
FVS	free vortex separator
GCNR	gas-core nuclear rocket
GCR	galactic cosmic radiation
GRC	Glenn Research Center, NASA (formerly Lewis Research Center)

HEDS	Human Exploration and Development of Space, a NASA enterprise
HZE	high atomic number and energy
ISRU	in situ resource utilization
ISS	International Space Station
JPL	Jet Propulsion Laboratory, jointly operated by NASA and the California Institute of Technology
JSC	Johnson Space Center, NASA
LES	large eddy simulation
LHP	loop heat pipe
LHTES	latent heat thermal energy storage
LHV	lower heating value
LPS	liquid-phase sintering
M-C	Mohr-Coulomb
MCFC	molten carbonate fuel cell
MMC	metal matrix composite
MRD	Microgravity Research Division
MSAD	Microgravity Science and Applications Division
NASA	National Aeronautics and Space Administration
NASDA	National Space Development Agency (Japan)
NEA	near-Earth asteroid
NEP	nuclear electric propulsion
NERVA	nuclear engine for rocket vehicle application
NRA	NASA Research Announcement
NRC	National Research Council
NTR	nuclear thermal rocket
PAFC	phosphoric acid fuel cell
PCM	phase-change material
PEMFC	proton-exchange membrane fuel cell
PMC	polymer matrix composite
PRA	probabilistic risk assessment
PV	photovoltaic
R&D	research and development
RF	radio frequency
RFMD	rotary fluid management device
RTG	radioisotope thermoelectric generator
RWGS	reverse water gas shift
SAND	solid amine-water desorption
SBR	soil-bed reactor
SFWE	static feed water electrolysis
SLS	selective laser sintering
SOFC	solid oxide fuel cell

SPFC	solid polymer fuel cells
SPWE	solid polymer water electrolysis
SWV	solar wind volatile
TE	thermoelectric
TI	thermionic
TIMES	thermoelectric integrated membrane evaporator
TPV	thermophotovoltaic
TRL	technical readiness level
UV	ultraviolet
VAPCAR	vapor-phase catalytic ammonia removal
VCD	vapor-compression distillation
VHT	vertical handgun tool
VOF	volume of fluid
VPL	vapor-pressure pumped loop

E

Biographies of Committee Members

Raymond Viskanta, *Chair,* is the W.F.M. Goss Distinguished Professor of Engineering in the School of Mechanical Engineering at Purdue University. His areas of specialty include radiation transfer in gases and solids, applied thermodynamics, heat transfer in combustion systems, and heat transfer in materials processing. Dr. Viskanta's honors include the Senior U.S. Scientist Award presented by the Alexander von Humboldt Foundation, Germany (1975), the American Society of Mechanical Engineers (ASME) Heat Transfer Memorial Award (1976), the American Institute of Aeronautics and Astronautics (AIAA) Thermophysics Award (1979), the American Institute of Chemical Engineers/ASME Max Jakob Memorial Award (1986), the ASME Melville Medal (1988), and the Japan Society of Mechanical Engineers Thermal Engineering Award for International Activity (1994). He was named an ASME fellow (1976), a Japan Society for the Promotion of Science fellow (1983), and an AIAA fellow (1988). He was elected to the National Academy of Engineering (1987) and as a foreign member of the Academy of Engineering Sciences of the Russian Federation (1995). He also received an honorary doctor of engineering degree (Doctor Honoris Causa) from the Technical University of Munich (1994).

Robert A. Altenkirch is the vice president for research at Mississippi State University. His areas of expertise are radiative heat transfer in flames, flame spreading over solid and liquid fuels, buoyancy effects on flames, low-gravity combustion modeling, microgravity experimentation, and the chemistry of pulverized-coal combustion. He has published a variety of articles on these subjects. He is also a fellow in the American Society of Mechanical Engineers and was awarded its Gustus L. Larson Memorial Award.

Robert L. Ash is professor of engineering and associate vice president for research at Old Dominion University. His current research is focused on developing systems to produce oxygen from the Martian atmosphere using glow-discharge techniques, accurate methods for including atmospheric effects in modeling commercial aircraft wake vortex hazards at congested airports, and methods for incorporating relaxation effects in simple fluids. He has also held positions as chairman of the Mechanical Engineering and Mechanics Department at Old Dominion and as a visiting distinguished research engineer at NASA's Langley Research Center.

Robert J. Bayuzick is a professor of materials science and director of materials science at Vanderbilt University. His research for the last several years has been directed toward materials processing under microgravity conditions, with a particular emphasis on the structure and properties of alloys resulting from deep undercooling through containerless solidification. Dr. Bayuzick recently served on the lead team working on TEMPUS, an electromagnetic levitation facility that allowed containerless processing of metallic samples in microgravity aboard the space shuttle Columbia. Dr. Bayuzick also served as the director of the Center for the Space Processing of Engineering Materials, one of the NASA-supported centers for the commercial development of space. He was a visiting senior scientist at the NASA Office of Space Science and Applications and won NASA's Public Service Medal.

Charles W. Carter, Jr., is a professor of biochemistry and biophysics at the University of North Carolina at Chapel Hill. His study of protein crystallography focuses on structure determination by X-ray crystallography of electron transport proteins and macromolecules responsible for incorporating tryptophan into proteins. He is a member of the Biophysical Society, the American Crystallographic Association, the American Society of Biochemistry and Molecular Biology, and the American Cancer Society.

Gretchen J. Darlington is a professor in the Departments of Pathology and Cell Biology and Molecular Genetics at the Baylor College of Medicine. She is currently investigating the role of the C/EBP family of transcription factors in the acute phase response, in the differentiation of adipose tissue in vivo, and growth inhibition. Dr. Darlington is a member of the American Society of Human Genetics, the American Association for the Advancement of Science, the New York Academy of Sciences, Sigma Xi, the Society for Cell Biology, and the Society for In Vitro Biology. She has been the recipient of an American Heart Association Established Investigatorship and an awardee of the Irma T. Hirschl Trust.

Richard T. Lahey, Jr., is the Edward E. Hood, Jr., Professor of Engineering and was the former dean of engineering at Rensselaer Polytechnic Institute. He is an internationally recognized authority on multiphase flow and heat transfer technology. He also has extensive experience in nuclear reactor thermal hydraulics and safety, two-fluid modeling, and the evaluation of instability mechanisms in phase-change systems. Prior to joining Rensselaer in 1975, he held several technical and managerial positions with the General Electric Company, including overall responsibility for all domestic and foreign research and development programs associated with nuclear reactor thermal-hydraulics and safety technology. He has received numerous honors and awards, including the E.O. Lawrence Memorial Award of the U.S. Department of Energy, the American Nuclear Society's Seaborg Medal, the Technical Achievement Award, the Arthur Holly Compton Award, the Meritorious Service Award, and the American Institute of Chemical Engineering's Kern Award. Dr. Lahey has been a Fulbright fellow and an Alexander von Humboldt senior fellow. He is a member of the National Academy of Engineering, the Russian Academy of Science, and the New York Academy of Science and is a fellow of the American Nuclear Society and the American Society of Mechanical Engineers.

Ralph A. Logan is a distinguished member of the technical staff (retired) at AT&T Bell Laboratories, where he worked in the Solid State Research Department at Murray Hill, New Jersey. His work includes studies of dislocations, p-n junctions, electroluminescence, lasers, and crystal growth to form optical communication devices. He is a fellow of the American Physical Society and the Institute of Electrical and Electronic Engineers and a member of the Optical Society.

Franklin K. Moore is the Joseph C. Ford Professor of Mechanical Engineering (emeritus) at Cornell University. Dr. Moore pioneered fundamental research in fluid mechanics and has continued to contribute innovative engineering in the fields of gas turbine dynamics, thermal engineering, and boundary layer theory. In addition to his work at Cornell, Dr. Moore is a consultant for NASA and a variety of other public and private institutions. He is a past member of the Aeronautics and Space Engineering Board of the National Research Council. Dr. Moore's honors include NASA's Distinguished Scientific Achievement Medal and membership in the National Academy of Engineering. He is also a member of the National Physical Society and a fellow of the American Society of Mechanical Engineers, and the American Institute of Aeronautics and Astronautics.

William W. Mullins is a professor (emeritus) of applied science in the Department of Materials Science and Engineering at Carnegie Mellon University. He was employed at the Westinghouse Research Laboratories in

Pittsburgh (1955-1960) and then at Carnegie Mellon University (formerly, the Carnegie Institute of Technology), where he served as head of the Department of Metallurgical Engineering and Materials Science (1963-1966) and as dean of the then College of Engineering and Science (1966-1970). In 1985, he was appointed University Professor of Applied Science. His research has been concentrated in the areas of the morphology of phase transformations, the capillarity-induced evolution of surfaces, the thermodynamics of stressed solids and solid surfaces, and the mathematical theory of grain boundary motion, grain growth, and coarsening. He was awarded a Fulbright fellowship and a Guggenheim fellowship (1961), received the Mathewson Gold Medal (1963) and the Philip M. McKenna Memorial Award (1981), was elected to the National Academy of Sciences (1984), received a Professional Achievement Citation from the University of Chicago Alumni Association (1990), a Humboldt senior fellowship (1992), the Mehl Medal and Memorial Lecturership (1994), and the Von Hippel Award of the Materials Research Society (1995), and was elected a fellow of TMS (1995).

Rosalia N. Scripa is associate dean for academic and student affairs and professor of materials and mechanical engineering at the University of Alabama's (UAB's) School of Engineering. She is also professor of biomedical engineering at UAB's Center for Telecommunications, Education, and Research Biomedical Implant Center. Dr. Scripa's research interests include semiconductor crystal growth, solidification of semiconductors under low gravity, development and characterization of biocompatible coatings for orthopedic materials, properties and processing of ceramic/glass and ceramic composites, and fracture analysis of ceramics and glass. She has been the recipient of more than 25 awards and honors, including the NASA Marshall Space Flight Center (MSFC) Certificate of Appreciation for her work as co-investigator on a crystal growth experiment aboard a 1992 space shuttle mission, the 1995 NASA MSFC Certificate of Excellence, the 1997 Alabama Society of Professional Engineers Outstanding Engineering Faculty Award, and the 1997 Outstanding Alumni of Achievement Award from the University of Florida.

Forman A. Williams is a professor of engineering physics and combustion, director of the Center for Energy and Combustion Research, and chair of the Department of Applied Mechanics and Engineering Sciences at the University of California at San Diego. His areas of specialty are flame theory, combustion in turbulent flows, asymptotic methods in combustion, fire research, reactions in boundary layers, and other areas of combustion and fluid dynamics. His current research topics include the prediction of emissions of oxides of nitrogen from large diesels, theory of flames with real chemistry, high-pressure combustion of binary fuel sprays, droplet burning experiments in the space shuttle, experimental and theoretical studies of fuel droplets and flames subject to straining flows, stretched diffusion flames in von Karman swirl flows, catalytic combustion fundamentals, and the fundamentals of acoustic instability in liquid propelled rockets. He is the author of *Combustion Theory*, co-author of *Fundamental Aspects of Combustion*, and a member of the editorial advisory boards of *Combustion and Flame*, *Combustion Science and Technology*, *Progress in Energy and Combustion Science*, and *Archivium Combustionis*.